BIOCHEMISTRY RESEARCH TREND

NATURAL PRODUCTS

STRUCTURE, BIOACTIVITY AND APPLICATIONS

BIOCHEMISTRY RESEARCH TREND

Additional books in this series can be found on Nova's website under the Series tab.

Additional E-books in this series can be found on Nova's website under the E-book tab.

BIOCHEMISTRY RESEARCH TREND

NATURAL PRODUCTS

STRUCTURE, BIOACTIVITY AND APPLICATIONS

RAMIRO E. GONCALVES
AND
MARCOS CUNHA PINTO
EDITORS

New York

Library of Congress Cataloging-in-Publication Data

Library of Congress Control Number: 2012940133

ISBN: 978-1-62081-728-5

Published by Nova Science Publishers, Inc. †New York

Contents

Preface vii

Chapter 1 Pteridophytes: Ethnobotany and Pharmacologic Bioactivities 1
Maria Flaviana Bezerra Morais-Braga, Teógenes Matias de Souza, Irwin Rose Alencar de Menezes, José Galberto Martins da Costa, João Batista Teixeira da Rocha, Rogério de Aquino Saraiva, Pablo Nogara, Diones Bueno, Margareth Linde Athayde, Aline Augusti Boligon, Antonio Álamo Feitosa Saraiva and Henrique Douglas Melo Coutinho

Chapter 2 In-vitro assessment of Chromones, Alkaloids and other Natural Products from Caribbean Plants as Potential Anti-tuberculars and Chemopreventors 35
Sheena Francis, Damion Morris, Mario Shields, Helen Jacobs and Rupika Delgoda

Chapter 3 Natural Products as Inhibitors of the Ubiquitin-/Ubiquitin-Like Protein-Proteasome Pathway 55
Dik-Lung Ma, Victor Pui-Yan Ma, Daniel Shiu-Hin Chan, Ka-Ho Leung, Hai-Jing Zhong and Chung-Hang Leung

Chapter 4 A Review of the Plant Origins, Composition and Biological Activity of Red Propolis 75
Begoña Gimenez-Cassina López and Alexandra Christine Helena Frankland Sawaya

Chapter 5 Nanoencapsulation of Natural Products with Multifunctional Properties in Silica and Hybrid Organic – Silica Host Matrices 91
Ioana Lacatusu, Nicoleta Badea and Aurelia Meghea

Chapter 6 Natural Products as a Source of Potential Drugs for the Treatment of Fungal Infections 167
Fernanda F. Campos, Susana Johann and Vera L. dos Santos

Index 203

PREFACE

A natural product is a chemical compound or substance produced by a living organism - found in nature that usually has a pharmacological or biological activity for use in pharmaceutical drug discovery and drug design. A natural product can be considered as such even if it can be prepared by total synthesis. In this book, the authors present current research in the study of the structure, bioactivity and applications of natural products. Topics discussed include the in-vitro assessment of chromones and alkaloids from Caribbean plants as potential anti-tuberculars and chemopreventors; natural products as inhibitors of ubiquitin and ubiquitin-like protein-proteasome pathways; the biological activities of red propolis; nanoencapsulation of natural products; and natural products as a source of potential drugs for the treatment of fungal infections.

Chapter 1 - To talk about one of the largest groups of vascular plants, emphasizing its utility for well-being and improvemment of quality of life in human populations, it is essential that we first describe and characterize these plants to better understand the relation between the species of this group and humans. Next, before going into different reports on bioactivities, it is necessary to emphasize the medicinal use of plants for the treatment of various types of illnesses, through ethnobotanical approaches in terms of use (parts of plant, species utilized and purpose) in different types of cultures. Only then can we appreciate the various pharmacologic possibilities of this group which has inhabited this planet for millions of years, confirming the empirical and traditional knowledge that expresses the power of phytotherapeutic medicine. To enrich the discussion on the potential of pteridophytes, an interesting case study will be demonstrated on the investigation of the antioxidant activity of ferns belonging to different families. In this study, emphasis will be given to phenolic compounds, phytoconstituents found in plants, and their inter-relation with the capacity of scavenging free radicals.

Chapter 2 - Resurgence in the tuberculosis pathogen, *Mycobacterium tuberculosis*, which significantly reduces survival rates of persons co-infected with the immunodeficiency disease HIV, poses a major health issue particularly for the developing countries, which accounts for over 95 percent of reported global HIV cases. Compounding this issue is the emergence of multiple drug resistant strains, highlighting an urgent need for the search for novel and effective therapies against *M. tuberculosis*. Arylamine *N*- acetyltransferase (NAT), a drug metabolizing enzyme found expressed in *M. tuberculosis* and responsible for the metabolism of the frontline anti-TB drug isoniazid has also been identified to play a significant role in its cell wall lipid synthesis. Knowledge on NAT enzyme's essential role in the survival of *M.*

tuberculosis, garnered through recent *nat* gene knockout experiments in *M. bovis*, has identified it as a useful target in the search for anti-tubercular therapy. In this study we employed this molecular target in an *in-vitro* assay, to search for natural products with potential for future *in-vivo* examinations. Eleven novel/known compounds including tetranoterpinoids and quassinoids (at100 μM) previously extracted and purified from five rare endemic and/or indigenous Caribbean plants, *Spathelia sorbifolia*, *Esenbeckia pentaphylla*, *Peperomia amplexicaulis*, *Hortia regia* and *Clusia havestiodes* were examined for their inhibitory properties using heterologously expressed NAT from *M. Smegmatis*, a homologue of *M. tuberculosis* NAT. Greatest inhibition was obtained for anhydrosorbifolin (65%), a chromone whose extended linear side chain appear to contribute to the increased inhibition compared with its derivatives, alloptaeroxylin which has no side chains (displayed 25% inhibition) and spatheliabischromone which has a cyclicized side chain (displayed 42% inhibition). These compounds were also examined for their inhibition of human cytochrome P450 (CYP) 1, a class of enzymes comprising of CYPs1A1, 1A2 and 1B1, important in the activation of polyaromatic hydrocarbons to their carcinogenic precursors. Results revealed greatest inhibition by the alkaloid dictamnine of CYP1B1 activity with a potent IC_{50} value of 0.27μM. These studies show natural chromones and alkaloids as potential anti-tuberculars and chemopreventors worth further analysis.

Chapter 3 - The proteasome is the final player in the regulated degradation of intracellular proteins through in both ubiquitin-dependent and ubiquitin-independent proteolytic pathways. In eukaryotic cells, the ubiquitin-proteasome system coordinates the polyubiquitination and subsequent proteolysis of unwanted proteins, crucial to normal cellular homeostasis. The first-in-class proteasome inhibitor, bortezomib. promotes apoptosis and chemosensitization of cancer cells, and is used clinically as either a single-agent chemotherapeutic or in combination with other drugs. Natural products offer medicinal chemists with a cornucopia of diverse chemical scaffolds and bioactive substructures, and historically have represented an important source of new drugs. Salinosporamide A (NPI-0052), a second-generation proteasome inhibitor of marine microbial origin, has been effective against bortezomib-resistant cancers and has entered Phase I clinical trials. This review discusses the application of natural products as inhibitors of the ubiquitin-proteasome system through targeting of both conventional proteolytic pathways as well as those involving ubiquitin-like proteins, such as NEDD8, which have recently emerged as novel anti-cancer targets.

Chapter 4 - Propolis is a mixture of various amounts of beeswax and resins collected by bees from plants, particularly from buds and resinous exudates. The composition of propolis varies according to the kind of bee, geographic and plant origin of the samples. Propolis is important for the hive defense and has been used for its medicinal properties since ancient times. Red propolis has been found in the northeastern coast of Brazil, as well as in Cuba, Venezuela, Mexico and China. Among the most frequently cited plant sources are species of the Leguminosae and Clusiaceace families. The composition of red propolis varies both qualitatively and quantitatively, confirming different plant sources. The compounds that have been found in red propolis samples are listed herein. Red propolis has shown diverse biological activities: antimicrobial activity against different bacteria and yeast; against *Leishmania amazonensis* parasites; antioxidant, cytotoxic and potential antitumor activity; antipsoriatic, anti-inflammatory and analgesic activities; anti-obesity and hepatoprotective effects. Many new studies of red propolis are being developed due to the promising therapeutic activity; however, future studies must differentiate between red propolis from diverse geographic origins and chemical profiles.

Chapter 5 - In the large field of nanotechnology, the synthesis of organic – inorganic hybrid nanomaterials based on different polymer matrices have become a prominent area of current research and development. Sol-gel chemistry has been easily modified to the changing needs of society to produce fine-tuned sol-gel nanostructured materials for relevant applications. In this respect, there is an increasing need for natural and versatile raw materials, as well as biocompatible compounds that could be extensively used in the large field of biomedicine. The design flexibility of sol-gel technique and its ease of fabrication can create surfaces with structural and chemical features that could be compatible with biomaterials. Silicate and derivative silicate frameworks are the most abundant compounds in nature, their use in science, medicine and engineering has increased drastically in the last decade. Therefore, silica and organo-silsesquioxane matrices are the focus of this chapter because they seem to have the properties needed to encapsulate a variety of compounds with active principles. The chapter is intended to give an overview on exploitation of the sol-gel template preparation route in order to improve the main properties of some natural products (e.g. flavones, vegetal extracts and vitamins) such as fluorescence intensity and antioxidant capacity by physical entrapment in appropriate silica-derived matrices. Synthesis of some novel fluorescence nanomaterials loaded with bioactive polyphenols which are present in most plants (concentrated in seeds, fruit skin or peel) with a high spectrum of biological activity, by replacing synthetic chemicals, may open new opportunities for optical and bio-medical applications.

Chapter 6 - The increasing numbers of immunocompromised people and the advances in medical technology, such as organ and bone marrow transplants, have been accompanied by an increasing occurrence of invasive fungal infections. In addition, the indiscriminate use of antifungal agents has resulted in the development of resistance to the drugs currently in use, which emphasizes the need for the discovery of new classes of antifungal compounds to treat fungal infections. Thus, there is an urge need for the discovery of new antifungal drug therapies that are safe and effective. Moreover, the increase of azole-resistant strains emphasizes the importance of identifying lead compounds that act on novel targets. Microorganisms and plants are sources of various natural products with antimicrobial activity. Biomolecules, such as biosurfactants and secondary metabolites belonging to different chemical classes including terpenoids, saponins, alkaloids and flavonoids, have been evaluated for their anti-adhesive, fungistatic and fungicidal properties in vitro and in vivo. These molecules can be used directly or as precursors to the development of new antifungal drugs. Formerly, combinatorial chemistry appeared to be the future for drug discovery, but in the late 1990s, synthetic chemists realized that combinatorial libraries lacked the "complexity" usually associated with natural compounds. Thus, with the advent of the concept of diversity-oriented synthesis, research into biologically active natural products has had a reprise. This review attempts to summarize the current status of the importance of natural products of microorganisms and plants as sources of drugs to treat fungal infections that are more effective and less toxic.

In: Natural Products: Structure, Bioactivity and Applications ISBN 978-1-62081-728-5
Editors: Ramiro E. Goncalves and Marcos Cunha Pinto

Chapter 1

PTERIDOPHYTES: ETHNOBOTANY AND PHARMACOLOGIC BIOACTIVITIES

Maria Flaviana Bezerra Morais-Braga[1], Teógenes Matias de Souza[2], Irwin Rose Alencar de Menezes[2], José Galberto Martins da Costa[2], João Batista Teixeira da Rocha[4], Rogério de Aquino Saraiva[4], Pablo Nogara[4], Diones Bueno[4], Margareth Linde Athayde[3], Aline Augusti Boligon[3], Antonio Álamo Feitosa Saraiva[1] and Henrique Douglas Melo Coutinho[2,*]

[1]Departamento de Ciências Biológicas
[2]Departamento de Química Biológica,
Universidade Regional do Cariri, Crato, CE, Brasil
[3]Departamento de Farmácia Industrial
[4]Departamento de Química, Universidade Federal de Santa Maria,
RS, Brasil

ABSTRACT

To talk about one of the largest groups of vascular plants, emphasizing its utility for well-being and improvemment of quality of life in human populations, it is essential that we first describe and characterize these plants to better understand the relation between the species of this group and humans. Next, before going into different reports on bioactivities, it is necessary to emphasize the medicinal use of plants for the treatment of various types of illnesses, through ethnobotanical approaches in terms of use (parts of plant, species utilized and purpose) in different types of cultures. Only then can we appreciate the various pharmacologic possibilities of this group which has inhabited this planet for millions of years, confirming the empirical and traditional knowledge that

* Henrique Douglas Melo Coutinho, Laboratório de Microbiologia e Biologia Molecular, Universidade Regional do Cariri – URCA, Crato-CE, Brasil. Rua Cel. Antonio Luis 1161, Pimenta, 63105-000. Fone: +55(88)31021212; Fax +55(88) 31021291. E-mail: hdmcoutinho@gmail.com.

expresses the power of phytotherapeutic medicine. To enrich the discussion on the potential of pteridophytes, an interesting case study will be demonstrated on the investigation of the antioxidant activity of ferns belonging to different families. In this study, emphasis will be given to phenolic compounds, phytoconstituents found in plants, and their inter-relation with the capacity of scavenging free radicals.

1. Introduction

Pteridophytes are vascular cryptogams and form a neglected group rich in biodiversity. Its food and medicinal values are not very well known by the world population, although widely reported in the scientific literature.

The utilization of these plants by ancient to contemporary societies as therapeutic agents, among other uses, has provoked curiosity in knowing a little more about popular medicinal practice, making them protagonists in human history and, above all, in verifying their efficacy through science.

Relations between pteridophytes and humans, therefore, have been recorded over the years. The use of pteridophytes is reported in the Ayurvedic system of medicine, elaborated by Sushruta (ca 100 AD) and Charka (ca 100 AD), who recommended the use of some ferns in their *Samhitas*, the traditional medical literature. Pammel [1] compiled a manual of poisonous plants of eastern North America with brief citations on the medicinal and economic values of various plants, including some pteridophytes. Ferns are also used by healers in Unani system of medicine [2]. In traditional Chinese medicine, many ferns are prescribed by local doctors [3]. Much later, modern biological and pharmaceutical studies were conducted on pteridophytes by different investigators. Benerjee and Sen [4] conducted extensive research on the antibiotic activity of innumerable ferns, reporting on a hundred species with such property. Dixit and Vohra [5] reported that important species of pteridophytes of India are used as food and in popular medicine. Kaushik [6] emphasized the ethnobotanical importance of ferns of Rajasthan, India. Important studies opn the food and medicinal value of pteridophytes were conducted by Nayar [7], Hodge [8], and Dixit [9]. In 2004, Gosh *et al.* [10] reported on the use of pteridophytes as food and medicinal plants.

The ethnobotanical use of this group is of immense importance, and this is clearly understood by scientists as a necessary basis for the discovery of active principles capable of promoting well-being and health in people.

2. Pteridophytes: Characterization and Ecological Aspects

Pteridophytes are vascular plants without flowers and seeds, which reproduce through spores, whose heteromorphic life cycle is shown in two phases, the gametophytic and sporephytic, thereby sharing an alternation of generations [11]. For a long time, these plants were treated as a single group, where they were classified as belonging to the division called Pteridophyta [12], constituted by evolutionarily distinct classes, forming a paraphyletic taxon [13,14].

Since the 1990s, new studies involving cladistic analyses with morphologic data, ultrastructure of gametes and analyses of DNA sequences with the use of molecular markers of plastid, mitochondrial and nuclear genes culminated with the separation of pteridophytes into two distinct monophyletic lineages: the lycophytes and monilophytes (ferns) [14,15,16].

In the new classification, therefore, the term "pteridophytes," designated as representing a paraphyletic group, is disappearing with disuse and that known before as "pteridophytes," according to Prado and Sylvestre [11], today correspond to the monophyletic lineages lycophytes and ferns. Despite the similarity in life cycle, these groups differ phylogenetically, and the ferns are more related to seed plants than are lycophytes.

Ferns have megaphylls and a differentiated vascularization, where the protoxylem appears confined to lobes of a string of xylem. The lateral roots are formed from the endodermis, protoxylem and mesarch (first cells of xylem in the middle, with radial maturation) in sprouts, and the anterozoids have 30 to 1000 flagella.

This group includes the classes Psilotopsida, Equisetopsida, Marattiopsida and Polypodiopsida, where 37 families and approximately 12,240 species are recognized [13,14,17].

Lycophytes are characterized by having microphyll-type leaves (with only one central vein), arrangement of helicoidal or opposing microphylls, microphylls ligulate or not, sporangia borne in the microphyll axil (adaxial side) or sporangia generally organized in the strobili on the apex of branches [14,18]. It is estimated that today the lycophytes represent less than 1% of all vascular plants, where they are composed of the families Lycopodiaceae, Selaginellaceae and Isoetaceae [19, 20], with about 1300 species described [17].

Lycophytes and ferns occupy a great diversity of environments in varied ecosystems, showing different biologicalal forms, including almost all forms of growth and adaptation of angiosperms [21] and are represented by terrestrial plants, epiphytes, hemiepiphytes, rupiculous or aquatic [11]. According to Moram [17], the number of species increases in both hemispheres in approaching the equator, reaching an approximate total of 13,600 species, distributed in four centers of greatest diversity: Southeast Asia with 4500 spp., South America with 3500 spp., Mesoamerica with 1800 spp. and the region of the Antilles with 1200 spp.

Humid and mountainous tropical environments with different forms of hybrid availabity, such as precipitation, condensation and clouds, are locations of greater prevalence of this plant group.

Occurrence in areas with dry periods or in semiarid climates is also recorded, but as expected, to a lesser degree. Information in the Catalog of Plants and Fungi of Brazil [11], point out that this country, for example, with an area of 8,514,880 km^2, probably has the largest flora in the world, i.e., 8.8 to 12.8%. Of this percentage, 9.2 to 13.1% of its floristic diversity is distributed among ferns and lycophytes, where the greatest occurrence is found in the Atlantic Forest Phytogeographic Domains, followed by Amazonia, Cerrado, Caatinga, Pampa and Pantanal.

For didactic purposes, we utilize in this chapter the term pteridophyte, since none of the plants mentioned are appropriately classified in lycophytes and monilophytes by their investigators.

3. Ethnobotanical Aspects: Socio-Economic Importance of Second Largest Group of Vascular Plants

With a large richness of species, it was expected that humans would benefit from pteridophytes with the aim of improving their quality of life. Over the years, these plants have become widely utilized in various ways by peoples, and the importance of this group includes a variety of economic aspects, although they are not so well recognized worldwide. For example, there is ethnobotanical use as ornamental plants, as raw material for craftwork, household utensils and cosmetics, in mystical rituals, in foods and in popular medicine.

The utilization of ferns as ornamental plants, live or dried, has for a long time caught people's interest, and has been considered one of the principal economic activities for this group of vascular plants. For example, we can mention the species *Rumohra adiantiformis* (G. Forst) *Ching,* the black or green ferns, which through their direct harvest from natural environments, has been a source of income for about 3000 farmers along the Atlantic coast of Rio Grande do Sul, Brazil (Ribas and Miguel, 2003) [22].

Other species are cited by Santos and Sylvestre (2006) [23] as ornamental plants, namely *Polypodium catharinae* Langsd. and Fisch., *Polypodium triseriale* Sw., *Adiantopsis radiata* (L.) Fee, *Adiantum raddianum* C. Presl., *Pityrogramma calomelanos* (L.) Link, *Anemia collina* Raddi, *Thelypteris dentata* (Forssk.) E.P. St. John, much utilized in landscaping or interior decorating or in the preparation of flower arrangements.

Dicksonia sellowiana Hook, tree fern, despite imminent threat of eradication as a nonrenewable natural resource and even being on the official list of plants threatened with extinction, they have been extensively extracted in the state of Parana – Brazil, for making cups, plates, sticks and powder [24].

In Nicaragua, the climbing fern *Lygodium venustum* SW., locally called crespillo, is harvested for making baskets, where it is of great importance for families who live off this economic activity. This species is also utilized in the preparation of the ayashuasca holy drink by the indigenous tribes Sharanahua and Culina in the Peruvian Amazon [25, 26].

The mystical use of some species is also reported. *Selaginella bryopteris* (L.) Bak. is smoked together with tobacco in mystical rituals, causing hallucinations, while *Trichomanes vittaria and Selaginella amazonica* are used in baths, for calming and to bring good luck. *Asplenium formosum* and *Microgramma vaccinifolia* are used in spiritual baths [27].

Pteridophytes are also appreciated in foods, where the great majority are eaten after cooking, to remove the sour taste. We can cite for example such species as *Maratta salicina* (Sm.), *Diplazium esculentum* (Retz.) Sw., *Maxima diplazium* (D. Don) C. Chr., *Dicranopteris linearis* (Burm. f.) Und., *Angiopteris evecta* (Hoffm.) and *Pteridium aquilinum* (L.) Kuhn var. *arachnoidum* (Kaulf.) Brade, where the last can cause tumors in the gastrointestinal tract in rats and because of this effect and its carcinogenic and mutagenic action demonstrated for other varieties of this species, it is recommended that its use in foods be stopped [23, 28, 29, 30].

However, the most common secondary use of pteridophytes is for medicinal plants. In the last years, various works have been conducted in innumerable localities, and listings containing the scientific and popular names, parts utilized and forms of use by the people, besides their purpose, are reported in the scientific literature. Currently, ethnobotanical surveys are very important, because they furnish valuable information on popular therapeutic products, besides guiding research on the pharmacologic activities of species mentioned as being medicinal, in the incessant search for clinically useful compounds.

4. Ethnomedicine of Pteridophytes: Popular Wisdom and Health

Over the years, experience with medicinal plants has been passed on from generation to generation. The continuous use of pteridophytes by peoples throughout the world has been given little attention in relation to the angiosperms. Species of various families are utilized in traditional phytotherapeutic treatments, in which leaves, spines, rhizomes, roots or spores are utilized, in the form of teas, decoctions, powder, oinment or syrup, among others, in combating various types of illnesses.

In the scientific literature, there are ethnopharmacologic studies carried out in different parts of the world, but especially India where it is one of the countries with the most publications on the medicinal use of pteridophytes. Various ethnopharmacologic surveys have been carried out recording a considerable number of plants. Researchers including Karthik [31], Kumari [32], Upreti [28], Srivastava [29] and Benjamim and Manickam [33] conducted ethnobotanical surveys on the popular phytotherapy of this country, which totaled 91 different species, where only the species *Angiopteris evecta* (Forst.) Hoff. of the family Marattiaceae was cited in common in all the works. Such studies, despite their great importance, are not able to express or represent the world's ethnopharmacology of pteridophytes.

Many species utilized as phytotherapeutic products in different regions of the world have not yet been cited, or because they are not part of the Indian biodiversity or simply because they are used for such purposes in the localities studied.

A small sample of the popular medicine of the Indian pteridoflora is demonstrated here, in a selection of seven species commonly cited in at least four works (Table 1). The available information can be considered an interesting guide for the search of bioactive natural products.

Table 1. Medicinal uses of some pteridophytes

Botanical name	Ethnomedicinal use
Adiantum capillus-veneris Linn.	The decoction of leaves is used for acute bronchitis and fever. The fronds are used for cough, colds, and halitosis and as eye drops. The plant is even used as a demulcent, expectorant, diuretic, emmenagogue, tonic, astringent, depurative, and emetic. The whole plant is used in a gel mixed with aloe vera and applied to cuts and wounds and as a capillary tonic. The tea or juice is used for cough, bronchitis and throat infections. Internally, it is also used to treat alcoholism and worms. Externally, it is used for snake bite and of bee sting. This plant is also reported to have anticancer, hypoglycemic, aphrodisiac, antimicrobial and antiviral activities.
Angiopteris evecta Hoffm.	The powdered rhizome is utilized mixed with water for diarrhea. It is used as a food and to treat fractures. It is even utilized in traditional medicine as a calming agent and for pain, fever and lice. The essential oil is utilized in a local perfumery. The decoction of leaves is ingested with lemon juice for the treatment of stomach ache and ulcers. The extract of leaves is utilized in the treatment of dysentery. The spores are efficacious in the treatment of leprosy and other skin diseases.
Blechnum orientale Linn.	The rhizome has anthelmintic action and is effective against *S. typhimurium*. The juice of the leaves is used to cure intestinal ulcers and the paste of the rhizome is utilized for treating bladder problems.
Dicranopteris linearis (Burm. f.) Und.	The sprouts are eaten and the fronds are utilized as covering on roofs and walls. The decoction is utilized as a laxative. Fronds are also utilized against asthma and show antibacterial and anthelmintic activities. In traditional medicine, it is still utilized for treating female sterility, when mixed with cow's milk.
Helminthostachys zeylanica (L.) Hook.,	This plant is utilized for intoxication processes and for sciatic nerve pain. The fronds are used as an aphrodisiac. The decoction of the rhizome is utilized against impotence. The young leaves are cooked as vegetables. The pulverized rhizome mixed with cow's milk is used as a stimulant and cerebral tonic and appetite stimulant. In traditional medicine, this plant is utilized as an anesthetic and in the treatment of blisters, ulcers and dysentery.
Lygodium flexuosum (L.) Sw.	The pulverized rhizome is used against herpes and diseases of the skin, rheumatism, lice, eczema, cuts and wounds and as an expectorant. Fresh roots mixed with mustard oil are utilized to treat rheumatism, carbuncles and ulcers. The aqueous extract of rhizome is utilized against gonorrhea. The plant is even used as an expectorant. The juice of the plant suppresses fever and ovulation.
Pteris vittata Linn.	The leaves are utilized for a prevention of diseases. The extract of the plant is utilized as a demulcent, hypotensive, tonic, antiviral and antibacterial. The whole plant is utilized as a paste and applied on areas affected by wounds. The paste is mixed with pepper and ingested to fight colds, cough and fever.

If crossing the data described in cited works and those in other worldwide ethnobotanical suveys, we would see that some species are in fact utilized for the same purposes in different cultures. The common use of some species by different populations has stirred the attention and curiosity of scientists with respect to the investigation of the efficacy of bioactivities so much referred to by popular tradition in pre-selection tests with regard to therapeutic usefulness, applied continuously over several generations.

Therefore, various questions arise, for example: What is so special in this group of plants that explains their use since primitive to contemporary cultures? Are there, in fact, bioactive components that act on body systems reequilibrating their functions? What part of the chemodiversity of these plants is responsible for the cure of illnesses? What quantity of the product is necessary for the desired effect? On what cell type does it act best, and in what form? What consequences could there be to human health with continuous use of pteridophytes? In fact, the search for answers to these questions has prompted worldwide studies on the bioactivities of this old group of vascular plants, and though without boasting, the mysteries of the ethnomedicine of pteridoflora are being increasingly unveiled in laboratories all over the world. Phytoconstituents are identified and compared regarding their biological properties to those in plants of other groups of known importance, and assays are carried out *in vitro* and *in vivo* disclosing therapeutic bioactivities whether guided by ethnobotanical studies, or by the intuition of scientists with respect to secondary metabolites found in plants.

5. Proving Bioactivities: Elucidation of Ethnomedicine Mysteries

Biological assays conducted for the purpose of proving the efficacy of pteridophytes as phytotherapeutic products have been more often conducted since the end of the XX century. Various bioactivities were then explored, where their eventual possibility of clinical-therapeutic application was confirmed or contested, indicating interesting alternatives for the creation of new drugs or classifying the plants as ineffective for therapeutic purposes.

As examples of these possibilities, we listed some studies done that proved some biological activities demonstrated by pteridophytes. It is important to point out that the great majority of studies concern antimicrobial activities (mainly antibacterial), where natural products in the form of extracts, fractions or isolated substances of pteridophytes are assayed against different types of microrganisms of pathologic importance. However, other activities could also be covered here, such as the antifungal, antiviral and antiparasitic potential of some species and anticarcinogenic and antioxidant activity as well could be useful in the identification of starting material for the synthesis of bioactive substances potentially viable as new drugs.

Antibacterial Activity

The development of bacterial resistance has led a growing demand for new drugs that can someway eliminate or neutralize microbial defenses. In the last decades, synthetic antibiotics had their efficacy reduced beause of the versatility of bacteria, and natural products represent today an alternative to be explored regarding their pharmacologic properties. Various representatives of the pteridoflora were evaluated for their antimicrobial potential and showed very promising results.

Osmunda regalis was evaluated for its antimicrobial potential because it is a pteridophyte known in traditional medicine systems. Thomas [34] used the disk diffusion method to test extracts of different polarities against pathogenic bacterial strains. Antibacterial activity was confirmed by the minimal inhibitory concentration (MIC) and minimal bactericidal concentration (MBC). The acetone extract exhibited better results with MIC and MBC values of 12.5 mg/ml and 25 mg/ml, respectively, against *Pseudomonas aeruginosa*, while MIC and MBC values of 25 mg/ml and 50 mg/ml were observed against *Shigella sonnei*. The active compounds of *O. regalis*, therefore, are soluble in acetone and show the capacity of hindering the growth and multiplication of some strains of multiresistant bacteria.

Parihar *et al.* [35] developed a very inclusive study involving 12 species of pteridophytes: *Adiantum capillus-veneris* L., *Adiantum incisum* Forsk., *Adiantum lunulatum* Burm. F., *Actiniopteris radiata* (Swartz.), Link, *Araiostegia pseudocystopteris* Copel., *Athyrium pectinatum* (Wall Mett ex.) T. Moore, *Chelienthes albomarginata* Clarke, *Cyclosorus dentatus* (Forsk.) Ching, *Dryopteris cochleata* (Don.) C. Chr., *Hypodematium crenatum* (Forsk.) Kuhn, *Marsilea minuta* L. and *Tectaria coadunata* (J. Smith) C. Chr. against Gram-negative, Gram-positive and phytopathogenic bacteria: *Escherichia coli* MTCC 443, *Salmonella arizonae* MTCC 660, *Salmonella typhi* MTCC 734, *Staphylococcus aureus* MTCC 96 and *Agrobacterium tumefaciens* MTCC 431. Utilizng the disk diffusion method, they tested the aqueous and alcoholic (methanol) extracts of leaves, obtaining results that showed, with few exceptions, that all the natural products were effective in inhibiting bacterial growth, thus serving as an excellent source for the search of active substances with potential for the development of new antibacterial drugs.

Some species of the genus *Adiantum* were tested against 11 bacteria by the microdilution method by Singh *et al.* [36]. The methanolic extracts of the species *Adiantum capillus-veneris*, *Adiantum peruvianum* and *Adiantum venustum* inhibited the growth of all the Gram-positive bacterial strains used with MIC values that varied between 3.90 and 62.50 μg/mL. Of the species utilized in the study, *Adiantum venustum* was the only one that demonstrated activity against all the Gram-negative strains. It should be noted that the inhibition of *E. coli* by *A. capillus-veneris* was seen at a concentration of 0.48 μg/mL. All species of *Adiantum* tested demonstrated notable antibacterial activity, attributed by the authors to the presence of phenolic compounds found in the species.

Another screening of antibacterial agents that deserves mention because of the large number of species investigate and the results obtained was performed by Banerjee and Sen in 1980 [4]. In this pioneering study, 114 species were investigated. Aqueous and organic (methanol, 70% ethanol, ether and aceton) extracts of the different parts of the plants (rhizome, stipe and sterile and fertile leaves) were obtained, and the assays were performed by the agar cup method against the bacterial strains *Staphylococcus aureus* (senstive and resistant to penicillin, *Sarcina lutea, Bacillus subtilis, Mycobacterium phlei, Escherichia coli, Pseudomonas aeruginosa, Salmonella typhi, Klebsiella pneumoniae* and *Vibrio cholerae*. A total of 73 species (64%) showed antibiotic activity, where 33 inhibited the growth of Gram-positive bacteria, 9 inhibited Gram-negative bacteria and 15 inhibited both types of bacteria. Some species stood out with respect to antimicrobial potential, such as *Microsorium alternifolium, Leptodecurrens chillus, Polypodium irioides, Pyrrosia mannii, Phymatodes ebenipes*, and those of the genera *Dryopteris* and *Adiantum*, among others.

With the large number of reports on the antibacterial activity of pteridophytes evidenced in these and in many other works, we can get a glimpse that this truly represents the great focus of studies for bioactivities. Active phytoconstituents in extracts and fractions have demonstrated a great capacity for disarming defense mechanisms developed by bacteria. This is very important, considering our expectations that they be utilized as sources of new drugs, because despite the various a types of antibiotic launched in the market, unfortunately, these have not kept up with development of microbial resistance, making them clinically ineffective.

Antifungal Activity

Studies on the antifungal potential of pteridophytes are more scarce when compared to those involving antibacterial effects. Thus, the majority of reports found on this are from studies in which antibacterial activity was also examined in the species n question and not only antifungal. However, it is important to clarify that these plants have also demonstrated the capacity to inhibit the growth of various types of fungi and for this reason they should be given greater attention in this aspect.

As examples of studies that search for antibacterial and antifungal activities concomitantly, we can cite those by Singh *et al.* [36] and Banerjee and Sen [4] mentioned earlier. In the first study, species of *Adiantum* were tested against eight fungal strains (*Candida albicans*, *Cryptococcus albidus*, *Trichophyton rubrum*, *Aspergillus niger*, *Aspergillus flavus*, *Aspergillus spinulosus*, *Aspergillus terreus* and *Aspergillus nidulans*), and the authors found that the methanolic extract of *A. capillus veneris*, *A. peruvianum*, *A. venustum* and *A. caudatum* were effective in inhibiting the growth of *C. albicans* and that the last three were also active against *Aspergillus terreus*. In a study by Banerjee and Sen of 114 plants, only three showed potential against the phytopathogenic fungi *Curvularia lunata, Aspergillus Niger* and *Helminthosporium oryzae*, namely the species *Dryopteris cochleata, Pteris biaurita* and *Gleichenia glauca* .

In general, the pteridophytes grow in humid locations, where occasionally there is water, due to their limitations for reproductive conditions. Usually, they occur together with fungi, and thus, both end up sharing the same habitat. Therefore, it is natural that fungi developed different defense mechanisms as a form of adaptation and survival against secondary metabolites produced and released by the plants, when these also sought better conditions for existence free of phytopathogenic organisms.

The compounds formed by secondary metabolic pathways of plants play an important role in the perpetuation of the species. Chemical prospecting of the species *Psilotum nudum, biserrata Nephrolepis* and *cordifolia Nephrolepis* done by Hani *et al.* [37], revealed the presence of flavonoids, tannins, alkaloids, reducing sugars, triterpenoids and steroids, and the antimicrobial activity demonstrated by these species can be attributed to these phytoconstituents, which have already demonstrated an inhibitory effect against microorganisms. In this study, antimicrobial assays were performed utilizng aqueous and non-aqueous extracts of aerial parts of the species by the disk diffusion method against nine types of bacteria and three important dermatophytic fungi (*Microsporum gypseum*,

Trichophyton mentagrophytes and *Trichophyton rubrum*). The aqueous extract of three species exhibited different degrees of inhibition of the fungi. *Psilotum nudum* was the only species whose majority of types of extract (aqueous, chloroform and ethanolic) was active against all the fungi tested. The results showed that these fern species had active substances with antifungal and antibacterial properties, and also inhibited the growth of Gram-positive and Gram-negative bacteria.

The genus *Adiantum* besides being one of best indications among the ferns against bacteria, also demonstrated antifungal potential. Ghosh *et al.* [38] studied the effect against *Aspergillus niger* and *Rhizopus stolonifer* by the species *Adiantum capillus-veneris* L. and *Adiantum lunulatum.* The crude extract and phenols (total phenols, carbohydrates and amino acids) extracted from both the gametophytes and different parts of the sporephytes were tested. The tests were carried out using three different methods: agar disk method, liquid culture method and suspension culture medium, to minimize the error, and both extracts exhibited antifungal activity, but the gametophytes showed better results.

This character can be attributed, according to the authors, to the large accumulation of metabolites, especially phenolic compounds in gametophytes, which as precursors of the zygote need to be more metabolically active to be resistant to pathogenic agents.

Other studies can be succinctly reported. Dalli *et al.* [39] found that the species *Pteris biaurita* had a strong antimicrobial activity, principally antifungal and antibacterial, supposedly due to the presence of eicosanoids and heptadecanes in its leaves, and Lee *et al.* [40] demonstrated that *Selaginella tamariscina* contained the biflavonoid isocryptomerin, greatly used in traditional medicine, and that this compound showed a strong antifungal action by causing depolarization of the plasma membrane. Ruiz-Bustos [41] demonstrated that the methanolic extract of the fern *Jatropha cuneata* at a concentration of 90□μg/mL exhibited a strong antifungal activity against *Fusarium verticillioides* and *Aspergillus niger.*

The natural evolution of fungi is of great interest to the scientific community. Their cells are similar to human cells, because they are all eukaryotes, and for this reason, there are difficulties in finding effective drugs attacks only specific targets. If new defense mechanisms develop (which is absolutely natural), commercially utilized drugs will become obsolete making it difficult to treat fungal infections. These studies with natural products, in this case the pteridophytes, in experimental screenings, point to a path for the discovery of new active principles extremely necessary in the current context for assuaging the lack of chemotheraputic substances. We see that these plants contain promising phytoconstituents from the gametophyte up to the sporophyte, and these can be used for safeguarding humans from opportunistic infections caused by fungi.

Anticancer Activity

Cancer has shaken many families throughout the world. The great suffering caused by this disease, as well as its extensive incidence in the world population, has led to a growing search for active principles with the capacity to inhibit the proliferation of tumor cells. Scientists have investigated various possibilities in relation to natural products. Plants, from bryophytes to angiosperms, have been tested with respect to antineoplastic potential in preliminary investigations. In view of the results indicating the existence of this bioactivity, special attention is given to proceeding with the isolation of compounds and *in vitro* and *in vivo* tests until finally determining their efficacy in humans. In this case, a great concern is that in the action of natural products on healthy and neoplastic cells only altered cells are damaged and chemotherapeutic collateral effects are null or minimized.

Some researchers have conducted antineoplastic studies on plant tumors caused by *Agrobacterium tumefaciens* based on the similarity between mechanisms of tumor formation in animals by the bacteria *Bartonella henselae* and *Helicobacter pylori*. It is a very unusual screening that may or may not have a promising effect in humans, but it is required before going further to new investigational phases. Sarker *et al.* [42] tested the aerial parts of *Selaginella ciliaris* (Retz.), *Marsilea minuta* (L.) and *Thelypteris prolifera* (Retz.) in a potato disk bioassay exhibiting crown-gall tumors. Before, however, they verified that the species itself had no antibacterial effect (at 250,000 ppm) against *A. Tumefaciens*. The plant extracts inhibited the tumor by 80, 82 and 76%, respectively, at 1000 ppm. Studies revealed that there are active phytoconstituents in the plants capable of interrupting the cell cycle; however, it is still a long way before knowing which metabolites are responsible for this antitumor action.

However, in more specific studies *in vitro* conducted by Lai *et al.* [43], the human cell lines colon adenocarcinoma HT-29, colon carcinoma HCT-116, breast adenocarcinoma MCF-7, leukemia K562 and Chang liver cells were used in cytotoxicity assays, in which the fractions of the methanolic extract of the fern *Blechnum orientale* Linn were tested. An interesting cytotoxic activity was demonstrated by the ethyl acetate, butanolic and aqueous fractions, against HT-29 cells. Curcumin was used as the positive control because it is widely used in clinical trials for chemoprevention of colon cancer, and Chang cells were utilized in cytotoxicity studies to test the effects of drugs/agents on normal cells. The butanolic fraction did not exhibit any cytotoxic effect against the normal cells, which means that the compounds present in this fraction could be utilized in chemotherapy of colon cancer. Three phenolic compounds have been found to be present in the extracts, namely terpenoids, flavonoids and tannins, where the last ones are in greater abundance in the butanolic and aqueous fractions. The isolation of the components of the butanolic fraction of this fern in complementary tests could lead to the elucidation of substances active against tumor cells, which is of extreme importance in the fight against colon cancer.

Compounds isolated have been investigated with respect to cancer chemopreventive activity. The dried roots of the ornamental fern *Neocheiropteris palmatopedata* (Baker) were used to isolate six kaempferol glycosides (palmatosides A, B and C, multiflorins A and B and afzelin), and these were assay for their capacity to inhibit TNF-α induced by NF-kB activity, nitric oxide (NO), production of aromatase, quinone reductase 2 (QR-2) and COX-1/-2 activity, important factors for the initiation of carcinogenic processes. Palmatosides A and B showed inhibition of TNF-α induced by NF-kB activity, with IC_{50} values of 15.7 and 24.1 μM, respectively, and only multiflorin B exhibited over 50% inhibition of nitric oxide production. Palmatoside A was the only compound that showed inhibition of the COX enzymes of more than 50% at 10 μg/mL. In the inhibition of aromatase, a multiflorin A was more active with an IC_{50} of 15.5 μM and the best performance in the inhibition of the enzyme QR2 was demonstrated by afzelin at a concentration of 11.5 μg/mL. In this study, Yang *et al.* [44] also evaluated cytotoxicity against hepatoma Hepa1c1c7 cells and MCF-7 breast cancer cells, and these compounds did not show a significant growth-inhibitory effect. The results suggested that the fern exerts a chemopreventive effect through its compounds and that this should be investigated in greater depth in view of it demonstrated ability of achieving specific targets without causing cytotoxic effects.

Besides kaempferol, other constituents of the flavonoid class have shown chemopreventive anticarcinogenic activity. The fern *Thelypteris torresiana* has been exhaustively studied because it has anticarcinogenic activity when tested in the form of an extract. Thus, researchers have isolated and tested its compounds in search of the phytoconstituent responsible for this effect. The main target of studies here have been the flavonoids, because they have shown chemopreventive potential and chemotherapeutic value in anticancer treatments with activities that vary from antiproliferative to cell cycle interruption to induction of apoptosis. The flavonoid protoapigenone was tested by Chang *et al.* [45] in a prostate cancer cell line (LNCaP) and showed inhibition of cell proliferation, induction of apoptosis by annexin V-FITC (labeling phosphatidylserine) and cell cycle blockade, as well as inhibition of the activation of p38 MAPK and JNK1/2, which regulate cell growth processes and cell proliferation and differentiation. Protoapigenone suppressed the growth of prostate cancer cells both *in vitro* and *in vivo*, in the latter without significant hepatotoxicity, nephrotoxicity and hematologic toxicity. This compound has been tested by Lin *et al.* [46], in hepatic and breast cancer cells (Hep G2, Hep 3B, MCF-7, A549, and MDA-MB-231), exhibiting considerable anti-tumor activity. In fact, this flavonoid isolated from fern could be a great alternative in cancer chemotherapy.

Antiviral Activity

The study of new substances with antimicrobial potential nowadays is growing and necessary, especially due to the development of resistance of infectious agents to the synthetic drugs widely utilized. Thus, natural compounds, principally flavonoids, have been very important for the discovery of antiviral agents.

According to the literature, the genus *Selaginella* shows numerous medicinal properties, where its species contain various secondary metabolites such as: alkaloids, terpenoids and phenols. Besides, they also have a large variety of biflavonoids, dimeric forms of flavonoids, still very little studied. The biflavonoidis show various medicinal properties, mainly antioxidant, anticancer, antiinflammatory and antimicrobial (antiviral, antibacterial, antifungal, antiprotozoan). The genus *Selaginella* is widely used in traditional Chinese medicine (TCM) as a complementary and alternative form to conventional synthetic drugs, where *S. tamariscina* is the most utilized species. The genus *Selaginella* possesses 13 bioactive compounds studied to date, especially, amentoflavone and ginkgetin [47]. The amentoflavone was obtained for the first time from the species *Selaginella sinensis* and studied by Ma *et al.* [48]. In 2009, Hafidh *et al.* [49] reportded that amentoflavone showed potent antiviral activity against respiratory syncytial virus (RSV), exhiting an IC_{50} of 5.5 μg/mL. In this study, the quantity of amentoflavone in nine species of *Selaginella* was determined by reversed-phase HPLC, and the species *S. sinensis* displayed the highest level, i.e., 1.13% amentoflavone.

In another study on the genus *Selaginella*, Maly [50] isolated five compounds of the species *Selaginella uncinata.* Two of these compounds are new chromone glycosides, named 5-hydroxy-2,6,8-trimethylchromone 7-O-beta-D-glucopyranoside (uncinoside A) and 5-acetoxyl-2,6,8-trimethylchromone 7-O-beta-D-glucopyranoside (uncinoside B). Their structures were elucidated by spectroscopic methods including three-dimensional techniques such as NMR. The other three compounds were identified as 8-methyl eugenitol, amentoflavone and hinokiflavone. The glycosides, uncinoside A and B, showed potent antiviral activity against respiratory syncytial virus (RSV) exhibiting IC_{50} values of 6.9 and 1.3 μg/mL and moderate antiviral activity against the parainfluenza type 3 virus (PIV 3), exhibiting IC_{50} of values of 13.8 and 20.8 μg/mL, respectively.

Some works have focused on the discovery natural products against HIV virus. This virus is the causative agent of acquired immunodeficiency syndrome (AIDS), and has caused serious public health problems, because each day the number of persons infected by HIV and AIDS patients continues to increase in the world population, especially in developing countries. Recent studies have observed the enzyme reverse transcriptase (RT) of HIV has shown to be very important for viral replication, because each catalytic function of RT is involved in the production of the virus. This enzyme exhibits RNA-dependent DNA polymerase activity, DNA-dependent DNA polymerase and ribonucleases H [51].

To date, two classes of inhibitory drugs, analogues and non-analogues of nucleotideos, have been developed; however, their utilization for the treatment of patients with AIDS is limited due to their toxicity and emergence of increasingly resistant viruses. The development and production of inhibitory drugs selective for HIV RT is pertinent to this objective. Min *et al.* [51] carried out an *in vitro* screening of 50 species of Korean and Chinese medicinal plants in search of antiviral potential.

The rhizome of *Dryopteris crassirhizoma* Nakai (Aspidiaceae) was found to strongly inhibit RNase H activity of HIV-1 RT, where the methanolic extract exhibited an IC_{50} of 25 μg/mL. According to Min *et al.*, this rhizome is known in Chinese medicine to have taenicidal action. The species of the genus *Dryopteris* are generally characterized by the presence of derivatives of phloroglucinol, such as flavaspidic acids, triflavaspidic acids, dryocrassins and albaspidin and filixic acid, and besides, glycosides and kaempferol were isolated from species of the genus *Dryopteris*.

Antiparasitic Activity

Leishmaniasis is an noncontagious infectious disease, caused by protozoans of the genus *Leishmania*, which shows three types of manifestations. It is estimated that there are 1.6 million new cases of the disease annually, of which about 500,000 are visceral and 1.1 million cutaneous or mucocutaneous, according to information from the World Health Organization [52]. The distribution of leishmaniasis has spread over the years, reaching countries where its occurence had not been previously recorded, which is very worrisome, considering the few options for therapeutic agents and also the fact that *Leishmania* has shown potential resistance to drugs. Scientific works involving the study of leishmanicidal activity of pteridophytes, however, are still very scarce. One of the few studies conducted in this area is that of El-On [53], who in 2009 reported that the methanolic extract of *Pteris vittata* had moderate leishmanicidal activity (25%-50%). According to the World Health Organization [52], it is estimated that more than 10 million people are infected with the parasite *Trypanosoma cruzi* in the whole world.

This protozoan causes Chagas disease, which in initial phase shows little or no symptomology, up to the chronic phase when there are systemic disturbances, culminating in the progressive compromise of the heart muscle, which can lead to death.

In the last year, the activity against protozoans has been evaluated for various natural products, where the genus *Selaginella* has been extensively studied. Currently, it is known that more than 60 species of *Selaginella* occur in India, where they are little used in popular medicine. According to the literature, this genus is rich in biflavonoids, a dimeric form of flavonoids. In a study by Olaf *et al.* [54], species of *Selaginella* were evaluated for antiprotozoan activity. In this work, an ethanolic extract and fractions of *Selaginella bryopteris,* of different polarities, were obtained using toluene, ethyl acetate and butanol by liquid-liquid partition. These natural products were tested against *Trypanosoma brucei, Trypanosoma rhodesiense* STIB 900, *Trypanosoma cruzi* strain tulahuen C2C4, *Leishmania donovani* strain MHOM-ET-67 and *Plasmodium falciparum* K1. The ethyl acetate fraction showed a high activity and was selected for the isolation of 11 pure compounds.

A high antimalarial activity was found for the compound 7'4'7-tri-O-methylamentoflavone, which exhibited an IC_{50} of 0.26 μM. This compound did not show significant cytotoxic activity (IC_{50} □ 150 μM) against L-6 cells. To assess in vivo activity against the protozoan *Plasmodium berghei*, a partial synthesis was carried out starting with amentoflavone (biflavonoid) for the production of 7'4'7-tri-O-methylamentoflavone.

This semi-synthetic was tested at a concentration of 50 mg/kg and did not show activity against *Plasmodium berghei.* A strong leishmanicidal activity was detected with compound 2,3 dihydronokiflavone, showing an IC_{50} of 1.6 μM, but against *Trypanosoma,* no significant activity was observed (IC_{50} □ 12.5 μg/mL for the extract).

Del Olmo [55], in a preliminary study in search of new compounds of natural origin, found that isonotholaenic acid, a stilbenoid, was the main component of a dichloromethane extract of the fern *Notholaena nivea.* This extract produced very promising results against the epimastigote forms of *T. cruzi*, exhibiting an IC_{50} of 30 μg/mL, similar to benznidazole with an IC_{50} of 7.4 μg/mL. Curiously, the transformation of isonotholaenic acid to the piperidide doubled its potency in relation to benznidazole. Based on other assays in this study, it was seen that these compounds were not cytotoxic to normal human cells. Therefore, these compounds represent important structures for the development of new agents against Chagas disease.

Antioxidant Activity

The equilibrium of living systems can be characterized by the constant production of free radicals, which in turn are neutralized by enzymatic and non-enzymatic antioxidant defense mechanisms. When for some reason this equilibrium is disrupted and the production of free radicals supercedes antioxidant capacity, highly unstable reactive species of oxygen, nitrogen or sulfur, among others, can bind to biomolecules essential for life, such as DNA, proteins and lipids, producing consequences such as mutations and other types of structural and functional damages, which can lead to cell death. Degenerative diseases such as Parkinson's, Alzheimer's and multiple sclerosis, besides various types of diseases of the lung and immune system, diabetes, cardiopathies, atherosclerosis and cancer, are illnesses that can develop due to damage caused by oxidative stress [56].

Antioxidant substances are those that are present in low concentrations but in higher concentrations compared to oxidative compounds, retard or inhibit the action of the latter, preventing or reducing the extent of oxidative damage [57]. It is known that natural products possess compounds that exhibit this capacity and for this reason, foods, seasonings and medicinal plants have been the targets of studies on their protective properties. Notwithstanding, various studies have shown the antioxidant potential of pteridophytes as an important bioactivity of this group.

The medicinal ferns *Blechnum orientale* L., *Dicranopteris linearis* (Burm.), *Cibotium barometz* (L.) J. Sam, *Acrostichum Aureum* L. and *Asplenium nidus* L. were investigated by Lai *et al.* [58], and all demonstrated antioxidant potential.

In methods of free radical reduction 2,2–diphenyl–1–picrylhydrazil (DPPH) and ferric reducing power (FRP), the species demonstrated activity following the order in decreasing potential: *B. orientale* ≥ *D. linearis* > *C. barometz* > *A. aureum* > *A. nidus*. This same order was also observed with respect to the presence of phenolic compounds, and antioxidant capacity was attributed to these, already demonstrated in many other studies. Using the ß-carotene bleaching (BCB) method, the evaluation of *D. linearis* showed 99% inhibition of lipid peroxidation. In contrast to the tendency observed in these preceding assays, ferrous ion chelating activity (FIC) showed that these species had low activity, with the exception of *A. Aureum,* which exhibited 58% (6.7 mg/mL).

The results provide evidence that these plants have a notable protective potential, which can help living systems in defense against free radicals. Recently, a study conducted by Lai and Lim [59] screened fifteen ferns in search of natural antioxidant sources utilizing methods including FRP, BCB and reduction of DPPH radical, and found five species whose methanolic extracts exhibited a strong capacity for sequestering free radicals. Of the species *Cyathea latebrosa* (Wall. ex Hook) Copel, *Cibotium barometez* (L.) J. Sm., *Drynaria quercifolia* (L.) J. Sm., *Blechnum orientale* L. and *Dicranopteris linearis* (Burm.), the last demonstrated the highest antioxidant potential, 61% at 0.1 mg/mL and 99% at 0.7 mg/mL. These data therefore confirm the marked potential of *Blechnum orientale* and *Dicranopteris linearis,* previously investigated.

In studies of phenolic compounds, these five species stood out because of their high levels, indicating as in an earlier study, that the antioxidant effect of the species is probably related to the presence and action of these phytoconstituents.

In the search for active substances capable of binding to free radical species, some works have gone a little further besides conducting assays with extracts, and thus compounds isolated from species such as *Cheilanthes anceps* Swartz and *Salvinia natans* L. were evaluated with respect to antioxidant potential. Chowdhar [60] isolated six flavonoids of *C. anceps* from the butanolic fraction of an ethanolic extract of leaves: (1) kaempferol-3-O- δ-L-rhamnopyranosyl (1→2)-ß- D-glucopyranoside-7-O-ß -D-glucopyranoside, (2) quercetin 3-O- δ-L-rhamnopyranosyl (1→2)-ß-D-glucopyranoside-7-O-ß-D-glucopyranoside, (3) quercetin-3-O-ß-D-glucosyl (1→2)-ß -D-galactoside-7-O- ß -D-glucoside, (4) quercetin-3-methyl ether-5-O-glucoside, (5) kaempferol-3-O-glucoside and (6) quercetin-3-O-glucoside. Using thin-layer bioautography and DPPH, antioxidant potential was found according to the following order: 6 > 3 > 2> 4, showing that quercetin glycosides were active, while glycosides of kaempferol did not exhibit activity. The antioxidant effect of *S. natans,* an aquatic fern, was observed by Srilaxmi [61] using an *in vivo* assay in Wistar albino rats with hepatic lesion induced by carbon tetrachloride (CCl_4).

This substance is an industrial solvent known for its hepatotoxicity, causing oxidative stress and cellular degeneration. Column chromatography was used to isolate the compound natansnin, a dibenzoyl glycoside, which was then assayed by the DPPH method, demonstrating a high antioxidant potential (60.6%) compared to the positive control BHT (63.6%). In *in vivo* tests, the rats were intoxicated with CCl_4 and then treated with natansnin (20 mg/kg body weight).

The reading of the parameters demonstrated that both doses diminished the damage caused by induced oxidative stress and inhibited the expression of inflammatory proteins, besides apoptosis.

The protective effect found can be related to the antioxidant activity of the flavonoid natansnin. Thus, the finding of this compound having been isolated from a fern reinforces the importance of the biological activities of this group as sources of chemoconstituents of clinical interest.

6. A Case Study: Assessment of Antioxidant Activity and Characterization of Phenolic Compounds of *Lygodium Venustum* Sw. and *Pityrogramma Calomelanos* (L.) Link

Introduction

Lygodium venustum (Figure 1) is a fern of the family Lygodiaceae that is normally seen growing in clearings inside forests or along roads in perturbed areas, where it can show a ruderal behavior, as it is observed in vacant lots in urban areas. The species is characterized by having pinnulae of different sizes, where the closest are lobed and larger than the others.

Source: Flaviana Morais.

Figure 1. Lygodium venustum.

Source: Flaviana Morais.

Figure 2. Pityrogramma calomelanos.

Because it has changeable structure, with adaptive form of "cipo", it is specialized in the lianescent habit, where it is considered, after reaching 50 cm in height, a free climber, winding and supporting itself generally on gravestones, dead wood or shrubs, capable of reaching a height of more than 5 m [61].

The utilization of *L. venustum* as a medicinal plant has been recorded in Mesoamerica by indigenous populations, among others, where it possesses antiseptic, fungicidal and trichomonacidal activities and is indicated for the treatment of dermatosis, mycosis and infection [62].

It is also used in the treatment of gastrointestinal and gyneco-obstetric disorders and as a postpartum antiinflammatory [63]. It is utilized as one of the components for the traditional preparation of the hallucinogenic drink ayahuasca among the Sharanahua and the natives of the upper Purus River in the Peruvian Amazon [26]. In Brazil, it is utilized by Afro-Brazilians in mystic cults for cleansing baths, besides being indicated in popular medicine for nervosismo and emotional instability [64, 65]. *Pityrogramma calomelanos* (Figure 2) belongs to the family Pteridaceae and displays a subcosmopolitan distribution, where it is very common in the tropics. Normally, it grows in sandy-clayey soil of ravines, as soilborn in very waterlogged soils, often close to the banks of creeks or to dams, exposed to sun and with few individuals. *P. calomelanos* is known by the popular names feto-branco, avenca-branca or avenca-preta. The species differs from others of family by showing equilateral pinnae and ascending pinnulae and the presence of white, yellow or rosy powdery covering on the abaxial surface [66, 67, 68]. In popular medicine, it is utilized as an ornamental and medicinal plant, where it is indicated as an astringent, analgesic, anti-hemorrhagic, depurative, emmenagogue, anti-cold, anti-hypertensive, anti-fever, anti-cough and blood circulation stimulant, besides being indicated for the treatment of renal problems [69, 70, 71].

The aim of this study was to evaluate the antioxidant potential of extracts and fractions of the pteridophytes *L. venustum* and *P. calomelanos* and to investigate the existence of phenolic compounds (phenolic acids and flavonoids), determining their levels.

Material and Methods

Material Plant

Lygodium venustum and *Pityrogramma calomelanos* were collected in the municipality of Crato, Ceara – Brazil, on the slope of Chapada do Araripe in a locality called Grangeiro (Figures 3 and 4).

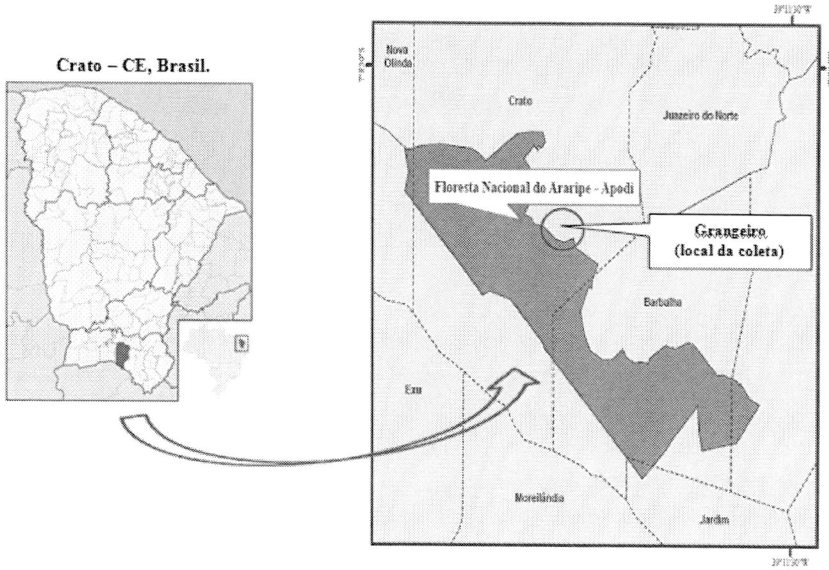

Source: http://commons.wikimedia.org/wiki/File:Ceara_Municip_Crato.svg , Access 15 november 2011; Centro de Processamento Remoto – Brasília – DF, Brasil.

Figure 3. Collection área of the assayed ferns *Lygodium venustum* and *Pityrogramma calomelanos.*

Source: Flaviana Morais.

Figure 4. Photo of the Collection área of the assayed ferns *Lygodium venustum* and *Pityrogramma calomelanos* (Araripe-Apodi Plateau, County of Crato, Ceará, Brasil.

Ethanolic extracts were prepared, from which hexane, dichloromethane, ethyl acetate and methanolic fractions were obtained for *L. venustum* and hexane, chloroform, ethyl acetate and methanolic fractions for *P. calomelanos.*

Qualitative Chemical Prospecting

A preliminary qualitative phytochemical analysis to detect the presence of classes of secondary metabolites such as tannins, flavonoids and alkaloids was carried out according to a phytochemical prospecting method previously described [72].

Quantification of Compounds Phenolic Por HPLC-DAD

The reverse-phase chromatographic analyes were performed with gradient elution using a C_{18} column (4.6 mm x 250 mm) packed with particle of 5μm diameter; the mobile phase was water containing 2% acetic acid (A) and methanol (B), and the composition of the gradiente was: 5% B for 2 minutes with change to 25%, 40%, 50%, 60%, 70% and 100% B, at 10, 20, 30, 40, 50 and 80 minutes, respectively, following the method described by Laghari [73] with minor modifications. The samples (extracts and fractions) were analyzed, by first dissolving them in ethanol at a concentration of 3 mg/mL. The presence of six phenolic compounds was investigated: gallic, chlorogenic and caffeic acids and the flavonoids quercetin, rutin and kaempferol. The identification of these compounds was performed by comparing their retention time and UV absorption spectrum with reference standards. The flow rate was 0.6 mL/min, injection volume 40 μL and wavelength 254 nm for gallic acid, 325 nm for caffeic and chlorogenic acids, and 365 nm for quercetin, rutin and lampferol. All the samples and mobile phase were filtered with 0.45-μm membrane (Millipore) and then degassed in ultrasound bath before use. Stock solutions of reference standards were prepared in HPLC mobile phase in a concentration range of 0.020 – 0.200 mg/mL for kaempferol, quercetin and rutin, and 0.050 – 0.250 mg/ml for gallic, caffeic and chlorogenic acids. The chromatogram peaks were confirmed by comparing their retention time with the reference standards and DAD spectra (200 to 400 nm). The calibration curves for the standards were: Gallic acid, $Y = 10523x + 1478.8$ ($r = 0.9999$); caffeic acid, $Y = 12765x + 1381.7$ ($r = 0.9995$); rutin, $Y = 12691 - 1165.0$ ($r = 0.9998$); quercetin, $Y = 13495x - 1092.6$ ($r =0.9999$); and kaempferol, $Y = 15692x - 1218.1$ ($r = 0.9997$). All chromatographic runs were at ambient temperature and in triplicate.

Assessment of Antioxidant Activity– DPPH

Antioxidant activity was determined utilizing the colorimetric method with DPPH, according to Choi *et al.* [74]. The crude extract was utilized and the fractions at concentrations of 1 to 500 μg/mL in ethanol (2.5 mL) were tested. A 2.5-mL aliquot of each sample was mixed with 1 mL of 0.3 nM DPPH. The reaction solutions were kept in the dark at ambient temperature for 30 minutes, and the absorbance was then read in a spectrophotometer (Shimadzu- UV-1201) at 518 nm, where the DPPH radical shows an absorption maximum. A solution of DPPH (1 mL; 0.3 nM) in ethanol (2.5 mL) was used as the negative control and a preparation of ascorbic acid as the standard (positve control), at concentrations varying from 1 to 100 μg/mL. Ethanol was used to zero the spectrophotometer, using as blanks the test solutions of each sample (without addition of DPPH), aiming to minimize the interference from the components of the samples in the reading. The assay was performed in triplicate and the calculation of antioxidant activity was according to the equation:

$$\%\text{inhibition} = 100 - \frac{\text{Abs}_{\text{sample}} - \text{Abs}_{\text{blank}} \times 100}{\text{Abs}_{\text{control}}}$$

where: $\text{Abs}_{\text{sample}}$ is the absorbance of fraction or crude extract; $\text{Abs}_{\text{blank}}$ is the absorbance of fraction or crude extract without addition of DPPH and $\text{Abs}_{\text{control}}$ is the absorbance of the DPPH solution in ethanol. The tests were performed in duplicate with three repetitions. The percentage of inhibition of the DPPH radical calculated and percentage of inhibition was plotted *versus* the concentration of the extract or fraction.

Statistical Analysis

All analyses were carried out in triplicate. The data are expressed as mean ± standard deviation (SD). Differences were evaluated by analysis of variance (ANOVA) followed by the minimum significant difference (MSD) test. Probability values less than 0.05 were considered statistically significant.

RESULTS AND DISCUSSION

The preliminary phytochemical evaluation (screening) of *L. venustum* indicated the presence of phenols, tannins, flavonoids and alkaloids. For Pityrogramma calomelanos, the screening of the extract revealed the presence of alkaloids, catequinas, chalcones, saponins, flavonoids and phenols. These compounds exert important functions in defense and conservation of species in the environment. High performance liquid chromatography (HPLC) was used to detect and quantify the phenolic constituents of the species. Extracts and fractions of ferns revealed the presence of gallic acid ^{A1} (RT =17.83 min; peak 1), chlorogenic acid ^{A2} (RT = 28.14 min; peak 2), caffeic acid ^{A3} (RT = 34.09 min; peak 3), rutin^{F} (RT = 42.11 min; peak 4), quercetin^{F} (RT = 49.78 min; peak 5) and kaempferol^{F} (RT = 58.96 min; peak of 6) (Figures 5 and 6), thereby demonstrating the the natural products analyzed contain flavonoids of the group of flavonols^{F} and phenolic acids (derivate of benzoic acid ^{A1}, derivate of phenylacrylic acid ^{A2} and derivate of cinnamic acid ^{A3}, whose percentages are displayed in Tables 2 and 3. Chlorogenic and caffeic acids, as well as the flavonoid quercetin, were found extracts as well as fractions of both plants. Gallic acid and rutin, however, were not found in the hexane fraction of *P. calomelanos* and rutin was absent in the dichloromethanol fraction of *L. venustum*. Kaempferol was also not in the methanolic fraction or the ethanolic extract of this species. Phenolic compounds and some of their derivates are considered effective agents in the prevention of oxidation, inhibiting the formation of or eliminating free radicals, which compromise living systems. They are very reactive chemically and the majority are found in the form of esters or of heterosides, making them soluble in water and polar organic solvents [75]. Based on our findings, the level of total phenolic compounds present in *P. calomelanos* was higher than that found in *L. venustum*. Quercetin, the flavonoid found in greatest quantity in *L. venustum* showed higher affinity with the solvent ethyl acetate, where the same occurred with the flavonoid kaemferol, the major compound in *P. calomelanos.* The ethyl acetate fraction of all the natural products tested showed the highest phenolic content, and the lowest levels were found in the hexane fraction both plants.

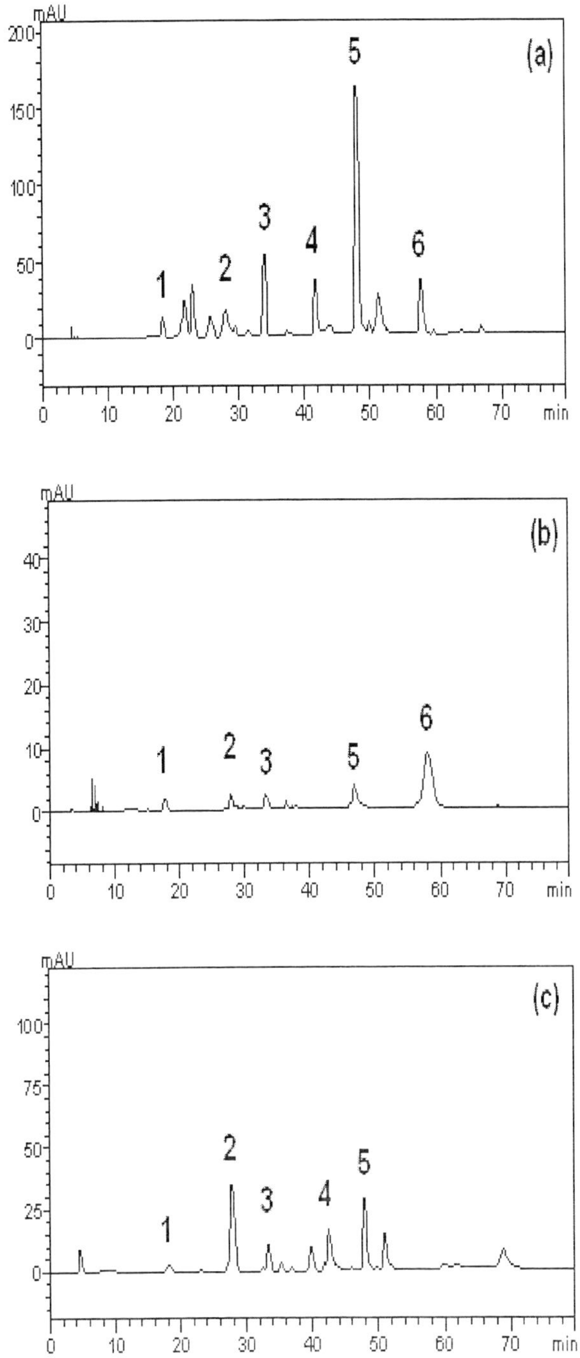

Figure 5. (Continued)

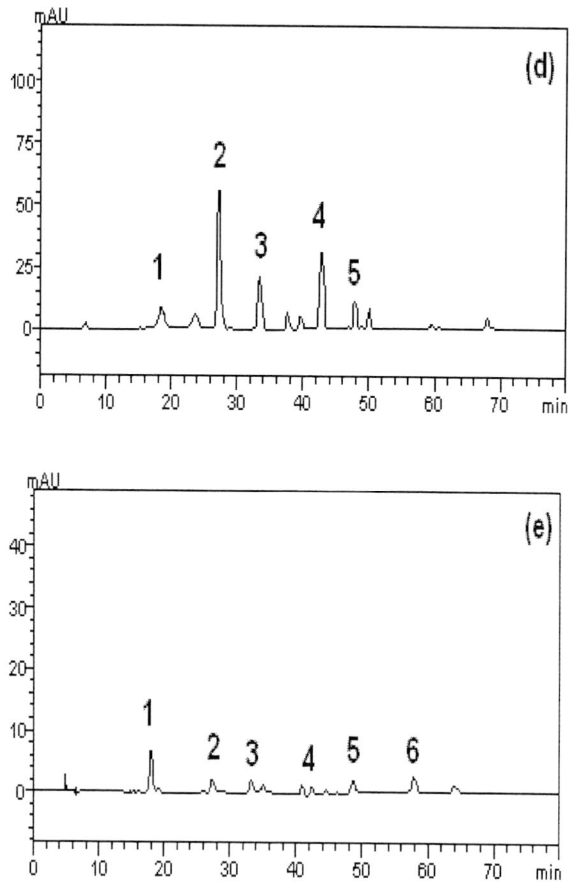

Figure 5. HPLC profile of the Ethyl Acetate Fraction of Lygodium venustum (AEFLV) (a); Dichoromethane fraction of L. venustum (DMFLV) (b); Ethanol Extract of L. venustum (EELV) (c); Methanol Fraction of L. venustum (MFLV) (d) and Hexane Fraction of L. venustum (HFLV) (e). The lecture was obtained using a wavelength of 327 nm. Gallic acid (peak 1), Chlorogenic acid (peak 2), Caffeic acid (peak 3), rutine (peak 4), quercetin (peak 5) and kampferol (peak 6).

This finding was in line with a study by Simões *et al.* [75], which showed that flavonoids are preferentially extracted by ethyl acetate. In screening for antioxidant activity and phenolic compounds carried out by por Lai and Lim [59], *P. calomelanos* was one of the fifteen species evaluated, and their data showed that this species contained a moderate quantity of phenolic compounds in the methanolic extract, but the authors were not able to specify the compounds and their percentages. It should be noted that these assays were performed utilizng methanolic extracts, while our study evaluated, besides the extract (ethanolic), fractions with different polarities.

This was the first screening for phenolic compounds in *L. venustum*. Jeetendra and Manish [76] analyzed different types of extracts of another species of the same genus, *L. flexuosum,* and found substantial quantities of compounds, which varied depending on the extraction solvent utilized, finding higher levels when using the solvent methanol. In *L. venustum* and *P. calomelanos,* the methanolic fraction was the second best extract fraction, behind ethyl acetate, also considered polar.

The capacity to inhibit the formation of free radicals can be measured by, among other methods, the decoloration of an ethanolic solution by DPPH. According to the method described by Choi, [74] A DPPH solution has an absorption band of 518 nm, at which the intensity of of violet color is measured. In the presence of good activity against free radical, there is a decoloration [77].

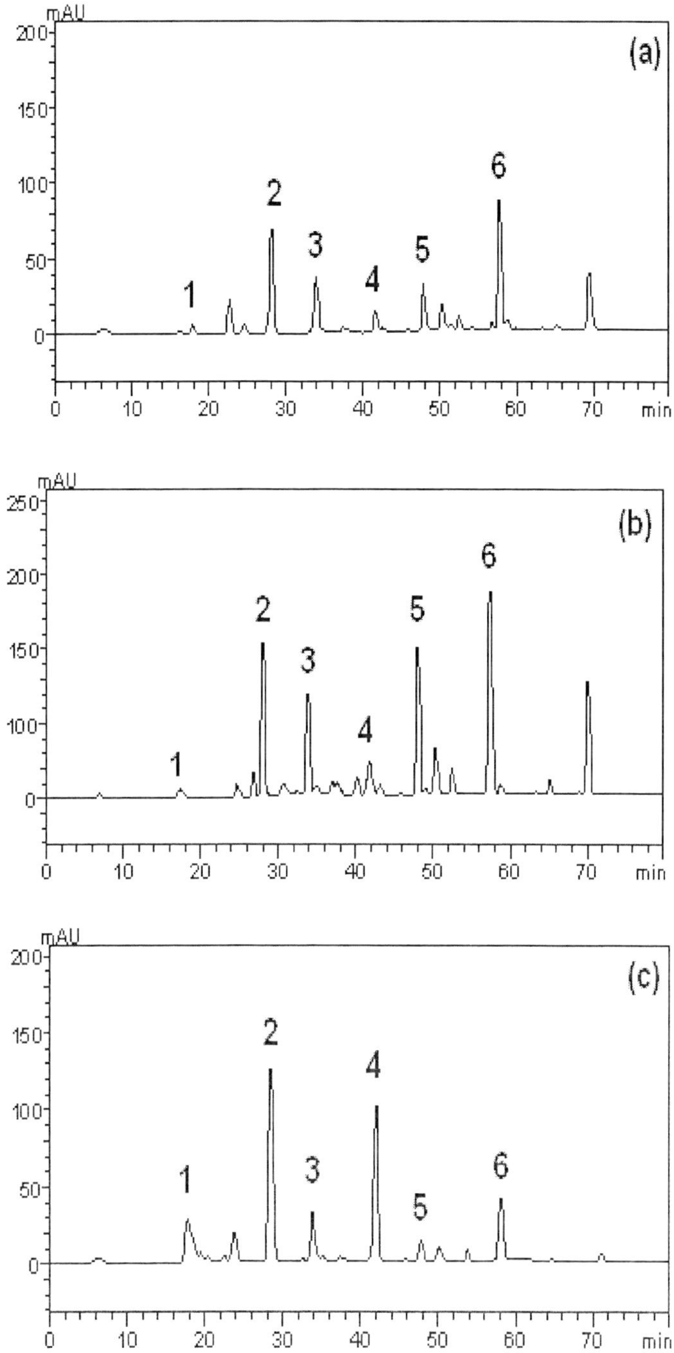

Figure 6. (Continued)

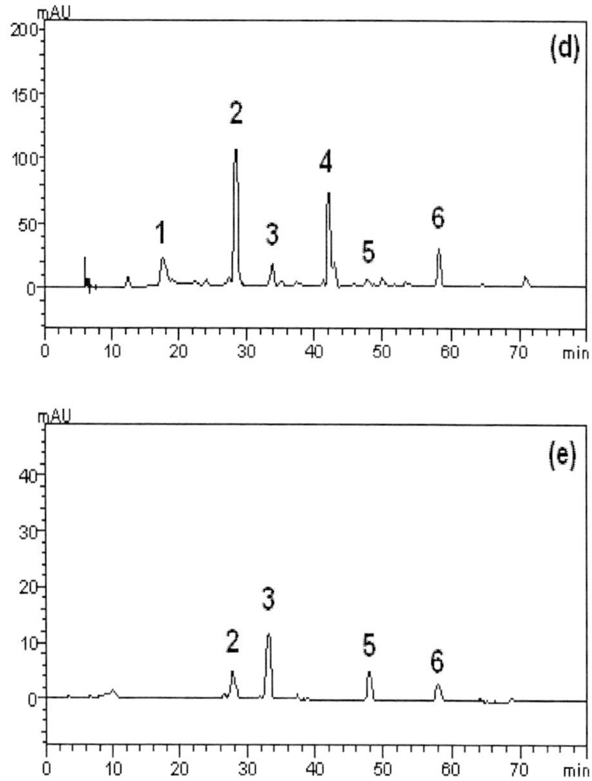

Figure 6. HPLC profile of the Ethyl Acetate Fraction of *Pityrogramma calomelanos* (AEFPC) (a); Chloroform fraction of *P. calomelanos* (CFPC) (b); Ethanol Extract of *P. calomelanos* (EEPC) (c); Methanol Fraction of *P. calomelanos* (MFPC) (d) and Hexane Fraction of *P. calomelanos* FHPC (e). The lecture was obtained using a wavelength of 327 nm. Gallic acid (peak 1), Chlorogenic acid (peak 2), Caffeic acid (peak 3), rutine (peak 4), quercetin (peak 5) and kampferol (peak 6).

Table 2. Phenols and flavonoids of Lygodium venustum SW

Natural Product	Gallic Acid	Chlorogenic acid	Caffeic acid	Rutine	Quercetin	Kampferol
FAELV mg/g	3.07 ± 0.01a	6.65 ± 0.03b	13.02 ± 0.04c	10.74 ± 0.02d	59.83 ± 0.01e	11.09 ± 0.01cd
FAELV %	0.30	0.66	1.30	1.07	5.98	1.10
FDCMLV mg/g	0.62 ± 0.03a	0.57 ± 0.01b	0.71 ± 0.01ab	-	0.80 ± 0.02b	3.93 ± 0.01c
FDCMLV %	0.06	0.05	0.07	-	0.08	0.39
EELV mg/g	1.76 ± 0.01a	6.12 ± 0.01b	2.95 ± 0.03c	3.28 ± 0.01c	6.05 ± 0.05b	-
EELV %	0.17	0.51	0.29	0.32	0.60	-
FMLV mg/g	2.02 ± 0.03a	11.39 ± 0.02b	3.91 ± 0.03c	6.16 ± 0.01c	1.88 ± 0.01a	-
FMLV %	0.20	1.13	0.39	0.61	0.18	-
FHLV mg/g	1.53 ± 0.02a	0.70 ± 0.03b	0.64 ± 0.07b	0.44 ± 0.01c	0.75 ± 0.01b	0.86 ± 0.03d
FHLV %	0.15	0.07	0.06	0.04	0.08	0.09

The results are expressed as mean ± standard deviation (SD) of three means, follow by different letters, according the Tukey test ($p < 0.005$).

Table 3. Phenols and flavonoids of Pityrogramma calomelanos (L.) Link

Natural Product	Gallic Acid	Chlorogenic acid	Caffeic acid	Rutine	Quercetin	Kampferol
FCPC mg/g	2.65 ± 0.01a	18.09 ± 0.04b	11.48 ± 0.09c	5.73 ± 0.01a	9.38 ± 0.05c	22.52 ± 0.01d
FCPC %	0.26	1.70	1.14	0.57	0.93	2.25
EAFPC mg/g	2.08 ± 0.03a	36.49 ± 0.11b	27.13 ± 0.02c	9.22 ± 0.01d	36.89 ± 0.03b	48.03 ± 0.03e
EAFPC %	0.20	3.64	2.71	0.92	3.68	4.80
EEPC mg/g	10.07 ± 0.03a	33.52 ± 0.01b	8.19 ± 0.04a	28.03 ± 0.01c	5.44 ± 0.02d	12.75 ± 0.09a
EEPC %	1.00	3.35	0.81	2.80	0.54	1.27
MFPC mg/g	9.37 ± 0.05a	30.15 ± 0.02b	6.84 ± 0.01c	20.63 ± 0.01d	1.19 ± 0.03e	9.96 ± 0.01a
MFPC %	0.93	3.01	0.68	2.06	0.11	0.99
HFPC mg/g	-	1.05 ± 0.01a	4.59 ± 0.03b	-	1.14 ± 0.02a	0.93 ± 0.01a
HFPC %	-	0.10	0.45	-	0.11	0.09

The results are expressed as mean ± standard deviation (SD) of three means, follow by different letters, according the Tukey test ($p < 0.005$).

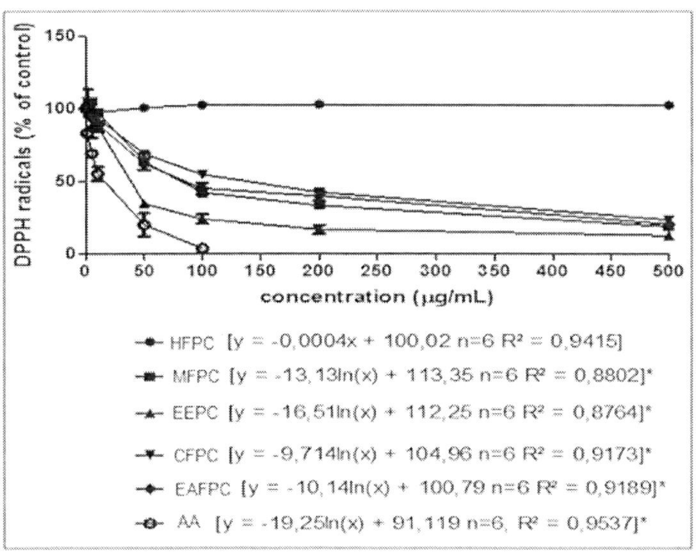

Figure 7. Antioxidant activity of Ascorbic acid (standard); Ethanol Extract of *Pityrogramma calomelanos* (EEPC); Ethyl acetate fraction of *Pityrogramma calomelanos* (EAFPC); Chloroform Fraction of *P. calomelanos* (CFPC); Methanol Fraction of *P. calomelanos* (MFPC); Hexane Fraction of *P. calomelanos* (HFPC).

The binding of ascorbic acid (known antioxidant substance) to the free radical DPPH is often used as a parameter for evaluating antioxidant potential of extracts and fractions of plants *in vitro* [78].

The IC_{50} values of the ethanolic extracts and fractions were obtained by the use of straight line equations: $y = -0.7104x + 98.01$, $n = 6$ ($R = 0.9953$) and $y = -16.51\ln(x) + 112.25$, $n = 6$ ($R^2 = 0.8764$), respectively (Figures 7 and 9).

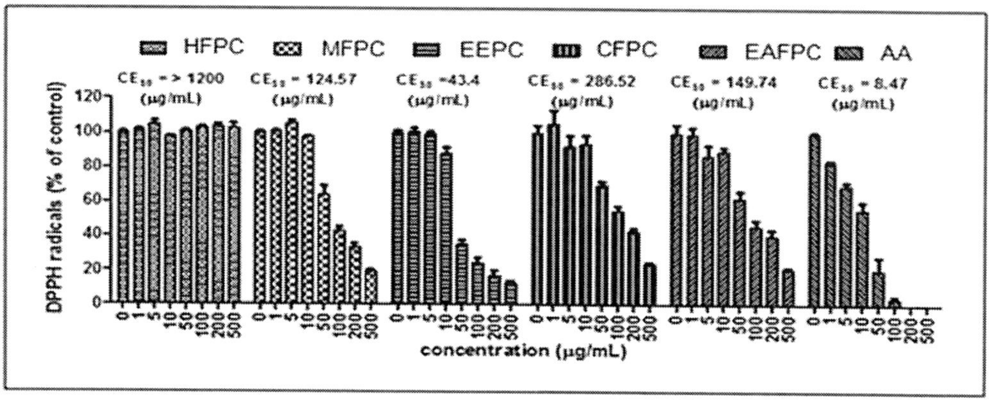

Figure 8. Free radical scavenging graphic (DPPH) of ascorbic acid (standard); ethanol extract of *Pityrogramma calomelanos* (EEPC); ethyl acetate fraction of *Pityrogramma calomelanos* (EAFPC); Choroform fraction of *P. calomelanos* (CFPC); Methanol fraction of *P. calomelanos* (MFPC); Hexane fraction of *P. calomelanos* (HFPC) and EC_{50} values.

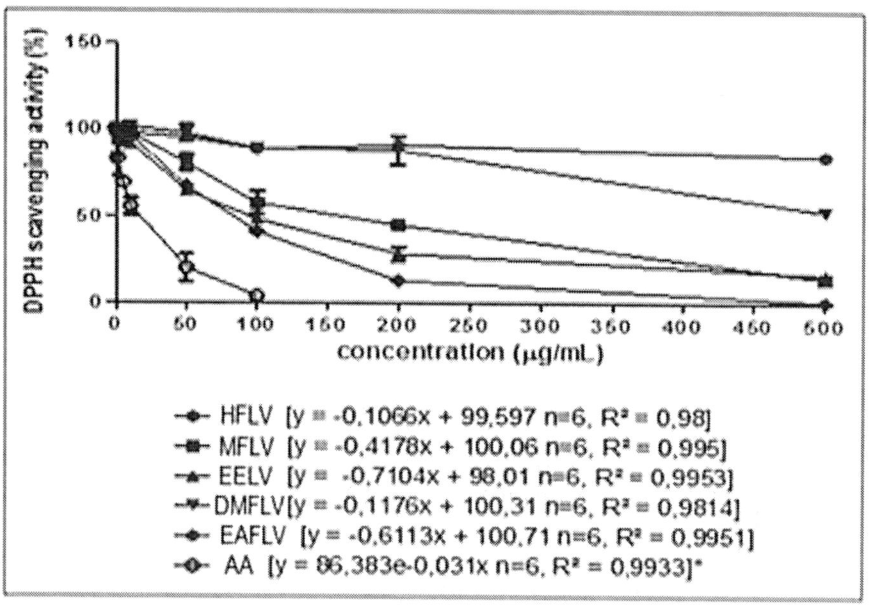

Figura 9. Antioxidant activity of Ascorbic acid (standard); Ethanol Extract of Lygodium venustum (EELV); Ethyl acetate fraction of Lygodium venustum (EAFLV); Dichloromethane fraction of L. venustum (DMFLV); Methanol Fraction of L. venustum (MFLV) ; Hexane Fraction of L. venustum (HFLV).

The first four points of the curve were considered, resulting in an IC_{50} value for both pteridophytes. For ascorbic acid, the IC_{50} was estimated mathematically based on the first concentration utilized (7.81 µg/mL), which demonstrated antioxidant activity close to or over 50%. For ascorbic acid, it was not possible to demonstrate linearity with the other concentrations tested. Ascorbic acid, whose antioxidant activity is well documented, showed an IC_{50} of 17.64 and 8.47 µg/mL.

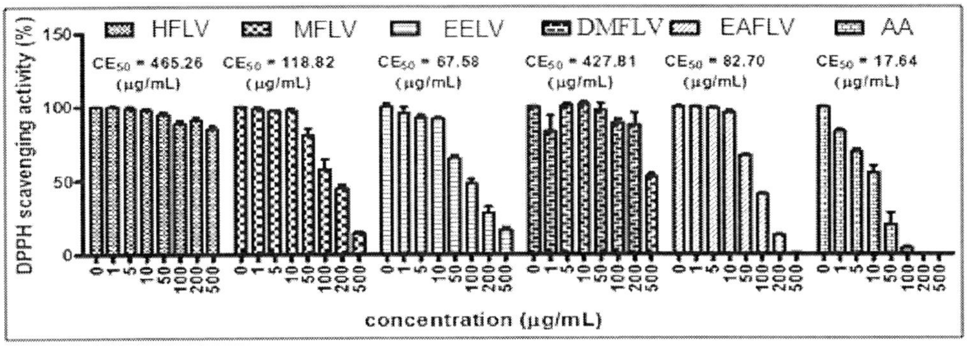

Figure 10. Free radical scavenging graphic (DPPH) of ascorbic acid (standard); ethanol extract of Lygodium venustum (EELV); ethyl acetate fraction of Lygodium venustum (EAFLV); Dichloromethane fraction of L. venustum (DMFLV); Methanol fraction of L. venustum (MFLV) ; Hexane fraction of L. venustum (FHLV) and EC_{50} values.

In our study, we found that among all the products analyzed, the ethanolic extracts of P. calomelanos and L. venustum exhibited the best results for antioxidant activity, showing respectively IC_{50} values of 43.4 µg/mL and 67.58 µg/mL (Figures 8 and 10). However, with respect to the fractions, the plants diverged in activity potential, since the that showed the best peformance in *P. calomelanos* was the methanolic extract, while that of ethyl acetate displayed the highest capacity of sequestering free radicals, among all fractions of *L. venustum*.

This behavior can be explained by the presence of compounds with polar properties in extracts, as is the case of phenolic acids and flavonoids, which possess provedn antioxidant activity. Various studies have shown the presence of quercetin and kaempferol in ferns, and some of them have demonstrated that these compounds, some already isolated and appropriately described, contribute effectively in capturing free radicals and help in the protection of molecules essential for life, as as the case in the study carried out by Chowdhary, previously described. The antioxidant potential of the phenolic acids caffeic and chlorogenic acids is already known [79], possessing good antioxidant activity in lipid systems and in inhibition of cellular peroxidation.

The results obtained from this study indicate that the species analyzed possess chemical compounds capable of capturing free radicals, and that these substances can be considered promising in the search for antioxidant drugs that prevent diseases as a consequence of oxidative. It should be noted that the DPPH test does not allow a precise definition of antioxidant effects because it is an *in vitro* method [80]. It is known that the activity of extracts and fractions of plants cannot be evaluated by only one method [74]; therefore, *in vivo* studies are needed to determine if these medicinal species can be utilized on an industrial scale.

Conclusion

Ethnobiological studies have demonstrated over the years that species of plants are the natural products most utilized by human populations. Without any doubt, the wide biodiversity of the world pteridoflora furnishes varieties of species rich in biologically active compounds, which in turn, are capable of reestablishing the homeostasis affected by microorganisms, viruses and oxidative stress, among other factors.

Living systems show relative fragility when faced with emerging and reemerging threats to health. Thus, new challenges emerge each time that we are exposed to them. The constant struggle for life, ours and that of other beings, is a fight where both sides launch various manoevers and skillful mechanisms of survival. In each battle, natural products with their defense strategies, are purposely used in our favor, and from these, "magical formulas," medicines, are created with the aim of dismantling the defenses of aggressors or to reequilibrate molecules and damaged systems, in a constant attempt to reestablish human well-being.

The study conducted for the preparation of this chapter allowed us to recognize how the pteridophytes are necessary to people, effectively contributing to the cure or treatment of illnesses in loctions where health systems are precarious or nonexistent. However, we also see that the potential of this group of plants is still far from being fully understood. Empirical knowledge, in fact, has been the principal contributor in the elucidation of biologicalal properties of pteridophytes; however, rational and scientific knowledge is without a doubt a decisive aspect for the safety of their use in therapies.

In summary, in view of the many biological activities exhibited by different and numerous species of pteridophytes, a greater quantity of studies are still necessary. A special and critical look should be given to this group in investigations more in-depth and also more inclusive, considering its abundance in biodiversity, worldwide distribution in propitious regions and, evidently, its use by populations since ancient times where they could only rely on the richness of crude extracts.

Acknowledgements

To all members of the Laboratório de Bioquímica Toxicológica da Universidade Federal de Santa Maria – RS, Brasil, with special mention to Rogério de Aquino Saraiva, Diones Caeran Bueno, Pablo Andrei Nogara and Aline Augusto Boligon by the realization of the assays.

References

[1] Pammel, LH. A Manual of Poisonous Plants - Chiefly of Eastern North America with Brief Notes on Economic and Medical Plants and Numerous illustrations. *The Torch Press Cedar Rapdis,* 1911, p. 323-325.

[2] Uddin, MG; Mirza, MM; Pasha, MK. The medicinal uses of pteridophytes of Bangladesh. *Journal Plant Taxonomy*, 1998, v. 5, n. 2, p. 29-41.

[3] Kimura, K; Noro, Y. Pharmacognostical studies on Chinese drug "Gu-sui-bu". I. consideration on "gu-sui-bu" in old herbals (Pharmacognostical studies on fern drugs XI). Syoy - akugaku Zasshi, 1965, v. 19, p. 25-31.

[4] Benerjee, RD; Sen, SP. Antibiotic activities of Pteridophytes. Ec. Botany, 1980, v. 34, n. 2, p. 284-298.

[5] Dixit, RD; Vohra, JN. A Dictionary of the Pteridophytes of India (Flora of India Series 4) Botanical Survey of India Publication, Department of Environment, Government of India, Botanical Garden, Howrah, 1984, p. 1-177.

[6] Kaushik, P. Ethnobotanical Importance of Ferns of Rajsthan: Indigenous Medicinal Plants. Today and Tommorrow Printers and Publication, New Delhi, 1998, p. 61-66.

[7] Nayar, BK. Medicinal ferns of India. Bulletin of the National Botanic Gardens, 1957, n. 58, p. 1-38.

[8] Hodge, WH. Fern food of Japan and the problem of toxicity. American Fern Journal, 1973, n. 63, p. 77-80.

[9] Dixit, RD. Ferns - a much neglected group of medicinal plants. III. *Journal Research India Medical*, 1975, v. 10, n. 2, p. 74-90.

[10] Ghosh, SR; Ghosh, B; Biswas, A and Ghosh, RK. The Pteridophytic Flora of Eastern India. Flora of India Series 4. Botanical Survey of India, 2004, v. 1, p. 1-591.

[11] Prado, J; Sylvestre, LS. As samambaias e licófitas do Brasil. In: Forzza, RC. *et al.* Catálogo de plantas e fungos do Brasil. Rio de Janeiro: Andrea Jakobsson Estúdio: Instituto de Pesquisas Jardim Botânico do Rio de Janeiro, v. 1, 2010; p. 69-74.

[12] Tryon, RM; Tryon, AF. Ferns and Allied Plants, with Special Reference to Tropical America. Berlin: Springer-Verlag; 1982.

[13] Pryer, KM; Schuettpelz, E.; Wolf, PG; Schneider, H; Smith, AR; Cranfill, R. Phylogeny and evolution of ferns (monilophytes) with a focus on the early leptosporangiate divergences. *American Journal of Botany,* 2004, v. 91, n. 10, p. 1582–1598.

[14] Smith, AR.; Pryer, KM.; Schuettpelz, E.; Korall, P.; Schneider, H.; Wolf, PG. A Classification for Extant Ferns. *Taxon*, 2006, v. 55, n. 3, p. 705-731.

[15] Pryer, KM; Schneider, H; Smith, AR; Cranfill, R; Wolf, PG; Hunt, JS; Sipes, SD. Horsetails and ferns are a monophyletic group and the closest living relatives to seed plants. *Nature*, 2001, v. 409, p. 618-622.

[16] Smith, AR; Pryer, KM; Schuettpelz, E; Korall, P; Schneider, H.; Wolf, PG. Fern Classification. In: T.A. Ranker; C.H. Haufler. Biology and Evolution of Ferns and Lycophytes. Cambridge: Cambridge University Press, 2008; p. 417-467.

[17] Moran, RC. Diversity, biogeography, and floristics. In: Ranker, TA; Haufler, CH. Biology and evolution of ferns and lycophytes. Cambridge: Cambridge University Press, 2008; p. 367-394.

[18] Pietrobom, MR; Maciel, S; Costa, JM; Souza, MGC; Trindade, MJ; Fonseca, MSS. Licófitas ocorrentes na Floresta Nacional de Caxiuanã, estado do Pará, Brasil: Lycopodiaceae e Selaginellacea. Boletim Museu Paraense Emílio Goeldi. *Ciências Naturais*, Belém, 2009, v. 4, n. 1, p. 37-45.

[19] Bateman, RM. An overview of Licophyte phylogeny. In: Camus, JM.; Gibby, M.; Johns, RN. Pteridology in perspective. Kew: *Royal Botanical Gardens*, 1996; p. 405-415.

[20] Kenrick, P; Crane, PR. The origin and early evolution of plants on Land. *Nature*, 1997, v. 389, p. 33-39.

[21] Holttum, RE. The ecology of tropical pteridophytes. In: Veerdorn, F. Manual of Pteridology. Amsterdan: The Hague Martinus Nijhoff, p. 420-450, 1938.

[22] Ribas, RP; Miguel, LA. Extração e comercialização de folhagens ornamentais da Mata Atlântica: o caso da verdes (*Rumohra adiantiformis*) no RS. RER, Rio de Janeiro, v. 42, n. 4, 2004, p. 575-596.

[23] Santos, MG; Sylvestre, LS. Aspectos florísticos e econômicos das pteridófitas de um afloramento rochoso do Estado do Rio de Janeiro, Brasil. *Acta botânica brasílica*, 2006, v. 20, v. 1, p. 115-124.

[24] Schmitt, JL; Schneider, PH; Windisch, PG. Crescimento do cáudice e fenologia de *Dicksonia sellowiana* Hook. (Dicksoniaceae) no sul do Brasil. *Acta botânica brasílica*, 2009, v. 23, n. 1, p. 282-291.

[25] FUNDECI - Gendered Access: Commodity Chain Analysis of Non-timber Forest Products from Laguna de Apoyo Nature Reserve, Nicaragua, Managua, Nicaragua: WIDTech/USAID; 2002.

[26] Rivier, L; Lidgren, JE. "Ayahuasca: the South American hallucinogenic drink. An ethnobotanical and chemical investigation", *Economic Botany*, 1972, v. 26, n.2, p. 101-129.

[27] Silva, VA; Andrade, LHC. Etnobotânica Xucuru: espécies místicas. *Biotemas*, 2002, v. 15, n. 1, p. 45-57.

[28] Upreti, K; Jalal, JS; Tewari, LM; Joshi, GC; Pangtey, YPS, Tewari, Geeta. Ethnomedicinal uses of Pteridophytes of Kumaun Himalaya, Uttarakhand, India. *Journal of American Science*, 2009, v. 5, n. 4, p. 167-170.

[29] Srivastava, K. Ethnobotanical Studies of Some Important Ferns. *Ethnobotanical L*eaflets, 2007, v. 11, p. 164-172.

[30] Fenwick, GR. Bracken (Pteridium aquilinum) - toxic effects and toxic constituents. Journal of the Science of Food and Agriculture, 1989, v. 46, n. 2, p. 147–173.

[31] Karthik, V; Raju, K; Ayyanar, M; Gowrishankar, K; Sekar, T. Ethnomedicinal uses of pteridophytes in Kolli Hills, Eastern Ghats of Tamil Nadu, India. *Journal of Natural Product Plant Resourse*, 2011, v. 1, n. 2, p. 50-55.

[32] Kumari, P; Otaghvari, AM; Govindapyari, H; Bahuguna, YM; Uniyal, PL. Some ethno-medicinally important pteridiphytes of India. *International Journal of Medicinal and Aromatic Plants* , 2011, v. 1, n. 1, p. 18-22.

[33] Benjamin, A; Manickam, VS. Medicinal pteridophytes from the Western Ghats. *Indian Journal of Traditional Knowledge*, 2007, v. 6, n. 4, p. 611-618.

[34] Thomas, T. Preliminary Antibacterial and Phytochemical Assessment of *Osmunda regalis* L. *International Journal of Pharmaceutical and Biological Archives,* 2011, v. 2, n. 1, p. 559-562.

[35] Parihar, P.; Parihar, L.; Bohra, A. *In vitro* antibacterial activity of fronds (leaves) of some important pteridophytes. *Journal of Microbiology and Antimicrobials*, 2010, v. 2, n. 2, p. 19-22.

[36] Singh, M; Singh, N; Khare, PB; Rawat, AKS. Antimicrobial activity of some important Adiantum species used traditionally in indigenous systems of medicine. *Journal of Ethnopharmacology*, 2008, v. 115, n. 2, p. 327-329.

[37] Rani, D; Khare, PB; Dantu, PK. *In vitro* antibacterial and antifungal properties of aqueous and non-aqueous frond extracts of *Psilotum nudum*, *Nephrolepis biserrata* and *Nephrolepis cordifolia.* Indian *Journal of pharmaceutical sciences*, 2010, v. 72, n. 6, p. 818-822.

[38] Ghosh, PG; Mukhopadhyay, R; Gupta, K. Antifungal activity of the crude extracts and extracted phenols from gametophytes and sporophytes of two species of *Adiantum.* Taiwania, 2005, v. 50, n. 4, p. 272-283.

[39] Dalli, AK; Saha, G; Chakraborty, U. Characterization of antimicrobial compounds from a common fern, *Pteris biaurita. Indian Journal of Experimental Biology*, 2007, v. 45, p. 285-290.

[40] Lee, J; Choi, Y; Woo, E-R, Lee, DG. Isocryptomerin, a novel membrane-active antifungal compound from *Selaginella tamariscina. Biochemical and Biophysical Research Communications*, 2009, v. 379, n. 3, p. 676–680.

[41] Ruiz-Bustos, E; Velazquez, C; Garibay-Escobar, A; García, Z; Plascencia-Jatomea, M; Cortez-Rocha, MO; Hernandez-Martínez, J; Robles-Zepeda, RE. Antibacterial and antifungal activities of some Mexican medicinal plants. *Journal of Medicinal Food*, 2009, v. 12, n. 6, p. 1398-1402.

[42] Sarker, AQ; Mondol, PC; Alam, J; Parvez, MS; Alam, F. Comparative study on antitumor activity of three pteridophytes ethanol extracts. Journal of Agricultural Technology, 2011, v. 7, n. 6, p. 1661-1671.

[43] Lai, HY; Lim, YY; Kim, KH. *Blechnum Orientale* Linn - a fern with potential as antioxidant, anticancer and antibacterial agent. *BMC Complementary and Alternative Medicine*, 2010, 10:15.

[44] Yang, JH; Kondratyuk, TP; Marler, LE; Qiu, X, Choi, Y; Cao, H; Yu, R; Sturdy, M; Pegan, S; Liu, Y; Wang, L; Mesecar, AD; Breemen, RBV; Pezzuto, JM, Fong, HHS; Chen, Y; Zhang, H. Isolation and evaluation of kaempferol glycosides from the fern *Neocheiropteris palmatopedata. Phytochemistry*, 2010, v. 71, n. 5-6, p. 641–647.

[45] Chang, HL; Wu, YC; Su, JH; Yeh, YT; Yuan, SSF. "Protoapigenone, a novel flavonoid, induces apoptosis in human prostate cancer cells through activation of p38 mitogen-activated protein kinase and c-Jun NH2-terminal kinase 1/2," *Journal of Pharmacology and Experimental Therapeutics*, 2008, v. 325, n. 3, p. 841–849.

[46] Lin, AS; Chang, FR; Wu, CC; Liaw, CC; Wu, YC. New Cytotoxic Flavonoids from Thelypteris torresiana. *Planta Med,* 2005, v. 71, p. 867-870.

[47] Ahmad, DS. Review: Natural products from genus *Selaginella* (Selaginellaceae). *Bioscience*, 2001, v. 3, n. 1, p. 44-58.

[48] Ma, SC; But, PP; Ooi, VE; He, YH; Lee, SF; Lin, RC. Antiviral amentoflavone from *Selaginela sinensis*, *Biological and Pharmaceutical Bulletin*, 2011, v. 24, p. 311.

[49] Hafidh, RR; Abdulamir, AS; Jahanshiri, F; Abas, F; Bakar, FA; Sekawi, Z. Asia is the Mine of Natural Antiviral Products for Public Health. *The Open Complementary Medicine Journal*, 2009, v. 1, p. 58-68.

[50] Ma, LY; Ma, SC; Wei, F; Lin, RC; But, PP; Lee, SH; Lee, SF. Uncinoside A and B, two new antiviral chromone glycosides from *Selaginella uncinata. Chemical and Pharmaceutical Bulletin*, 2003, v. 11, n. 51, p. 1264-7.

[51] Min, BS; Tomiyama, M; Ma, CM; Nakamura, N; Hattor, M. Kaempferol Acetylrhamnosides from the Rhizome of *Dryopteris crassirhizoma* and Their Inhibitory Effects on Three Different Activities of Human Immunodeficiency Virus-1 Reverse Transcriptase. *Chemical and Pharmaceutical Bulletin*, 2001, v. 5, n. 49, p. 546-550.

[52] World Health Organization. First WHO report on neglected tropical diseases. *Working to overcome the global impact of neglected tropical diseases*, 2010.

[53] El-On, J; Ozer, L; Gopas, J; Sneir, R; Enav, H; Luft, N; Davidov, G; Golan-Goldhirsh, A. Antileishmanial activity in Israeli plants. *Annals of Tropical Medicine and Parasitology*, 2009, v. 4, n. 103, p. 297-306.

[54] Olaf, K; Rumalla, CS; Marcel, K; Armin, P; Silke,B; Appa, R; Wolfgang, S. Antiplasmodial and leishmanicidal and activity of biflavonoids from India *Selaginella bryopteris*. *Phytochemistry Letters,* 2008, v. 1, p. 171-174.

[55] Esther, DO; Marlon, GA; Jose, LLP; Victoria, M; Eric, D; Arturo, SF. Leishmanicidal Activity of Some Stilbenoids and Related Heterocyclic Compounds. *Bioorganic and Medicinal Chemistry Letters*, 2001, v. 11, p. 2123–2126.

[56] Bianchi, MLP; Antunes, LMG. Radicais livres e os principais antioxidantes da dieta. Revista de Nutrição, Campinas, 1999, v. 12, n. 2, p. 123-130.

[57] Halliwell, B. Free radical and antioxidant a personal view. *Nutrition Reviews,* 1994, v. 52, n.8, p. 253-261.

[58] Lai, HY; Lim, YY; Tan, SP. Antioxidative, tyrosinase inhibiting and antibacterial activities of leaf extracts from medicinal ferns. *Bioscience, Bioteechnology and Biochemistry*, 2009, v. 73, n. 6, p. 1362-1366.

[59] Lai, HY; Lim, YY. Antioxidant Properties of Some Malaysian Ferns. 3rd International Conference on Chemical, Biological and Environmental Engineering IPCBEE. Singapore: IACSIT Press, 2011, v. 20.

[60] Chowdhary, S; Verma, DL; Pande, R; Kumar, H. Antioxidative properties of flavonoids from *Cheilanthes anceps* Swartz. *Journal of American Science*, 2010, v. 6, n.5, p. 203-207.

[61] Mehltreter K. Leaf phenology of the climbing fern *Lygodium venustum* in a Semideciduous Lowland Forest on the Gulf of Mexico, *American Fern Journal*, v. 96, n. 1, p. 21–30, 2006.

[62] Duke, JA. Duke's Handbook of Medicinal plants of Latin America. New York: CRC Press Taylor and Francis group, 2008; 832 p.

[63] Argueta, A; Cano, L; Rodarte, M. Atlas de las plantas de la medicina tradicional mexicana, vol. I–III. Instituto Nacional Indigenista, Mexico City, 1994.

[64] Rodrigues, E; Tabach, R; Galduróz, JCF; Negri, G. Plants with possible anxiolytic and/or hypnotic effects indicated by three Brazilian cultures — Indians, Afro-Brazilians, and river-dwellers. *Studies in Natural Products Chemistry*, 2008, v. 35 (C), p. 549-595.

[65] Albuquerque, U; Barros, ICL; Chiapetta, AA. Pteridófitas utilizadas nos cultos afro-brasileiros em Recife – PE: um estudo etnobotânico, *Biológica Brasílica*, 1997, v. 7, p. 23-30.

[66] Moran, RC. *Pityrogramma* Link. In: G. Davidse *et al.* Flora Mesoamericana. México: Universidad Nacional Autónoma de México, 1995, p. 137-140.

[67] Prado, J. Flora da Reserva Ducke, Amazonia, Brasil: Pteridophyta - Blechnaceae, *Rodriguesia*, 2005, v. 56, n. 86, p. 33-34.

[68] Pietrobom, MR; Barros, ICL. Pteridófitas de um remanescente de Floresta Atlântica em São Vicente Férrer, Pernambuco, Brasil: Pteridaceae. *Acta botânica brasílica*, 2002, v. 16, n. 4, p. 457-479.

[69] 69 May, LW. The economic uses and associated folklore of ferns and fern allies. *The Botanical Review*, 1978, v. 4, p. 491-528.

[70] Corrêa, MP. Dicionário das plantas úteis do Brasil e das exóticas cultivadas. Rio de Janeiro, Instituto Brasileiro de Desenvolvimento Florestal, 1984.

[71] Barros, ICL; Andrade, LHC. Pteridófitas medicinais (samambaias, avencas e plantas afins). Recife: Ed. Universitária da Universidade Federal de Pernambuco, 1997; 213 p.

[72] Matos, FJA. Introdução à fitoquímica experimental. Fortaleza: UFC Edições, 1997; 141 p.

[73] Laghari, AH; Memon, S; Nelofar, A; Khan, K.M; Yasmin, A. Determination of free phenolic acids and antioxidant activity of methanolic extracts obtained from fruits and leaves of *Chenopodium album*. *Food Chemistry*, 2011, v. 126, p. 1850–1855.

[74] Choi, CW; Kim, SC; Hwang, SS; Choi, BK; Ahn, HJ; Lee, MY; Park, SH; Kim, SK. Antioxidant activity and free radical scavenging capacity between Korean medicinal plants and flavonoids by assay-guided comparison. *Plant Science*, 2002, v. 163, p. 1161-1168.

[75] Simões, CMO; Schenkel, EP; Gosmann, G; Mello, JCP; Mentz, LA; Petrovick, PR. Farmacognosia: da planta ao medicamento. 6ªed. Porto Alegre/Florianópolis: Editora da UFRGS/Editora da UFSC, 2010; 1102 p.

[76] Jeetendra, N; Manish, B. Correlation of antioxidant activity with phenolic content and isolation of antioxidant compound from *Lygodium flexuosum* (L.) SW. Extracts. *International Journal of Pharmacy and Pharmaceutical Sciences*, 2011, v. 3, n. 2, p. 48-52.

[77] Kulisic, T; Radonic, A; Katanilic, V; Milos, M. Use of different methods for testing antioxidative activity of oregano essential oil. *Food Chemistry*, 2004, v. 85, p. 633-640.

[78] Rice-Evans, CA; Miller, NJ; Paganga, G. Structure – antioxidant activity relationships of flavonoids and phenolic acids. *Free Radical Biology and Medicine*, 1996, v. 20, n. 7, p. 933-956.

[79] Soares, SE. Ácidos fenólicos como antioxidantes. *Revista de Nutrição, Campinas,* 2002, v. 15, n. 1, p, 71-81.

[80] Ursini, F; Maiorino, M; Marazzoni, P; Roveri, A; Pifferi, G. A novel antioxidant flavonoid (idB 1031) affecting molecular mechanisms of cellular activation. *Free Radical Biology and Medicine*, 1994, v. 16, n. 5, p. 547-553.

In: Natural Products: Structure, Bioactivity and Applications ISBN 978-1-62081-728-5
Editors: Ramiro E. Goncalves and Marcos Cunha Pinto ©2012 Nova Science Publishers, Inc.

Chapter 2

IN-VITRO ASSESSMENT OF CHROMONES, ALKALOIDS AND OTHER NATURAL PRODUCTS FROM CARIBBEAN PLANTS AS POTENTIAL ANTI-TUBERCULARS AND CHEMOPREVENTORS

Sheena Francis[1], Damion Morris[1], Mario Shields[1], Helen Jacobs[2] and Rupika Delgoda[1,*]

[1]Natural Products Institute, University of the West Indies, Mona , Kingston, Jamaica
[2]Department of Chemistry, University of the West Indies, Mona, Kingston, Jamaica

ABSTRACT

Resurgence in the tuberculosis pathogen, *Mycobacterium tuberculosis*, which significantly reduces survival rates of persons co-infected with the immunodeficiency disease HIV, poses a major health issue particularly for the developing countries, which accounts for over 95 percent of reported global HIV cases. Compounding this issue is the emergence of multiple drug resistant strains, highlighting an urgent need for the search for novel and effective therapies against *M. tuberculosis*. Arylamine *N*- acetyltransferase (NAT), a drug metabolizing enzyme found expressed in *M. tuberculosis* and responsible for the metabolism of the frontline anti-TB drug isoniazid has also been identified to play a significant role in its cell wall lipid synthesis. Knowledge on NAT enzyme's essential role in the survival of *M. tuberculosis*, garnered through recent *nat* gene knockout experiments in *M. bovis*, has identified it as a useful target in the search for anti-tubercular therapy. In this study we employed this molecular target in an *in-vitro* assay, to search for natural products with potential for future *in-vivo* examinations. Eleven novel/known compounds including tetranoterpinoids and quassinoids (at100 μM) previously extracted and purified from five rare endemic and/or indigenous Caribbean plants, *Spathelia sorbifolia*, *Esenbeckia pentaphylla*, *Peperomia amplexicaulis*, *Hortia regia* and *Clusia havestiodes* were examined for their inhibitory properties using heterologously expressed NAT from *M. Smegmatis*, a homologue of *M. tuberculosis* NAT. Greatest inhibition was obtained for anhydrosorbifolin (65%), a chromone whose

[*] Corresponding author: email: thejani.delgoda@uwimona.edu.jm.

extended linear side chain appear to contribute to the increased inhibition compared with its derivatives, alloptaeroxylin which has no side chains (displayed 25% inhibition) and spatheliabischromone which has a cyclicized side chain (displayed 42% inhibition). These compounds were also examined for their inhibition of human cytochrome P450 (CYP) 1, a class of enzymes comprising of CYPs1A1, 1A2 and 1B1, important in the activation of polyaromatic hydrocarbons to their carcinogenic precursors. Results revealed greatest inhibition by the alkaloid dictamnine of CYP1B1 activity with a potent IC_{50} value of 0.27μM. These studies show natural chromones and alkaloids as potential anti-tuberculars and chemopreventors worth further analysis.

INTRODUCTION

Status of Tuberculosis

Tuberculosis (TB), primarily a respiratory disease caused by the pathogen *Mycobacterium tuberculosis* has being listed as a global emergency since 1993 by the World Health Organization (WHO) (Chaisson and Martinson 2008; Adeniyi *et al.,* 2004). The control and management of the disease has faced multiple challenges, including that of funding, awareness and more recently the emergence of multidrug resistant strains of mycobacteria (Boccia and Evans, 2011). Lengthy and expensive therapy (Blumberg *et al.,* 2010) and drug toxicities associated with drug-drug interactions, particularly by patients co-infected with HIV/AIDS, (Tripathi *et al.,* 2005) have led to patient non compliance and suboptimal therapeutic regimes. Such treatment practices coupled with the ease of migration of individuals (Blumberg *et al.,* 2010) are thought as probable causes in the evolution and spread of new strains showing both moderate (MDR) to extreme (XDR) drug resistance (Van den Boogart *et al.,* 2009; WHO 2010; Pablos-Méndez *et al.,* 1998; Cole and Telenti 1995).

Though methods of early detection are available (Van den Wijngaert *et al.,* 2004; Kamerbeek *et al.,* 1997), and the use of decades old treatments, such as short course chemotherapy, Bacille Calmette-Gue´rin (BCG) vaccine, Isoniazid (INH) and its hydrazine derivatives, used on its own or in combination (Furin and Johnson 2005) remain the frontline defense against this deadly disease. In a mammoth effort to combat the spread of TB and evolution of *M.tuberculosis*, WHO (2007) launched directly observed treatment short-course (DOTS), a programme that holds health-care workers accountable for the administering and treatment of TB in their region. This labour intensive programme, met with some reluctance by health care professionals (Arata 1991) due to the social stigma associated with the disease and relies on second line drugs, which are not as effective as the first line drugs, further prolonging the already lengthy therapy (CCDC 2005; Mitchison 2005; Gillespie *et al.,* 2001), consequently increasing the overall treatment costs. Regardless of DOTS moderate success, underlying issues such as the length and expense of therapy remains unaddressed, especially in poverty stricken areas where resources and man power are limited (Mitnick *et al.,* 2003). These have contributed to the emmergence of drug resistant strains of mycobacteria posing the greatest challenge to scientists at present.

Mycobacteria have a complex and hydrophobic cell envelope (Belisle *et al.,* 1997; Kolattukudy *et al.,* 1997) which withstands treatment by most antibiotics (Brennan and Draper 1994). The bacteria is slow growing and has both a dormant and a virulent stage

(Rivera-Marrero *et al.,* 1998) enabling it to thrive in its host under adverse conditions (Balganesh *et al.,* 2004; Barry *et al.,* 1998). Investigations on its genome (Brosch *et al.,* 1998; Cole et al., 1998; Cole 1999; Philipp *et al.,* 1996), cell wall structure (Tripathi *et al.,* 2005; Brennan 2003; Crick *et al.,* 2010) and cellular functions including that of xenobiotic metabolizing enzymes (Glenn *et al.,* 2010; Upton *et al.,* 2001a), all aimed at elucidating its pathogenicity, have also identified new targets for drug development.

Arylamine N-acetyltransferase in *M. tuberculosis*

Arylamine *N*-acetyltransferase (NAT), a polymorphic phase II enzyme, initially discovered in humans due to varying metabolism of the tuberculosis drug isoniazid (Peters *et al.,* 1965), has also unsurprisingly been found in *M .tuberculosis* and other prokaryotes (Sim *et al.*, 2000). Like all NATs, NAT from *M.tuberculosis* (TBNAT) catalyses the transfer of an acetyl group from acetyl-CoA to the terminal nitrogen of arylamines and hydrazines, including INH (Upton *et al.,* 2001b). The active site catalytic triad first identified in NAT from *Salmonella typhimurium* (Sinclair *et al.*, 2000) and found common across multiple organisms (Sim *et al.,* 2000) is found in mycobacterial NATs. The discovery of NAT in *M .tuberculosis* and its capability to metabolise INH, led to the assumption that it may be involved in drug resistance. Although initial overexpression work were implicating such suspicions (Payton *et al.,* 1999), subsequent knock out studies of the *nat* gene in *M. tuberculosis* (Payton et al., 2001) and *M. bovis* (Bhakta *et al.,* 2004) identified NAT as a critical agent in cell wall integrity. The cell wall is a complex multi-layered structure (Crick *et al.,* 2010), which is the initial barrier for drugs aimed at eradicating the bacteria, and the tough cell wall is critical to the success of *M. tuberculosis* with approximately 250 genes coding for cell wall lipid biosynthesis (Cole *et al.,* 1998). NAT plays an essential role in regulating the production of mycolic acid (Bhakta *et al.,* 2004), a key component in the architecture of the cell wall (Crick *et al.,* 2010; Brennan 2003). NAT's role in mycobacterial cell wall biosynthesis and survival of the organisms within macrophage has identified it is as a useful target for drug development with small molecule inhibitors of NAT being potential drug leads (Westwood *et al.,* 2010; Fullam *et al.,* 2008; Sim *et al.,* 2008; 2003). Useful high throughput assays has now been developed to assist in rapid screening of small molecule inhibitors (Westwood *et al.*, 2011) and in the current paper we have utilized such methods to test several natural small molecules as potential NAT inhibitors.

Although NATs bioactive role is clearly important, its precise endogenous function remains unknown for most organisms. Recent evidence has shown that the expression of human NAT1 is up-regulated in estrogen receptor positive breast cancer (Adam *et al.* ,2003). Identification of human NAT1 specific small molecules (Russell *et al.,* 2009), including colourimetric indicators (Laurieri *et al.,* 2010) has been invaluable as useful molecular probes for studying the function of NATs in diseased and normal cells. The search for small molecules as inhibitors of important molecular targets therefore continue to play an important role in the quest to garner a total understanding of the role these targets play in human and pathogenic physiology.

Cytochrome P450s as a Drug Target

Recent geneome sequence, genetic and protein characterisation studies of *M. tuberculosis* have helped identifiy multiple cytochrome P450 (CYP) enzymes encoded in the genome, which have demonstated their potential as antitubercular targets as well as revealing novel aspects of CYP form and function. CYP enzymes are an ubiquitous superfamily of enzymes found expressed in a range of eucaryotes and procaryotes and whose functions include xenobiotic and endogenous substrate metabolism. The pathogenic *M. tuberculosis* encodes 20 CYP enzymes, and CYPs 128, 121 and 125 have been demonstrated for essentiality in viability, virulence or persistence in the host (McKlean *et al.*, 2010). Various azole-class drugs bind with high affinity to the CYP from *M. tuberculosis* heme and are mycobacterial antibiotics.

In humans, CYP enzymes play a major role in drug metabolism with over 90% of drugs in the market being subject to metabolism by at least one CYP enzyme. It is mandotory that new drug entities are characterised for their CYP interaction profile, as governed by the regulatory authority, the U.S. Food and Drug Administration (FDA). In the quest for the search for new drug leads active against any human pathogen, it is prudent then to investigate their inhibitory features on human CYP enzymes, to identify those seletive and potent against the pathogenic target. Simultaneous investigation of several molecular targets could offer a new route to effective antibiotics and minimize the development of drug resistance.

The human CYP1 enzymes (CYPs1A1, 1A2 and 1B1) which are under the transcriptional regulation of AhR receptor, are known for their induction by and metabolism of polyaromatic hydrocarbons (PAHs) present in cigarette smoke, industrial dyes and agricultural pesticides. These enzymes, particularly CYPs1A1 and 1B1 catalyse critical conversions to form the ultimate carcinogen that bind to cellular macromolecules such as DNA to form DNA-PAH adducts associated with neoplastic diseases (Zhang *et al.,* 2004). The conversion of benzo[a]pyrene, a prototypic PAH carcinogen, to the ultimate carcinogenic metabolite, (+)B[a]P-7, 8-diol-9,10-epoxide2, capable of producing tumours in new born mice (Levin *et al.*, 1976), involve all three CYP1 enzymes. Given the direct involvement with the formation of ultimate carcinogens, it has been well accepted that the inhibition of CYP1 enzymes is a cancer chemopreventive strategy. A well known group of natural chemopreventors include isothiocyantes found in cruciferous vegetables which are CYP1 inhibitors that drecrease DNA-PAH adducts in animal models (Hecht *et al.,* 2002). The search for CYP1 inhibitors have also gained strength by the evidence of a likely role in cancer progression and drug resistance (Martinez *et al.,* 2008) due to overexpression of CYP1B1 enzymes in certain tumour tissue compared with the surrounding normal cells. In work described in this paper, we therefore extended the biological activity characterization of small, natural compounds to their inhibitory properties against the activities of CYP1A1, 1A2 and 1B1 enzymes.

Natural Products as Drug Leads and Caribbean Biodiversity

For years, natural products and its derivatives have been used for their pharmacological properties and continue to play a vital role in drug discovery and its development (Newman and Cragg 2007; Atul Bhattaram *et al.,* 2002). Of the 1010 medicinal drugs reported, 25% (Atul Bhattaram *et al.,* 2002, Newman and Cragg 2007) are derived from natural products and used without chemical modification such as β-lactaman or its pharmacophores having a natural compound base, as exemplified by the first line anti-TB drug, rifampicin (Sensi 1983; Tribuddharat & Fennelward 1999) and second line treatment, streptomycin, kanamycin and capreomycin (Zhang *et al.,* 2006; Pauli *et al.,* 2005; Copp 2003; Tribuddharat and Fennelward 1999; Shu 1998). Although due to the modern *in-silico* aids in lead compound searches, a decline in the interest in natural compounds were observed in the 1990s, (Cragg and Newman 2005) the continued threat by drug resistant strains makes essential the need to hunt for unusual structural motifs hidden within nature.

The Caribbean region consists of an archipelago of islands with a rich biodiversity (Mittermeier *et al.,* 1998) and high endemism. In Jamaica, 27.2% of the 2888 known species of flowering plants that are native are endemic to the island (Adams, 1972). The practice of folklore medicine to treat ailments is a common tradition (Asprey *et al.,* 1953), with least 334 plants species been identified as having medicinal qualities (Mitchell and Ahmad, 2006). However, full scientific investigations on bioactive isolates lag behind, with only 44 of those plants having undergone phytochemical isolation and bioactivity characterisation. It is in this context that this research was undertaken, with the aim of extending knowledge on the bioactive value of natural isolates from several Caribbean plants.

Five endemic and/or indigenous plants used in ethnomedicine were identified for biological activity characterisation in this study. *Esenbeckia pentaphylla*, a rare, endemic Jamaican plant (Adams 1972) is known to produce alkaloids, chromones, and aldehydes (Simpson *et al.*, 2005) with insecticidal activity (Jagadeesh *et al,.* 2001) and germination stimulant activity (Mann, 1986). *Spathelia sorbifolia*, a native to Jamaica (Taylor *et al,.* 1977; Adams *et al.,* 1973) with bioactivities that include anti-inflammatory, antiviral and anticancer (Cassady *et al.*, 1990) is mainly found to accumulate chromones (Simpson *et al.,* 2010). *Hortia regia* is endemic to Guyana and is known for its chromones (Jacobs *et al.*, 1986; 1987) and alkaloid constituents (Tinto *et al.*, 1992) while *Peperomia amplexicaulis*, indigenous to the Caribbean (Boufford 1982; Mitchell 2011), is known for its chromane constituents (Burke *et al.*, 2003). *Clusia havestiodes* is a complex of endemic forms where three varieties have been identified (Adams 1972). *Clausia* are known mainly for their naturally occurring benzophenones (Christian *et al.,* 2001) and have numerous biological activities ranging from antimicrobial, antifungal and anti-HIV (Baggett *et al.*, 2005).

Evaluation of Biological Activity

Working with the hypothesis that natural compounds may hold multiple therapeutic benefits, eleven compounds isolated from the 5 plants mentioned above were characterized for their inhibitory properties against the NAT from *M. smegmatis* (MSNAT), a homologue of *M. tuberculosis* and heterologously expressed human CYP enzymes, in particular the CYP1 enzymes (CYPs1A1, 1A2 and 1B1). Since NAT has been shown to play a critical role in cell wall integrity and macrophage stability in mycobacteria, and CYP and NAT involvement in cancers, it is hoped these natural compounds may provide valuable information in the search for effective therapies for human diseases. Examples of natural products with multiple applicability is not uncommon and include antibiotics with antitumour, hypocholesterolomic, immunosuppressant activity (Demain and Zhang, 2005) and curcuminoids with antibacterial, antiviral and antifungal activities (Kawamori *et al.*, 1999: Babu *et al.*, 2006: Subramaniam *et al.*, 2012) .

METHODS AND MATERIALS

Chemicals

All CYP substrates and metabolites were purchased from Gentest Corporation (Worburn, MA, USA). All other chemicals for the CYP inhibition assays and for the Arylamine N-acetyltransferase (NAT) – 5, 5' dithobis 2-Nitrobenzate (DTNB) colourimetric assay along with reagents were purchased from Sigma-Aldrich (St.Louis, MO).

Plant Material: Extraction and Isolation

The aerial parts of Caribbean endemic and/ or indigenous plants were collected identified and samples were deposited in the Herbarium at the University of the West Indies, Mona, Jamaica. The heartwood of *Spathelia sorbifolia* (Simpson *et al.,* 2010) and *Esenbeckia pentaphylla* (Simpson and Jacobs 2005), roots of *Hortia regia* (Jacobs *et al.,* 1986) fruits of of *Clusia havetiodes* var. *stenocarpa* (Christian *et al.,* 2001) and whole plant of *Peperomia amplexicaulis* (Burke *et al.,* 2003) were air-dried, milled then defatted with hexanes and exhaustively extracted by cold percolated hexane, acetone and/or methanol. The resulting residues were subjected to repeated silica gel column chromatography. The identity and purity of the compounds were determined using NMR spectroscopic data, IR, UV, HREIMS, EIMS, ^{1}H, and ^{13}C. For a detailed extraction and isolation methodologies see Simpson *et al.,* 2010; Christian *et al.*, 2001; Burke *et al.*, 2003; Jacobs *et al.*, 1986.

Expression and Enzymatic Assay for Pure Recombinant NAT from *Mycobacterium Smegmatis*

The open reading frame for the gene for *Mycobacterium smegmatis* Arylamine N-acetyltransferase (MSNAT) and an N-terminal hexa histidine tag was expressed in *E. Coli* BL21 (DE3)pLysS and purified using Ni-NTA agrose as described previously (Payton *et al.,* 1999). An MSNAT enzymatic assay that exploits the NAT catalyse hydrolysis of acetylCoA in the presence of isoniazid to produce CoA was employed as described elsewhere (Brooke *et al.,* 2003). Briefly, pure recombinant MSNAT (0.6 μg/100μl) was incubated at 37^{0}C with isoniazid (450μM) for 10 mins in the presence or absence of test compound (at 100μM). The reation was initiated with the addition of acetylCoA (400μM) and the amount CoA formed was measured at 405nm using a μQuant universal microplate spectrophotometer, following complex formation with 5,5-dithiobis (2-nitrobenzoic acid), DTNB. Test compounds were dissolved in DMSO and less than 0.1 % DMSO used in the final assay.

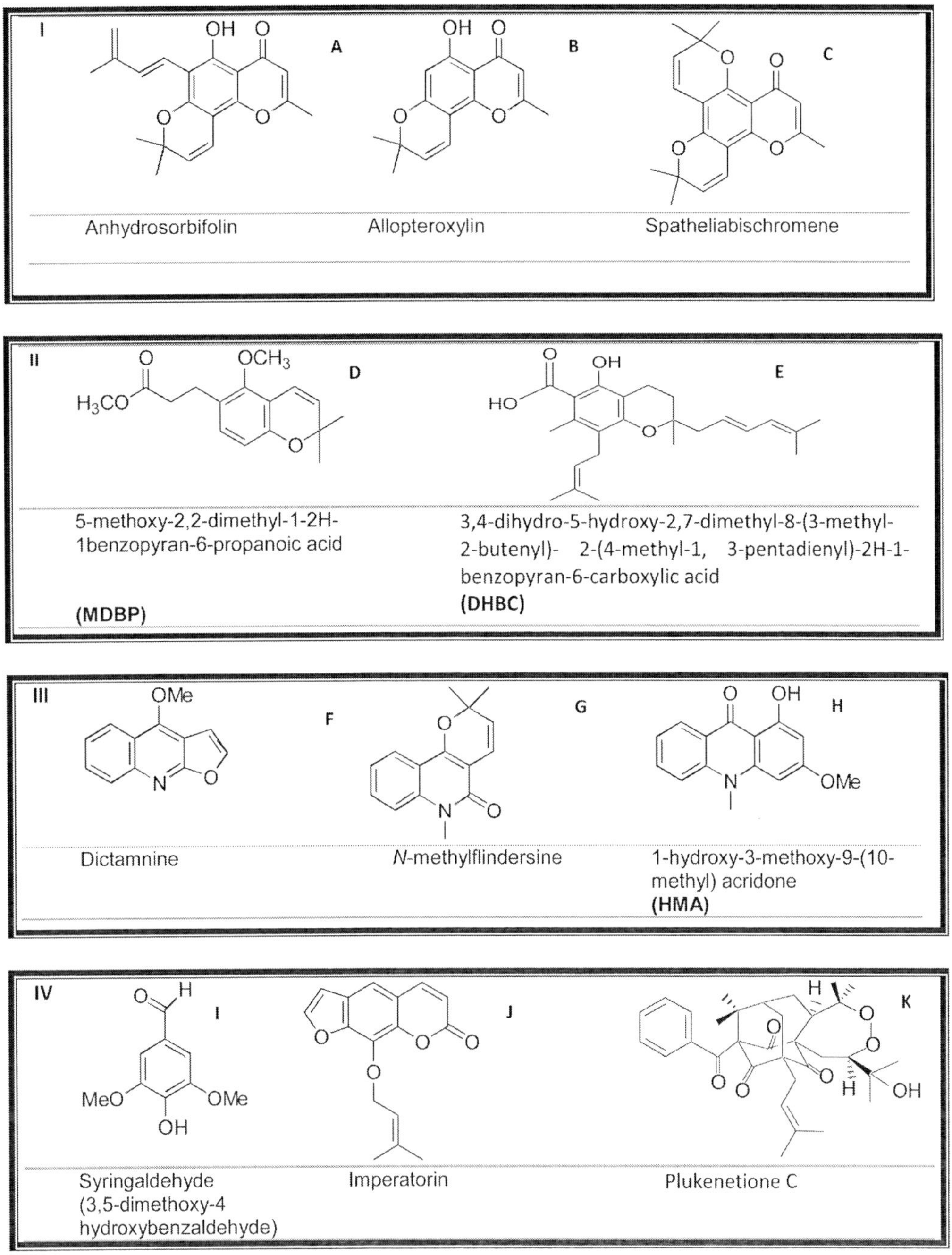

Figure 1. Phytochemicals evaluated for biological activity.
I Related *Chromones*.; Compounds A– C were extracted from *Spathelia Sorbifolia* (Simpson *et al*., 2010). II : Related *Chromanes*. Compounds D and E were extracted from *Hortia regia*, and *Peperomia amplexicaulis* respectively (Burke *et al*., 2003; Henry *et al*., 2001; Jacobs *et al*., 1986). III. *Alkaloids:* Compounds F – H were extracted from *Esenbeckia pentaphylla*; (Simpson and Jacobs 2005); Group IV Others: I, an *Aldehyde* was extracted from *Esenbeckia pentaphylla*; (Simpson and Jacobs 2005); a *Coumarin* – J extracted from *Esenbeckia pentaphylla*; (Simpson and Jacobs 2005) and K a *Benzophenone* extracted from *Clausia havetiodes* (Christian *et al*., 2001).

CYP Microsomes and Inhibition Assays

Escherichia coli membranes expressing human CYP1A1, CYP1A2, CYP1B1 co-expressed with CYP reductase were purchased from Cypex Ltd, (Dundee, U.K).

The test compounds were evaluated for their ability to inhibit the catalytic activity of recombinant CYP1 enzymes by means of high throughput fluorometric detection assays conducted in 96 well microtitre plates as described elsewhere (Badal *et al.,* 2011; Crepsi *et al.,* 1997). 7-ethoxy-3-cyanocoumarin (CEC) was used as a substrate for both CYP1A1 and CYP1A2 and 7-ethoxyresorufin (ERes) was used as a substrate for detecting activity of CYP1B1. The reactions were monitored fluorometrically at 37 °C, using a Varian Cary Eclipse fluorescence spectrophotometer. All the test compounds were dissolved in DMSO and less than 0.3% of DMSO was used in the final assay.

Data Analysis

IC50 values were determined by fitting the data in Sigma Plot (version 10.0) and enzyme kinetics module version 1.3, using non linear regression analysis. The data listed represent the average values from three different determinations.

Results

Eight compounds each belonging to a chromane, chromone, or alkaloid family, and three others classed as either a coumarin, aldehyde, or prenylated benzophenol, displayed in figure 1, were extracted and purified from four endemic and/or indigenous Caribbean plants. Groups I-III display structural isomers of each class of compounds while group IV displays the others, except for imperatorin (J), a coumarin, despite being a general isomer of the chromones have been placed in group IV in this paper, away from the more directly related chromones in group I in order to help in structure activity relationship analysis. Two of these compounds, 5-methoxy-2,2-dimethyl-1-2H-1benzopyran-6-propanoic acid (MDBP) and Spatheliabischromene extracted from *H.regia* and *S. sorbifolia* respectively, were reported as novel from the region (Jacobs *et al.*, 1986, and Taylor *et al.,* 1977). Full biological worth of these, in particular of MDBP, remains unexplored.

The impact of these natural compounds on the activity of MSNAT catalyzed metabolism of isoniazid was evaluated and results are reported as percentage inhibition compared to the activity in the absence of an inhibitor in figure 2. As seen therein, the highest inhibition (63%) was displayed by anhydrosorbiforlin. The alkaloids, dictamnine, N-methylflindersine, and 1-hydroxy-3-methoxy-9-(10-methyl) acridone (HMA) all displayed >50% inhibition. All other compounds displayed fairly moderate to poor inhibition of MSNAT.

All compounds were then also evaluated for their inhibition of human CYP1 enzyme activities (CYPs1A1, 1A2 and 1B1). To verify the accuracy of experimental techniques employed to detect CYP inhibition, assays with known inhibitors were carried out with furafylline (against CYP1A2), and ketoconazole (against CYP1A1, CYP1B1) and the obtained IC_{50} values compared well with published values (Badal *et al.,* 2011). Michaelis constant, Km, was determined for each marker substrate under the specified experimental conditions, in order to determine suitable substrate concentrations for assessing inhibitory potential of test compounds. All natural compounds underwent control experiments that gauaged intrinsic fluoroscence and the impact on the fluoroscence of the metabolite standard expected to be produced in the assay. Results of such control experiments eliminated the assessment of inhibitory impact by N-methylflindersine, and HMA using this method of activity detection.

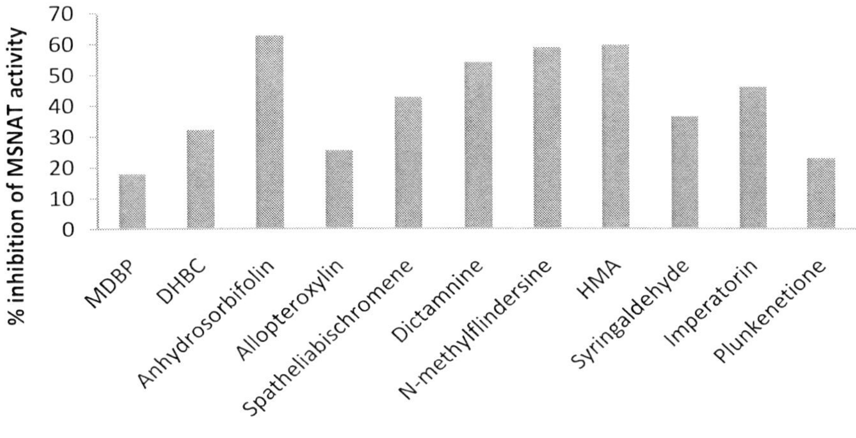

Figure 2. Inhibition of MSNAT activity by natural products.
Inhibition by the eleven natural compounds (at 100μM) on NAT activity was assessed using heterologously expressed NAT from *M. smegmatis* (MSNAT), as described in methods. In each case, results are shown as the mean of triplicate determinations of percentage inhibition of hydrolysis of acetylCoA in the presence of isoniazid. Compounds as listed in figure 1. MDBP, DHBC and HMA refers to 5-methoxy-2,2-dimethyl-1-2H-1benzopyran-6-propanoic acid, 3,4-dihydro-5-hydroxy-2,7-dimethyl-8-(3-methyl-2-butenyl)- 2-(4-methyl-1, 3-pentadienyl)-2**H**-1-benzopyran-6-carboxylic acid, and 1-hydroxy-3-methoxy-9-(10-methyl) acridone respectively.

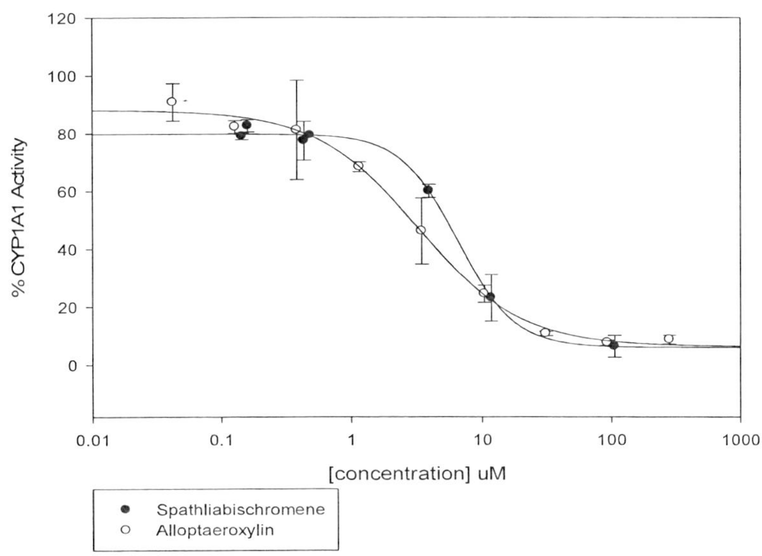

A

Figure 3. (Continued)

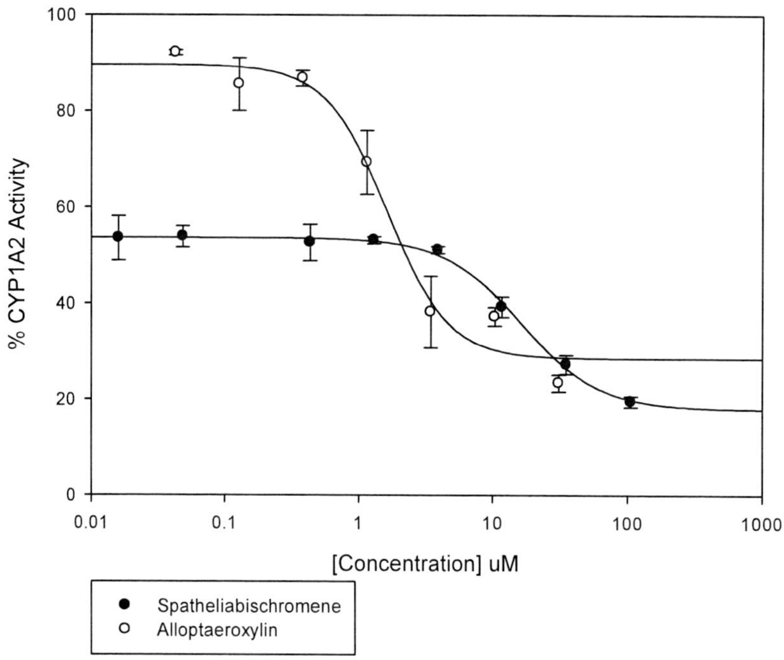

B

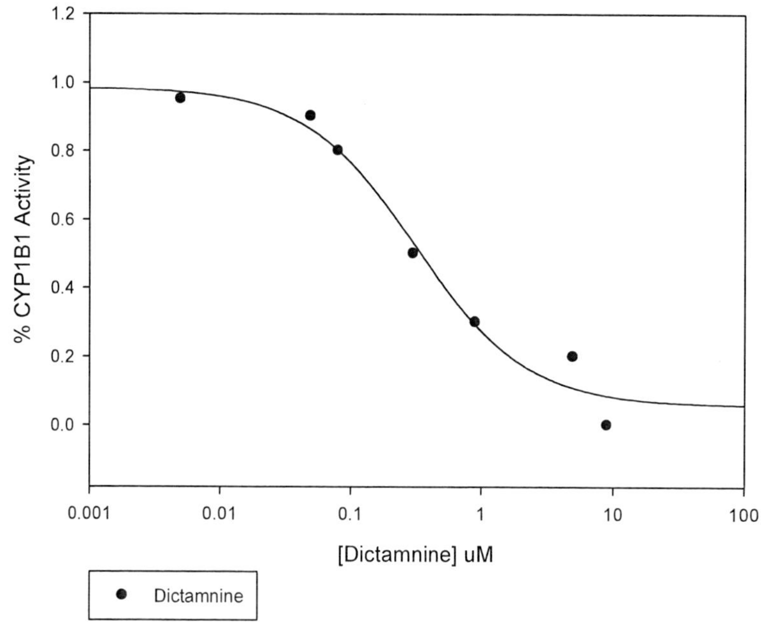

C

Figure 3. Inhibition of human CYP1 enzymes by natural compounds.
Inhibition of activities of human recombinant CYP1 enzymes by natural products. CYP1A1 (A), CYP1A2 (B), catalyzed 7-ethoxy-3-cyanocoumarin (0.5 and 5 μM respectively) and CYP1B1 (C) catalyzed 7-ethoxyresorufin (0.37 μM), activities were determined in the presence of varying concentrations of the chromone ranging between 0 and 1000 μM, as described in methods and materials. Control enzyme activity (mean ± SEM) for CYP1A1, CYP1A2 and CYP1B1 were 1.165 ± 0.1708, 1.8 ± 0.07 and 0.34 ± 0.08 μM/min/pmol respectively. Data are expressed as mean percentage of control enzyme activity for three independent experiments.

The inhibitions of natural compounds were evaluated following plots of percent activity vs concentration as exemplified by dictamnine, anhydrosorbifolin, and alloptaeroxylin (see figure 3). The obtained IC_{50} values are summarized in Table 2. Strikingly noticeable potency was displayed by dictamnine against the activity of CYP1B1, with an IC_{50} value of 0.27μM. Moderate inhibition was also displayed by anhydrosorbifolin, alloptaeroxylin and spatheliabischromene, on at least one CYP1 activity.

DISCUSSION

The inhibition of NAT enzyme activity in mycobacteria has been employed in the past as a useful target in the search for new anti-tuberculars, given its importance in mycobacterial cell wall synthesis and macrophage stability. In this paper, we investigated novel and known compounds emanating from Caribbean plant biodiversity for their mycobacterial NAT inhibitory potential. Of the eleven compounds examined, four were identified as having greater than 50% inhibition at 100μM. Although synthetic compounds with much greater potency than these values have been reported previously (Westwood *et al*, 2010), structure-activity relationships identified in related families of compounds in this study, can point to structural motifs key for improving inhibition and those worthy of further explorations including impact on growth of *M. tuberculosis*.

Greatest inhibition was obtained for anhydrosorbifolin (63%), a chromone whose extended linear side chain appear to contribute to the increased inhibition compared with its derivatives, alloptaeroxylin which has no side chains (displayed 26% inhibition) and spatheliabischromone which has a cyclicized side chain (displayed 43% inhibition). Linear, side chain presence extending from the benzene ring, appear to yield higher NAT inhibition in this chromone family and design of structural variants for further experiementation may prove to be worth the effort. This family of compounds also displayed moderate inhibition against the human CYP1 enzyme activities. Of particular interest is the loss of potency in CYP1A2 inhibition with cyclization of the side chain where nearly a ten fold difference is observed for spatheliabischromone compared with anydrosorbifolin. It is also interesting however; that the very same structural feature appears to be important in its binding between the two closely related CYP1A1 and CYP1A2 enzymes. Spatheliabischromone displayed an IC_{50} of 6.5μM against the activity of CYP1A1 and 16.5μM against CYP1A2, while no such significant difference was observed in the IC50s of the other two compounds across the two enzymes. Such selectivity for these closely related enzymes may be of diagnostic value in the pharmaceutical industry. Other natural products with distinct selectivity have been previously observed in our laboratory within the quassinoid family and the quassinoid bound CYP1A1 active site of model aided in identification of features key for selectivity and potency (Shields *et al.*, 2009).

The three alkaloids examined, dictamnine, N-methylflindersine and HMA, which share a basic quinolone core, displayed over 50% NAT inhibition. The second line anti-tuberculosis drug floroquinolone, bear resemblance to these quinolones. Dictamnine and N-methyflindersine, are also known antibiotics (Severino *et al.*, 2009, O'Donell *et al.*, 2010). Noticeably, dictamnine also demonstrated potent and selective inhibition (IC_{50}=0.27μM) against the activity of human CYP1B1 enzyme, an enzyme drawing keen interest as a target for novel and anticancer therapeutics. The discovery of the overexpression of CYP1B1 in tumour tissues compared with normal surrounding cells (McFadyen *et al.,* 1999; Murray *et al.,* 1997), have led to the search for prodrugs reliant on CYP1B1 metabolism for the conversion into cytotoxic therapeutics. The role of such overexpression is yet to be fully understood, however, modification in its expression has shown to modulate tumour progression (Castro *et al.,* 2008) and thus clearly identified as a target for therapy. Combined with the role that CYP1B1 play in activating arylhydrocarbon carcinogens, inhibitors of CYP1B1 enzymes are considered to carry chemopreventive and treatment benefits. Dictamine has been shown to display moderate cytotoxicity, (Akhmedzhanova *et al.,* 2010) and it's CYP1B1 inhibition reported here warrants full characterisation as a chemopreventive lead. N-methylflindersine (Jansen *et al.,* 2006) has been shown to have weak cytotoxicity on several cell lines, although its full chemopreventive potential remains unexplored.

Table 1. Summary of IC50 values reflecting inhibition of human CYP1 activity by natural products

Codes and Compound Names	CYP1A1	CYP1A2	CYP1B1
Anhydrosorbifolin	4.9□M[a]	1.9□M[a]	1.4□M[a]
Alloptaeroxylin	3.7 □M ± 0.27	2.5□M± 0.11	NI
Spatheliabischromene	6.5 □M ±0.5	16.5□M ± 1.7	NI
3,4-dihydro-5-hydroxy-2,7-dimethyl-8-(3-methyl-2-butenyl)- 2-(4-methyl-1, 3-pentadienyl)-2**H**-1-benzopyran-6-carboxylic acid (DHBC)	2.1□M[a]	5.8□M[a]	5.6□M[a]
Dictamnine	2.67□M ± 0.66	35.09□M	0.27□M
N-methyflindersine	ND	ND	ND
1-hydroxy-3-methoxy-9-(10-methyl) acridone (HMA)	ND	ND	ND
Syringaldehyde	NI	NI	NI
Imperatorin	NI	NI	NI

ND = not determined due to interferences with the assay; NI = No Impact; [a] values from Badal *et al.*, 2008.

MDBP, a novel compound and its structural isomer DHBC, displayed fairly weak inhibition on NAT activity, while DHBC displayed moderate inhibition on CYP1 activities. Imperatorin, a coumarin and an isomer of the chromones, displayed near 50% inhibition on NAT activity. Syringaldehyde, imparted only 37% inhibition on NAT activity, and is in line with growth suppressive characteristics on *M. tuberculosis* in an *in-vitro* assay which was also observed to be weak (Chiang *et al.,* 2010).

In conclusion this paper highlights, chromones with linear side chain arms for investigation of the impact on the growth of *M. tuberculosis*. While quinolone alkaloids were also found to have potential anti-tubercular properties, one in particular (dictamnine) was identified as having potent inhibition against CYP1B1enzyme, thus warranting future investigation as a potential chemopreventor. Natural products derived from Caribbean plants extracts thus display valuable leads in the search for antituberculars and chemopreventors.

ACKNOWLEDGEMENTS

We are grateful to Professor Edith Sim, University of Oxford, U.K. for the supply of NAT protein and for editorial input into the manuscript. We are also grateful for financial support received from the Forest Conservation Fund (Jamaica), International Foundation for Science (Sweden) and the University of the West Indies (Mona, Jamaica).

REFERENCES

Adam, P., Berry, B., Loader, J., Tyson, K., Craggs, G., Smith, P., De Belin, J., Steers, G., Pezzella, F., Sachsenmeir, K., Stamps, A., Herath, A., Sim E., O'Hare, M., Harris, A., and Terrett, J. (2003). Arylamine N-acetyltransferase-1 is highly expressed in breast cancers and conveys enhanced growth and resistance to etoposide *in vitro*. *Molecular Cancer Research, 1,* 826-835.

Adams, C.D. (1972). Flowering Plants of Jamaica, MacLehose and Co. Ltd, University Press, Glasgow. 848.

Adams, C. D., Taylor, D. R., and Warner, J. M. (1973). N-methylflindersine from spathelia sorbifolia. *Phytochemistry, 12*(6), 1359-1360.

Adeniyi, B. A., Groves, M. J., and Gangadharam, P. R. J. (2004). In vitro anti-mycobacterial activities of three species of cola plant extracts (sterculiaceae). *Phytotherapy Research, 18*(5), 414-418.

Akhmedzhanova, V., Angenot, L., and Shakirov, R. (2010). Alkaloids from *haplophyllum leptomerum*. *Chemistry of Natural Compounds, 46*(3), 502-503.

Arata, K. (1991). The global tuberculosis situation and the new control strategy of the world health organization. *Tubercle, 72*(1), 1-6.

Arunrattiyakorn, P., Suksamrarn, S., Suwannasai, N., Kanzaki, H. (2011). Microbial metabolism of α-mangostin isolated from *Garcinia mangostana* L. *Phytochemistry, 72*, 730–734.

Asprey, G.F., Thornton, P. (1953).Medicinal plants of Jamaica Part I. *West Indian Medical Journal*, *2*, 233–52.

Atul Bhattaram, V., Graefe, U., Kohlert, C., Veit, M., and Derendorf, H. (2002). Pharmacokinetics and bioavailability of herbal medicinal products. *Phytomedicine, 9, Supplement 3*, 1-33.

Babu, K., Shanmugam, V., Ravindranath, S.D., and Joshi, V.P. (2006). Comparison of chemical composition and antifungal activity of *Curcuma longa* L. leaf oils produced by different water distillation techniques. *Flav Frag J., 22(3)*: 191-196.

Badal, S., Williams, S., Huang, G., Francis, S., Vendantam, P., Dunbar, O., Jacobs, H., Tzeng, T.J., Gangemi, J., and Delgoda, R. (2011). Cytochrome P450 1 enzyme inhibition and anticancer potential of chromene amides from *Amyris plumier. Fitoterapia, 82*, 230 – 236.

Badal, S., Shields M., Niazi U, Jacobs H, Sutcliffe MJ and Delgoda R. (2008) Screening Natural Products for CYP1 Inhibitors, Proceedings of the 17th International symposium on Microsomes and Drug Oxidations, Saratoga Springs, New York, p63-67.

Baggett, S., Mazzola, E., Kennelly, E. J. (2005). The benzophenones: Isolation, structural elucidation and biological activities: *In* Studies in Natural Products Chemistry (Atta ur-Rahman) *Elsevier Press 32*, Part L 721-771

Balganesh, T., Balasubramanian, V., Kumar, S. (2004). Drug discovery for tuberculosis: Bottlenecks and path forward. *Current Science* 86, 167 – 176.

Barry (III), C. E., Lee, R. E., Mdluli, K., Sampson, A. E., Schroeder, B. G., Slayden, R. A., and Yuan, Y. (1998). Mycolic acids: Structure, biosynthesis and physiological functions. *Progress in Lipid Research, 37*(2–3), 143-179.

Belisle, J. T., Vissa, V. D., Sievert, T., Takayama, K., Brennan, P. J., and Besra, G. S. (1997). Role of the major antigen of mycobacterium tuberculosis in cell wall biogenesis. *Science, 276*(5317), 1420-1422.

Bhakta, S., Besra, G. S., Upton, A. M., Parish, T., Sholto-Douglas-Vernon, C., Gibson, K. J. C., Knutton, S., Gordon, S., daSilva, R. P., Anderton, M. C., and Sim, E. (2004). Arylamine n-acetyltransferase is required for synthesis of mycolic acids and complex lipids in mycobacterium bovis bcg and represents a novel drug target. *The Journal of Experimental Medicine, 199*(9), 1191-1199.

Blumberg, H. M., Migliori, G. B., Ponomarenko, O., and Heldal, E. (2010). Tuberculosis on the move. *The Lancet, 375*(9732), 2127-2129.

Boccia, D., and Evans, C. A. (2011). A new era for global tuberculosis control. *The Lancet, 378*(9799), 1293.

Boufford, D. (1982). Notes on Peperomia (Piperaceae) in the southeastern United States. *Journal of the Arnold Arboretum 63*, 317-325

Brennan, P. J. (2003). Structure, function, and biogenesis of the cell wall of mycobacterium tuberculosis. *Tuberculosis, 83*(1–3), 91-97.

Brennan, P. J. and Draper, P. 1994. Ultrastructure of Mycobacterium Tuberculosis. In: Tuberculosis: Pathogenesis, Protection, and Control (Bloom, B. R.). ASM Press, p271-284.

Brooke, E. W., Davies, S. G., Mulvaney, A. W., Pompeo, F., Sim, E., and Vickers, R. J. (2003). An approach to identifying novel substrates of bacterial arylamine N-acetyltransferases. *Bioorganic & Medicinal Chemistry, 11*(7), 1227-1234.

Brosch, R., Gordon, S. V., Billault, A., Garnier, T., Eiglmeier, K., Soravito, C., Barrell, B. G., and Cole, S. T. (1998). Use of a mycobacterium tuberculosis h37rv bacterial artificial chromosome library for genome mapping, sequencing, and comparative genomics. *Infection and Immunity, 66*(5), 2221-2229.

Burke, S. J., Jacobs, H., McLean, S., and Reynolds, W. F. (2003). Structural and spectral assignment by 2d nmr of a new prenylated benzopyrancarboxylic acid and structural reassignment of a related compound. *Magnetic Resonance in Chemistry, 41*(2), 145-146.

Cassady, J. M., Baird, W. M., and Chang, C.-J. (1990). Natural products as a source of potential cancer chemotherapeutic and chemopreventive agents. *Journal of Natural Products, 53*(1), 23-41.

Castro, D.J, Baird, W.M., Pereira, C.B., Giovanni, J., Löhr, C., Fischer, K., Yu, Z., Gonzalez, F.J., Krueger, S.K., Williams, D.E. (2008). Fetal mouse cyp1b1 and transplacental carcinogenesis from maternal exposure to Dibenzo[a,l]pyrene. *Cancer Prevention Research, 1*, 128-34.

Centre for Disease Control(CDC) (2005). Worldwide emergence of *Mycobacterium tuberculosis* with extensive resistance to second-line drugs. *Morbidity and Mortality Weekly Report*, 55, 250-253

Chaisson, R. E., and Martinson, N. A. (2008). Tuberculosis in africa — combating an hiv-driven crisis. *New England Journal of Medicine, 358*(11), 1089-1092.

Chiang, C.-C., Cheng, M.-J., Peng, C.-F., Huang, H.-Y., and Chen, I.-S. (2010). A novel dimeric coumarin analog and antimycobacterial constituents from fatoua pilosa. *Chemistry & Biodiversity, 7*(7), 1728-1736.

Christian, O. E., Henry, G. E., Jacobs, H., McLean, S., and Reynolds, W. F. (2000). Prenylated benzophenone derivatives from clusia havetiodes var. Stenocarpa. *Journal of Natural Products, 64*(1), 23-25.

Cragg, G., Newan, D. (2005). Plants as a source of anti-cancer agents. *Journal of Ethnopharmacology, 100,* 72–79.

Cole, S. T. (1999). Learning from the genome sequence of mycobacterium tuberculosis h37rv. *FEBS Letters, 452*(1–2), 7-10.

Cole, S. T. and Telenti, A. (1995). Drug resistance in *Mycobacterium tuberculosis. Eur. Resp. Rev.* 8, 701S–713S

Cole, S. T., Brosch, R., Parkhill, J., Garnier, T., Churcher, C., Harris, D., Gordon, S. V., Eiglmeier, K., Gas, S., Barry, C. E., Tekaia, F., Badcock, K., Basham, D., Brown, D., Chillingworth, T., Connor, R., Davies, R., Devlin, K., Feltwell, T., Gentles, S., Hamlin, N., Holroyd, S., Hornsby, T., Jagels, K., Krogh, A., McLean, J., Moule, S., Murphy, L., Oliver, K., Osborne, J., Quail, M. A., Rajandream, M. A., Rogers, J., Rutter, S., Seeger, K., Skelton, J., Squares, R., Squares, S., Sulston, J. E., Taylor, K., Whitehead, S., and Barrell, B. G. (1998). Deciphering the biology of mycobacterium tuberculosis from the complete genome sequence. [10.1038/31159]. *Nature, 393*(6685), 537-544.

Copp, B. R. (2003). Antimycobacterial natural products. *Natural Product Reports, 20*(6)., 535-557.

Crespi, C.L, Miller, V.P., and Penman, B.W. (1997). Microtitre plate assays for inhibition of human, drug metabolising cytochromes P450. *Analytical Biochemistry 248*, 188-90.

Crick, D. C., Chatterjee, D., Scherman, M. S., and McNeil, M. R. (2010). 6.13 - structure and biosynthesis of the mycobacterial cell wall. In M. Editors-in-Chief: Lew and L. Hung-Wen (Eds.), *Comprehensive natural products ii* (pp. 381-406). Oxford: Elsevier.

Fullam, E., Westwood, I. M., Anderton, M. C., Lowe, E. D., Sim, E., and Noble, M. E. M. (2008). Divergence of cofactor recognition across evolution: Coenzyme a binding in a prokaryotic arylamine n-acetyltransferase. *Journal of Molecular Biology, 375*(1), 178-191.

Furin, J. J., and Johnson, J. L. (2005). Recent advances in the diagnosis and management of tuberculosis. *Current Opinion in Pulmonary Medicine, 11*(3), 189-194.

Gillespie, S. H., Morrissey, I., And Everett, D. (2001). A comparison of the bactericidal activity of quinolone antibiotics in a mycobacterium fortuitum model. *Journal of Medical Microbiology, 50*(6), 565-570.

Glenn, A. E., Karagianni, E. P., Ulndreaj, A., and Boukouvala, S. (2010). Comparative genomic and phylogenetic investigation of the xenobiotic metabolizing arylamine n-acetyltransferase enzyme family. *FEBS Letters, 584*(14), 3158-3164.

Henry, G. E., and Jacobs, H. (2001). A short synthesis of 5-methoxy-2,2-dimethyl-2h-1-benzopyran-6-propanoic acid methyl ester. *Tetrahedron, 57*(25), 5335-5338.

Hecht, S., Kenny, P., Wang, M., and Upadhyaya, P. (2002). Benzyl isothiocyanate An effective inhibitor of polycyclic aromatic hydrocarbon tumorigenesis in A/J mouse lung. *Cancer Letters, 187*, 87-94.

Jacobs, H., Ramadayal, F., McLean, S., Perpick-Dumont, M., Puzzuoli, F., and Reynolds, W. F. (1987). Constituents of hortia regia: 6,7-dimethoxycoumarin, rutaecarpine, skimmianine, and (+)-methyl (e,e)-10,11-dihydroxy-3,7,11-trimethyl-2,6-dodecadienoate. *Journal of Natural Products, 50*(3), 507-509.

Jacobs, H., Ramdayal, F., Reynolds, W. F., Poplawski, J., and McLean, S. (1986). Isolation of 5-methoxy-2,2-dimethyl-1-2h-benzopyran-6-propanoic acid methyl ester and characterization by two-dimensional nuclear magnetic resonance spectroscopy. *Canadian Journal of Chemistry, 64*(3), 580-583.

Jagadeesh, S. G., Krupadanam, G. L. D., and Srimannarayana, G. (2001). Cheminform abstract: Antifeedant activity of the constituents of evodia lunu-ankenda. *ChemInform, 32*(3), 476.

Jansen, O., Akhmedjanova, V., Angenot, L., Balansard, G., Chariot, A., Ollivier, E., Tits, M., and Frédérich, M. (2006). Screening of 14 alkaloids isolated from haplophyllum a. Juss. For their cytotoxic properties. *Journal of Ethnopharmacology, 105*(1–2), 241-245.

Kamerbeek, J., Schouls, L., Kolk, A., van Agterveld, M., van Soolingen, D., Kuijper, S., Bunschoten, A., Molhuizen, H., Shaw, R., Goyal, M., and van Embden, J. (1997). Simultaneous detection and strain differentiation of mycobacterium tuberculosis for diagnosis and epidemiology. *Journal of Clinical Microbiology, 35*(4), 907-914.

Kawamori, T., Lubet, R., Steele, VE., Kelloff, G.J., Kaskey, R.B., Rao, C.V., and Reddy, B.S. (1999). Chemopreventive effect of curcumin a naturally occurring anti-inflammatory agent during the promotion progression stages of colon cancer. *Cancer Research, 59*, 597-601.

Kolattukudy, P. E., Fernandes, N. D., Azad, A. K., Fitzmaurice, A. M., and Sirakova, T. D. (1997). Biochemistry and molecular genetics of cell-wall lipid biosynthesis in mycobacteria. *Molecular Microbiology, 24*(2), 263-270.

Laurieri, N., Crawford, M., Kawamura, A., Westwood, I., Robinson, J., Fletcher, A., Davies, S., Sim, E., Russel, A. (2010). Small molecule colorimetric probes for specific detection of human arylamine N-acetyltransferase 1a potential breast cancer biomarker. *Journal of American Chemical Society*, 132, 3238 -3239.

Levin W, Wood A, Yari H, Jerina DM, and Conney AH. (1976). (±)-trans-7,8-dihydroxy-7,8-dihydrobenzo[*a*]pyrene: a potent skin carcinogen when applied topically to mice. *Proceeding of National Academy of Science USA*, 73, 3867-71.

Mann, J. (1987). Secondary Metabolism (2nd Edn.). 178 pp. Oxford University Press Inc.: New York.

Martinez, V., O' Connor, R., Liang, Y., and Clynes, M. (2008). CYP1B1 expression is induced by docetaxel: effect on cell vialbility and drug resistance. *British journal of cancer*; *98*, 564-70

McFadyen M.C., Breeman S., Payne S., Stirk, C., Miller, I.D., Melvin, W., Murray G. (1999). Immunohistochemical localization of cytochrome P450 CYP1B1 in breast cancer with monoclonal antibodies specific for CYP1B1. *Journal of Histochemistry and Cytochemistry* 47:1457–64.

McLean KJ, Belcher J, Driscoll MD, Fernandez CC, Le Van D, Bui S, Golovanova M, Munro AW, The Mycobacterium tuberculosis cytochromes P450: physiology, biochemistry & molecular intervention, *Future Med Chem.* 2010 Aug;2(8):1339-53.

Minchin, R. F., Hanna, P. E., Dupret, J.-M., Wagner, C. R., Rodrigues-Lima, F., and Butcher, N. J. (2007). Arylamine N-acetyltransferase i. *The International Journal of Biochemistry & Cell Biology, 39*(11), 1999-2005.

Mitchell, S. A. (2011). The Jamaican root tonics: A botanical reference. *Focus on Alternative and Complementary Therapies, 16*(4), 271-280.

Mitchell, S. A, and Ahmad, M. H. (2006). A review of medicinal plant research at the University of the West Indies, Jamaica, 1948-2001, *West Indian Medical Journal, 55(4),* 243-269.

Mitchison, D. A. (2005). Shortening the treatment of tuberculosis. *Nature Biotechnology, 23*(2), 187-188.

Mitnick, C., Bayona, J., Palacios, E., Shin, S., Furin, J., Alcántara, F., Sánchez, E., Sarria, M., Becerra, M., Fawzi, M. C. S., Kapiga, S., Neuberg, D., Maguire, J. H., Kim, J. Y., and Farmer, P. (2003). Community-based therapy for multidrug-resistant tuberculosis in lima, peru. *New England Journal of Medicine, 348*(2), 119-128.

Mittermeier, R. A., Myers, N., Thomsen, J. B., Da Fonseca, G. A. B., and Olivieri, S. (1998). Biodiversity hotspots and major tropical wilderness areas: Approaches to setting conservation priorities. *Conservation Biology, 12*(3), 516-520.

Mulholland, D. A., Parel, B., and Coombes, P. H. (2000). The chemistry of the meliaceae and ptaeroxylaceae of southern and eastern africa and madagascar. *Current Organic Chemistry, 4*(10), 1011-1054.

Murray GI, Taylor MC, McFadyen MC, McKay JA, Greenlee WF, Burke MD, Melvin WT (1997). Tumor specific expression of cytochrome P450 CYP1B1. *Cancer Research 57,* 3026–3031.

Newman, D. J., and Cragg, G. M. (2007). Natural products as sources of new drugs over the last 25 years. *Journal of Natural Products, 70*(3), 461-477.

O'Donnell, F., Smyth, T. J. P., Ramachandran, V. N., and Smyth, W. F. (2010). A study of the antimicrobial activity of selected synthetic and naturally occurring quinolines. *International Journal of Antimicrobial Agents, 35*(1), 30-38.

Okunade, A. L., Elvin-Lewis, M. P. F., and Lewis, W. H. (2004). Natural antimycobacterial metabolites: Current status. *Phytochemistry, 65*(8), 1017-1032.

Pablos-Méndez, A., Raviglione, M. C., Laszlo, A., Binkin, N., Rieder, H. L., Bustreo, F., Cohn, D. L., Lambregts-van Weezenbeek, C. S. B., Kim, S. J., Chaulet, P., and Nunn, P. (1998). Global surveillance for antituberculosis-drug resistance, 1994–1997. *New England Journal of Medicine, 338*(23), 1641-1649.

Pauli, G. F., Case, R. J., Inui, T., Wang, Y., Cho, S., Fischer, N. H., and Franzblau, S. G. (2005). New perspectives on natural products in tb drug research. *Life Sciences, 78*(5), 485-494.

Payton, M., Auty, R., Delgoda, R., Everett, M., and Sim, E. (1999). Cloning and characterization of arylamine n -acetyltransferase genes from mycobacterium smegmatis and mycobacterium tuberculosis: Increased expression results in isoniazid resistance. *Journal of Bacteriology, 181*(4), 1343-1347.

Payton, M., Gifford, C., Schartau, P., Hagemeier, C., Mushtaq, A., Lucas, S., Pinter, K., and Sim, E. (2001). Evidence towards the role of arylamine n-acetyltransferase in mycobacterium smegmatis and development of a specific antiserum against the homologous enzyme of mycobacterium tuberculosis. *Microbiology, 147*(12), 3295-3302.

Peters, J. H., Miller, K. S., and Brown, P. (1965). Studies on the metabolic basis for the genetically determined capacities for isoniazid inactivation in man. *Journal of Pharmacology and Experimental Therapeutics, 150*(2), 298-304.

Philipp, W. J., Poulet, S., Eiglmeier, K., Pascopella, L., Balasubramanian, V., Heym, B., Bergh, S., Bloom, B. R., Jacobs, W. R., and Cole, S. T. (1996). An integrated map of the genome of the tubercle bacillus, mycobacterium tuberculosis h37rv, and comparison with mycobacterium leprae. *Proceedings of the National Academy of Sciences, USA 93*(7), 3132-3137.

Rivera-Marrero, C. A., Burroughs, M. A., Masse, R. A., Vannberg, F. O., Leimbach, D. L., Roman, J., and Murtagh Jr, J. J. (1998). Identification of genes differentially expressed inmycobacterium tuberculosisby differential display pcr. *Microbial Pathogenesis, 25*(6), 307-316.

Russell, A., Westwood, I., Crawford, M., Robinson, J., Kawamura, A., Redfield, C., Laurieri, N., Lowe, E., Davies, S., Sim, E. (2009). Selective small molecule inhibitors of the potential breast cancer marker human arylamine N-acetyltransferase 1, and its murine homologue, mouse arylamine N-acetyltransferase 2. *Bioorganic & Medical Chemistry 17*, 905-918.

Sensi, P. (1983). History of the development of rifampin. *Review of Infectious Diseases, 5*(Supplement 3), S402-S406.

Severino, V. G. P., Silva, M. F. d. G. F. d., Lucarini, R., Montanari, L. B., Cunha, W. R., Vinholis, A. H. C., and Martins, C. H. G. (2009). Determination of the antibacterial activity of crude extracts and compounds isolated from hortia oreadica (rutaceae) against oral pathogens. *Brazilian Journal of Microbiology, 40*, 535-540.

Shields, M., Niazi, U., Badal, S., Yee, T., Sutcliffe, M. J., and Delgoda, R. (2009). Inhibition of cyp1a1 by quassinoids found in picrasma excelsa. *Planta Medica, 75*, 137,141.

Shu, Y.-Z. (1998). Recent natural products based drug development: A pharmaceutical industry perspective. *Journal of Natural Products, 61*(8), 1053-1071.

Sim, E., Payton, M., Noble, M., and Minchin, R. (2000). An update on genetic, structural and functional studies of arylamine n-acetyltransferases in eucaryotes and procaryotes. *Human Molecular Genetics, 9*(16), 2435-2441.

Sim, E., Sandy, J., Evangelopoulos, D., Fullam, E., Bhakta, S., Westwood, I., Krylova, A., Lack, N., and Noble, M. (2008). Arylamine N-acetyltransferases in mycobacteria. *Current Drug Metabolism, 9*(6), 510-519.

Sim, E., Pinter, K., Mushtaq, A., Upton, A., Sandy, J., Bhakta, S., and Noble, M. (2003). Arylamine N-acetyltransferases: a pharmacogenomic approach to drug metabolism and endogenous function. *Biochemical Society Transactions,* 31, 615–619.

Sim, E., Westwood, I., and Fullam, E. (2007). Arylamine n-acetyltransferases. *Expert Opinion on Drug Metabolism & Toxicology, 3*(2), 169-184.

Simpson, D. S., and Jacobs, H. (2005). Alkaloids and coumarins from esenbeckia pentaphylla (rutaceae). *Biochemical Systematics and Ecology, 33*(8), 841-844.

Simpson, D. S., McLean, S., Reynolds, W. F., and Jacobs, H. (2010). Tetranortriterpenoids from *spathelia sorbifolia* (rutaceae). *Natural Product Communications, 5*(6), 859 -862.

Subramaniam, D., Ponnurangam, S., Ramamoorthy P., Standing, D., Battafarano, R., Anant, S., and Sharma, P. (2012). Curcumin Induces Cell Death in Esophageal Cancer Cells through Modulating Notch Signaling. *PLos One, 7 (2)*: e30590.

Taylor, D. R., Warner, J. M., and Wright, J. A. (1977). New chromones from spathelia sorbifolia l. (rutaceae); synthesis of the benzo[1,2-b:3,4-b[prime or minute] dipyranone sorbifolin. *Journal of the Chemical Society, Perkin Transactions, 1*(4).

Tinto, W. F., McLean, S., and Reynolds, W. F. (1992). Hortiamide, a new tyramine alkaloid from hortia regia. *Journal of Natural Products, 55*(11), 1676-1678.

Tribuddharat, C., and Fennewald, M. (1999). Integron-mediated rifampin resistance inpseudomonas aeruginosa. *Antimicrobial Agents and Chemotherapy, 43*(4), 960-962.

Tripathi, R. P., Tewari, N., Dwivedi, N., and Tiwari, V. K. (2005). Fighting tuberculosis: An old disease with new challenges. *Medicinal Research Reviews, 25*(1), 93-131.

Upton, A., Johnson, N., Sandy, J., and Sim, E. (2001a). Arylamine n-acetyltransferases – of mice, men and microorganisms. *Trends in Pharmacological Sciences, 22*(3), 140-146.

Upton, A. M., Mushtaq, A., Victor, T. C., Sampson, S. L., Sandy, J., Smith, D. M., Van Helden, P. V., and Sim, E. (2001b). Arylamine N-acetyltransferase of mycobacterium tuberculosis is a polymorphic enzyme and a site of isoniazid metabolism. *Molecular Microbiology, 42*(2), 309-317.

van den Boogaard, J., Kibiki, G. S., Kisanga, E. R., Boeree, M. J., and Aarnoutse, R. E. (2009). New drugs against tuberculosis: Problems, progress, and evaluation of agents in clinical development. *Antimicrobial Agents and Chemotherapy, 53*(3), 849-862.

Van den Wijngaert, S., Dediste, A., VanLaethem, Y., Gerard, M., Vandenberg, O., and Zissis, G. (2004). Critical use of nucleic acid amplification techniques to test for mycobacterium tuberculosis in respiratory tract samples. *Journal of Clinical Microbiology, 42*(2), 837-838.

Westwood, I., Bhakta, S., Russell, A., Fullam, E., Anderton, M., Kawamura, A., Mulvaney, A., Vickers, R., Bhowruth, V., Besra, G., Lalvani, A., Davies, S., and Sim, E. (2010). Identification of arylamine *N* -acetyltransferase inhibitors as an approach towards novel anti-tuberculars. *Protein & Cell, 1*(1), 82-95.

World Health Organisation (2007). Global Tuberculosis Control Surveillance, Planning, Financing. URL October 2011. http://www.who.int/tb/ publications/global_ report/ 2007/pdf/full.pdf.

WHO (2010), WHO 2010. Global tuberculosis control 2010. URL October 2011. http://www.who.int/tb/publications/global_report/2010/gtbr10_ main.pdf.

Zhang, K., Ma, J., Chen, X., Sun, Y., Kong, Q.Y., Liu, J., Li, H. (2004). Frequent CYP1A1 expression in gastric cancers and their related lesions. *Oncology Report*; 12, 1335-40.

Zhang, Y., Post-Martens, K., and Denkin, S. (2006). New drug candidates and therapeutic targets for tuberculosis therapy. *Drug Discovery Today, 11*(1–2), 21-27.

Zhang, L., Demain, A. (2005). Natural Products and Drug discovery, In *Drug discovery and therapeutic medicine* (Zhang, L., Demain, A.), Humana press, Totowa, NJ. p3 – 32.

In: Natural Products: Structure, Bioactivity and Applications ISBN 978-1-62081-728-5
Editors: Ramiro E. Goncalves and Marcos Cunha Pinto

Chapter 3

NATURAL PRODUCTS AS INHIBITORS OF THE UBIQUITIN-/UBIQUITIN-LIKE PROTEIN-PROTEASOME PATHWAY

Dik-Lung Ma[1]*, Victor Pui-Yan Ma[1], Daniel Shiu-Hin Chan[1], Ka-Ho Leung[1], Hai-Jing Zhong[2,3] and Chung-Hang Leung[2,3,*]

[1]Department of Chemistry, Hong Kong Baptist University, Kowloon Tong, Hong Kong

[2]Institute of Chinese Medical Sciences, University of Macau, Macao SAR, China

[3]State Key Laboratory of Quality Research in Chinese Medicine, University of Macau, Macao SAR, China

ABSTRACT

The proteasome is the final player in the regulated degradation of intracellular proteins through in both ubiquitin-dependent and ubiquitin-independent proteolytic pathways. In eukaryotic cells, the ubiquitin-proteasome system coordinates the polyubiquitination and subsequent proteolysis of unwanted proteins, crucial to normal cellular homeostasis. The first-in-class proteasome inhibitor, bortezomib. promotes apoptosis and chemosensitization of cancer cells, and is used clinically as either a single-agent chemotherapeutic or in combination with other drugs. Natural products offer medicinal chemists with a cornucopia of diverse chemical scaffolds and bioactive substructures, and historically have represented an important source of new drugs. Salinosporamide A (NPI-0052), a second-generation proteasome inhibitor of marine microbial origin, has been effective against bortezomib-resistant cancers and has entered Phase I clinical trials. This review discusses the application of natural products as inhibitors of the ubiquitin-proteasome system through targeting of both conventional

* Correspondence to: Dik-Lung Ma, Department of Chemistry, Hong Kong Baptist University, Kowloon Tong, Hong Kong, Email: edmondma@hkbu.edu.hk; Chung-Hang Leung, Institute of Chinese Medical Sciences, University of Macau, Macao, China, Email: duncanleung@umac.mo.

proteolytic pathways as well as those involving ubiquitin-like proteins, such as NEDD8, which have recently emerged as novel anti-cancer targets.

INTRODUCTION

Protein homeostasis is a fundamental process of living cells. Regulating a balanced concentration of proteins is crucial for maintaining normal cellular physiology. Aberrant cellular protein levels can result in excessive cell proliferation and cycle abnormalities, leading to cancer development.[1] In eukaryotes, the ubiquitin-proteasome system (UPS) is the master cellular system for the regulated degradation of intracellular proteins having diverse cellular functions such as signal transduction, cell proliferation and apoptosis.[2] Ubiquitin, a 76 amino acid-long polypeptide widely expressed in tissues of eukaryotic organisms, is the central regulatory unit of the ubiquitin-proteasome pathway. Ubiquitin can be reversibly conjugated to target proteins, subsequently sent to proteasome for degradation.[2] Other ubiquitin-like proteins (Ubls), such as SUMO, NEDD8 and ISG-15, have also been discovered in recent years.[3]

Due to the biological importance of the ubiquitin and Ubl conjugation pathways, other therapeutic approaches targeting these systems have been investigated.[4] One of the most extensively studied targets of the UPS is the proteasome . As the furthest downstream agent of the UPS, the proteasome controls the degradation rate of most unwanted proteins. The first clinically evaluated proteasome inhibitor, bortezomib (Velcade; Millennium Pharmaceuticals), has shown good efficacy against cancer cell growth. The clinical success of bortezomib has validated the UPS as a biological target for the treatment of diseases. Targeting ubiquitin or ubiquitin-like enzymes has been shown to be another avenue for therapeutic intervention by small molecule inhibitors.[5]

Natural products (NPs) have been used as therapeutic agents over the centuries, and still are regarded as a major cornerstone of drug discovery and developmental research. Natural products provide medicinal chemists with many complex and interesting scaffolds, exhibiting unique interactions with biomolecules. A detailed survey by Newman and Cragg revealed that NPs, NP-derived small molecules or NP mimics represented 51% of marketed drugs in the years 1981–2006.[6] This promising figure clearly demonstrates that natural products still play an important role in the discovery of novel therapeutics for modulating the activity of biomolecular targets.

In this chapter, first we give an overview of the ubiquitin-proteasome system (UPS) and briefly describe the structure and function of the proteasome. Then we review representative examples of non-linear peptidic natural product inhibitors of the proteasome and their possible mechanisms of actions. Finally, we highlight successful cases of natural products that target other facets of the ubiquitin or Ubl conjugation pathways.

THE UBIQUITIN-PROTEASOME PATHWAY AND ITS HOMOLOGOUS SYSTEMS

The ubiquitin-proteasome pathway is the major protein degradation pathway in eukaryotic cells[7] and plays a vital role in the regulation of cellular protein levels. It is also involved in many cellular processes, such as transcription[8] and cell cycle progression.[9] The UPS basically functions as a housekeeper that removes unwanted proteins from the cell. Degradation of unwanted proteins by the ubiquitin-proteasome system involves two major stages: 1) covalent attachment of a polyubiquitin chain to the targeted proteins in an ATP-dependent process and 2) degradation of the ubiquitin-tagged proteins by the 26S proteasome. Three distinct enzymes, named E1, E2 and E3.[10] aid in the formation and attachment of the polyubiquitin tag The E1 (ubiquitin-activating) enzyme binds to ubiquitin and mediates the formation of a thioester linkage between E1 and ubiquitin using ATP. The activated ubiquitin is then transferred to the E2 (ubiquitin-conjugating) enzyme through the formation of an additional thioester bond. Finally, the E3 (ubiquitin ligase) enzyme mediates the transfer of ubiquitin to the lysine residues of the target protein. E3 enzymes can be broadly classified into three groups; these include HECT, RING finger and U-box domain-containing proteins.[4b] HECT E3s conjugate ubiquitin directly and mediate the transfer of ubiquitin to bound substrate. RING finger E3s do not react with ubiquitin, but rather act as templates for the reaction between ubiquitin-E2 and the protein substrate. U-box E3s are structurally similar to the RING finger E3 enzymes, but lack the metal-chelating residues which are present in the RING finger domain.[4b]

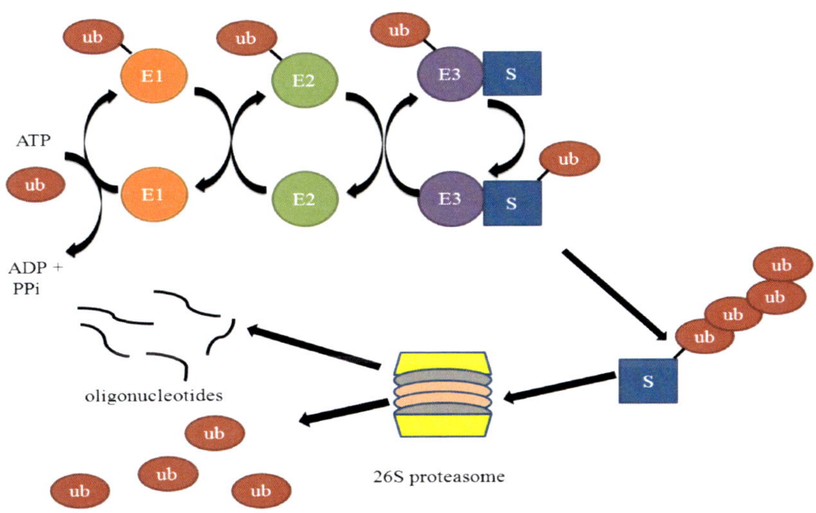

Scheme 1. The ubiquitin-proteasome pathway. Ubiquitin (ub) is first activated by E1 (ubiquitin-activating enzyme, UAE), and then is transferred to E2 (ubiquitin-conjugating enzyme). Ubiquitin is subsequently transferred to the target protein substrate (blue rectangle) with the aid of E3 (ubiquitin-ligase). The polyubiquitinated proteins are degraded by the 26S proteasome. The tagged proteins are first recognized by the 19S cap (yellow trapezoids) and are transported into the 20S catalytic core (grey and orange ovals) for degradation into short peptides. Ubiquitin is then released as monomers and recycled.

The polyubiquitin-tagged protein substrate is formed through successive attachment of ubiquitin residues to the substrate by the ubiquitin enzymes. The tagged proteins are recognized by the downstream 26S proteasome complex, which then degrades polyubiquitinated proteins into smaller peptide fragments. The ubiquitin residues are subsequently freed and can be recycled for further ubiquitination of other substrates.

Since the discovery of ubiquitin-conjugation pathway, other homologous pathways for Ubls have been discovered. These include NEDD8, SUMO and ISG15.[4b]

These Ubls resemble ubiquitin in terms of structure and sequence. Ubls are conjugated to a vast array of proteins via homologous pathways but are mechanistically distinct to the ubiquitin conjugation pathway. Each Ubl has its own dedicated set of E1, E2 or E3 enzymes. Compared to the ubiquitin conjugation pathway, the enzymes in Ubl conjugation pathways tend to be more specific towards their respective protein substrates. The Ubl conjugation pathways provide fascinating possibilities for the specific targeting of particular proteins involved in aberrant signaling activities.

STRUCTURE OF THE PROTEASOME

The 26S proteasome is a large multi-protein complex that is distributed in the nucleus and cytoplasm of eukaryotic cells. The structure and function of the proteasome is highly conserved from bacteria to eukaryotes. The proteasome consists of one 20S core catalytic unit capped with two 19S regulatory particles.[11] The 19S regulatory particles act as a recognition arm for the polyubiquitinated protein substrate. It cleaves the polyubiquitin chain from the tagged protein and is able to unfold the protein substrate before the latter enters the 20S core. Each regulatory particle is composed of six ATPase subunits, each responsible for energy-driven protein unfolding.[12] The unfolded protein substrate is then passed into the central 20S catalytic core. The 20S catalytic core is a cylindrical particle consisting of four stacked heptameric rings.[13] The outermost two rings comprise seven α-subunits, which contact the 19S regulatory particles and form a narrow channel that acts as a gateway for the transportation of the unfolded protein substrate. The inner rings consist of seven β-subunits, each containing three active catalytic sites which perform the proteolytic reactions. Although all three active sites share a common mechanism of proteolysis, the selectivity of the catalytic subunits is governed by the size of their binding pockets. The proteolytic activities of the proteasome can be termed chymotrypsin-like, trypsin-like, or peptidyl-glutamyl peptide-hydrolyzing (PGPH)-like.[14] The mechanism of proteolysis involves *N*-terminal threonine-dependent nucleophilic cleavage of the peptide bonds of the substrate, which is degraded into short peptide fragments. Small molecules which are able to recognize the proteasome binding pocket with high affinity, and antagonize the nucleophilic ability of the threonine moiety, can function as antagonists of the proteasome.

Natural Products Targeting the Proteasome

1. β-Lactones

The first natural product reported to target the proteasome was lactacystin, which was isolated from *Streptomyces lactacystinaeus*.[15] The story of lactacystin began in 1991 from a screening campaign aimed at identifying potential natural products which possess therapeutic values towards neurodegenerative diseases. Lactacystin was originally discovered as an inducer of neurite outgrowth in cancer cell lines.[15] The compound was later found to inhibit chymotrypsin-like proteasomal activity with a nanomolar IC_{50} value.

Investigation showed that lactacystin was not the active compound against the proteasome, but a prodrug that underwent spontaneous intramolecular lactonization in aqueous solution to give *clasto*-lactacystin β-lactone, the bioactive species inhibiting proteolytic activity.[16] The highly strained β-lactam ring in *clasto*-lactacystin β-lactone facilitates a favorable reaction with the hydroxyl group of the *N*-terminal threonine of the proteasome, therefore irreversibly blocking proteasome activity.

The discovery of lactacystin as a natural product inhibitor of proteasome has stimulated the search for novel β-lactone analogues as proteasome inhibitors. In 2003, Fenical *et al.* isolated salinosporamide A from *S.tropica* in the Bahamas, identifying the compound as a potent proteasome inhibitor.[17] Salinosporamide A contains the same molecular core scaffold as *clasto*-lactacystin β-lactone, but with significant differences in its substitution pattern.

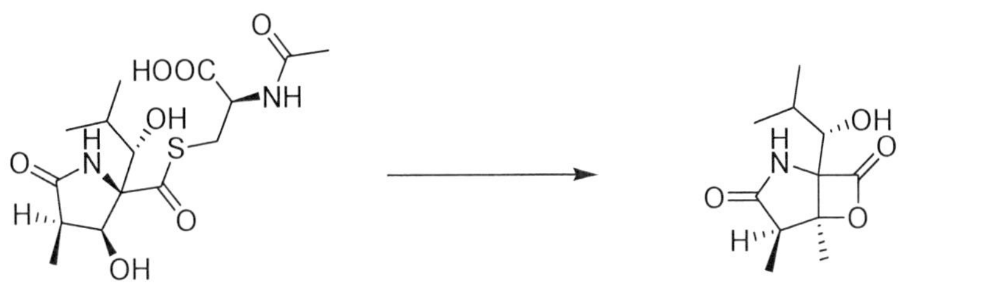

Scheme 2. Formation of *clasto*-lactacystin β-lactone from lactacystin through spontaneous intramolecular lactonization.

It was shown that salinosporamide A potently inhibited the chymotrypsin-like activity of the proteasome, with an IC_{50} value of 1.3 nM, presumably these are the different substituents in salinosporamide, greatly improving the potency of the compound when compared to *clasto*-lactacystin β-lactone. The pronounced activity of salinosporamide A has stimulated the examination of other salinosporamide family members as proteasome inhibitors, and at least eight salinosporamide derivatives have been discovered as potent proteasome inhibitors to date.[18] A closely related class of compounds to the β-lactone family are the cinnabaramides. Cinnabaramides A–G are isolated from terrestrial *Streptomyces* strains, and have been recognized as potent proteasome inhibitors.[19Belactosin A and C, metabolites of *Streptomyces*, have also been reported to be bioactive against the proteasome.[20]

Similar to *clasto*-lactacystin β-lactone, these compounds act as inhibitors of the proteasome by irreversibly blocking its proteolytic activity. The presence of the highly strained β-lactone ring is important to proteasome inhibition, as this structural feature facilitates the nucleophilic attack of the lactone ring by the *N*-terminal threonine of the proteasome, forming a covalent ester bridge. Since the proteolytic activity of the proteasome is highly dependent on threonine nucleophilic activity, the irreversible attachment of a small molecule causes a permanent loss of proteolytic activity.

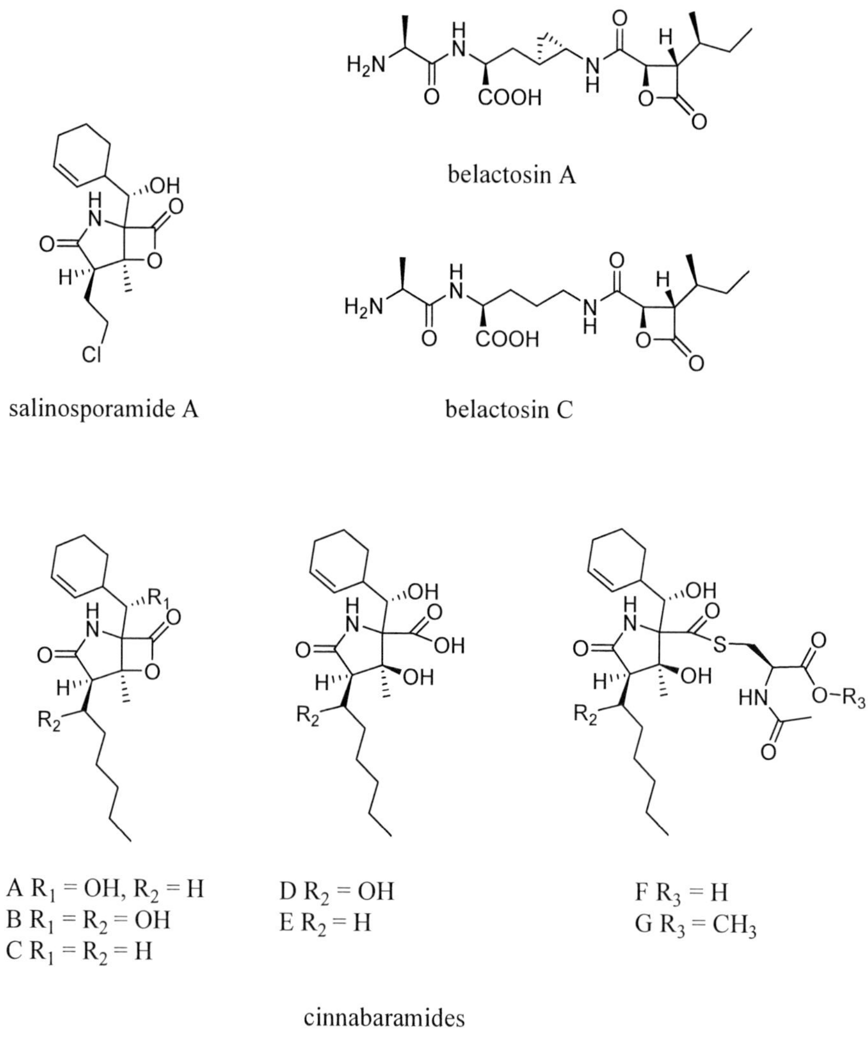

Figure 1. β-Lactone proteasome inhibitors.

2. Macrocyclic Compounds

Besides low molecular weight natural products, a number of large macrocyclic compounds have been reported to be potent proteasome inhibitors. These compounds can be classified into three sub-classes: mycalolides, syrbactins and macrocyclic alkanoids.

The mycalolides are macrolactones bearing tri-oxazole moieties isolated from sponges of the genus of *Mycale* and hard coral *Tubastrea faulkneri*. These mycalolides have been characterized to display cytotoxic and actin-depolymerizing properties. In 2005, Tsukamoto *et al.* isolated five structurally-related mycalolides from *Mycale*sponges that differed only in their oxidation state of C_{30}, or in the presence or absence of the third oxazole ring.[21] Three of the five compounds exhibited inhibitory effects against proteasome, with secomycalolide A showing the greatest activity with an IC_{50} value of 11 μg/mL. The common mechanism of action of these mycalolides involves inhibiting the chymotrypsin-like proteolytic activity of the proteasome.

Another class of macrocyclic compounds which display inhibitory effects against the proteasome are the syrbactins.[22] Syringolide A and its structural analogue glidobactin A. These were first reported to inhibit the proteolytic activity of yeast,[23] and later identified to suppress mammalian proteasome activity in leukaemic cell lines.[24] These compounds, like the β-lactones, react with the *N*-terminal threonine of the proteasome via Michael addition reactions. This results the formation of a covalent linkage and irreversibly inhibiting activity of the proteasome.

The third class of macrocyclic compounds which target the proteasome are classified as macrocyclic alkanoids. In 2000, Koguchi *et al.* isolated four structurally unusual alkanoids, TMC-95 A–D,[25] from *Apiospora montagnei*, screening the compounds for anti-proteasomal activity. TMC-95 A exhibited the most potent inhibition against the chymotrypsin-like, trypsin-like and PGPH-like activities of the proteasome.

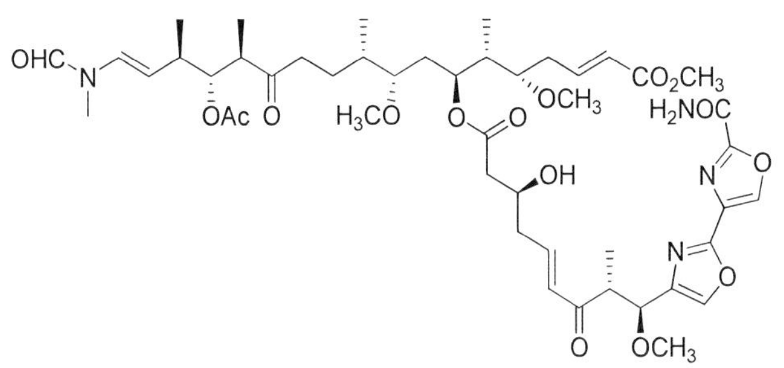

Secomycalolide A

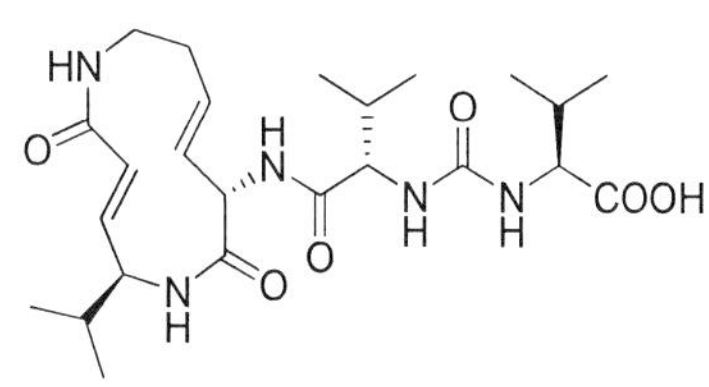

Syringolin A

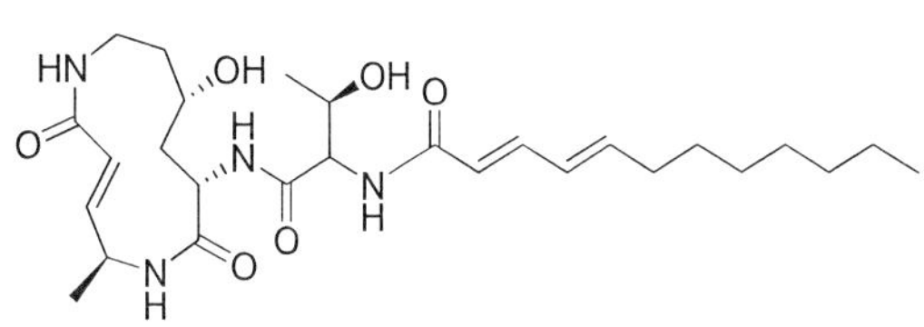

Glidobactin A

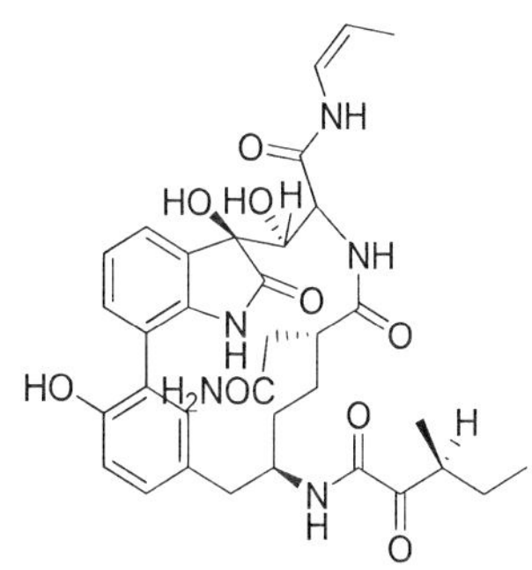

TMC-95 A

Figure 2. Representative examples of macrocyclic proteasome inhibitors.

In particular, TMC-95 A inhibits the chymotrypsin-like activity of the proteasome with an IC_{50} value in the nanomolar range. The mechanism of action of these alkaloids was remarkably different from those of the other macrocyclic proteasome inhibitors. The inhibitory effects of these compounds were found to be reversible, and highly selective towards the proteasome. Crystallographic analysis revealed that TMC-95 A bound to the core particle of the yeast proteasome with multiple hydrogen bonds, instead of through covalent modification of the *N*-terminal catalytic threonine unit of the proteasome.[26]

3. Phenolic Compounds

Polyphenolic compounds are common chemical constituents of the human diet and are mostly found in food sources such as fruits, beverages and vegetables.[27] Polyphenols have a wide range of beneficial effects, and they have been proposed to play important roles in prevention of human diseases, such as cancer and cardiovascular disease.[28] Reports have studied the mechanistic action of polyphenolic compounds in the body, and some have been shown to inhibit proteasomal activity.

Flavonoids are one of the major classes of polyphenolic compounds, and abundantly found in edible plants. They display a range of pharmacological properties including tumor suppression, anti-inflammation and anti-oxidant effects.[29] In 2001, Dou and co-workers reported that (–)-EGCG, a polyphenol found in green tea, inhibited the chymotrypsin-like, but not trypsin-like, proteasomal activity of purified 20S proteasome with nanomolar potency.[30] (–)-EGCG also exhibited promising proteolytic inhibitory effects against intact tumor cells with an IC_{50} value in the low micromolar range. Structure-activity relationship analysis of (–)-EGCG analogues attributed the inhibition of proteasomal activity to the presence of the gallate ester functionality.[31] Using computer modeling techniques, these ester-containing compounds inhibited the catalytic activity of the chymotrypsin-like domain by acylating its *N*-terminal threonine.[32]

Chen *et al.* reported that quercetin, kaempferol, myricetin and apigenin inhibited proteasomal activity *in* vitro.[33] Quercetin, myricetin and kaempferol are mainly found in grapes, whereas apigenin is found in chamomile flowers. Quercetin and apigenin showed the greatest inhibitory effects of both purified 20S proteasome and cultured leukemia cells, with IC_{50} values in the low micromolar range. Wu *et al.*[34] demonstrated that apigenin strongly inhibited the chymotrypsin-like and trypsin-like activities of the proteasome, but exhibited a weaker inhibitory effect towards PGPH-like proteasomal activity. Dosenko *et al.*[35] reported quercetin was potent against all three catalytic domains, but had the strongest inhibitory effect against chymotrypsin-like activity in purified 20S proteasome.

Chang investigated the pharmacological effects of a number of polyhydroxyflavones on the proteasome.[36] Superior activity against all three catalytic domains of the 20S core was exhibited by 5,6,4□-trihydroxy-7,3□-dimethoxyflavone, 5,6,3□,4□-tetrahydroxy-7-methoxyflavone and 6,7,4□-trihydroxyisoflavone . Emerging as the most potent inhibitor of ubiquitin-dependent proteasomal activity in *in vitro* assays was 5,6,4□-trihydroxy-7,3□-dimethoxyflavone. This suggests that hydroxylation at the C7 and C3□ positions are important for activity against the proteasome. Apigenin derivatives luteolin and chrysin have also been reported to possess anti-proteolytic activity. Luteolin is found in edible plants, while chrysin has been isolated from *Passifora caerulea*. Recently, Wu *et al.* have shown that luteolin and chrysin selectively inhibit chymotrypsin-like and trypsin-like proteolytic activity in cultured tumor cells. These, however, exhibited weak effects against PGPH activity. [34] Luteolin was found to be the strongest inhibitor of proteasomal activity, followed by apigenin and then chrysin. These results further suggest that the potency of flavone compounds against the proteasome may depend strongly on the hydroxylation pattern of the flavone backbone.

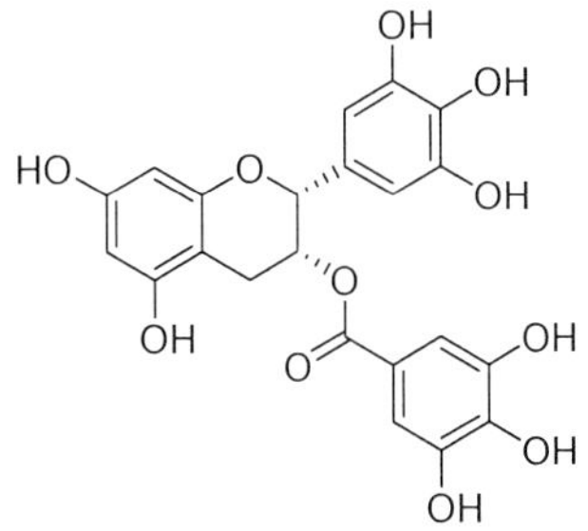

(-)-EGCG

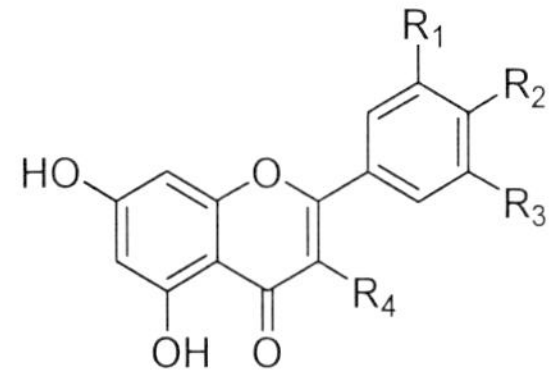

apigenin $R_1 = R_3 = R_4 = H$, $R_2 = OH$
kaempferol $R_1 = R_3 = H$, $R_2 = R_4 = OH$
quercetin $R_3 = H$, $R_1 = R2 = R_4 = OH$
myricetin $R_1 = R_2 = R_3 = R4 = OH$
chrysin $R_1 = R_2 = R_3 = R4 = H$
luteolin $R_1 = R_2 = OH$, $R_3 = R_4 = H$

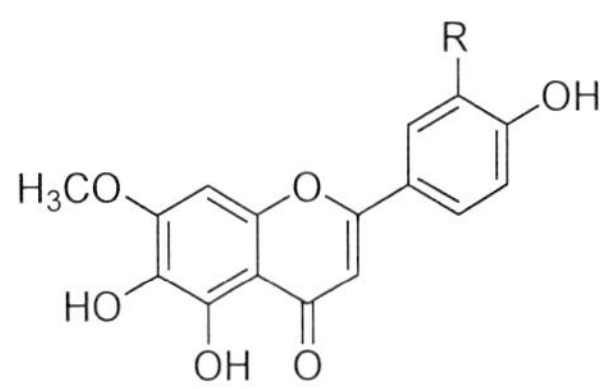

5,6,4'-trihydroxy-7,3'-dimethoxyflavone R = OCH_3
5,6,3',4'-tetrahydroxy-7-methoxyflavone R = OH

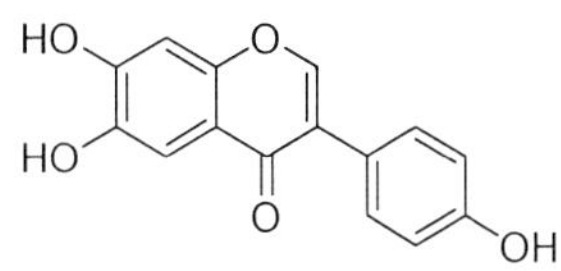

6,7,4'-trihydroxyisoflavone

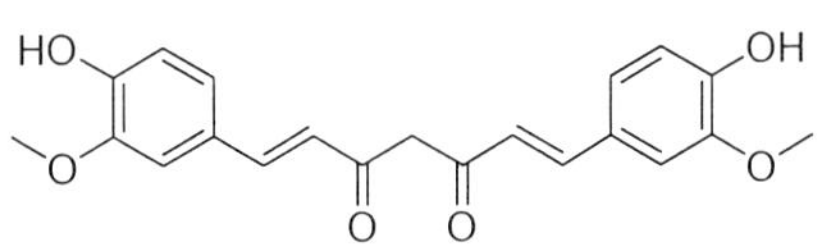

curcumin

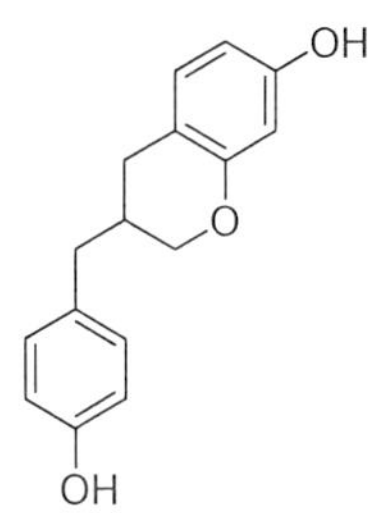

7-hydroxy-3-(4-hydroxybenzyl)-chroman

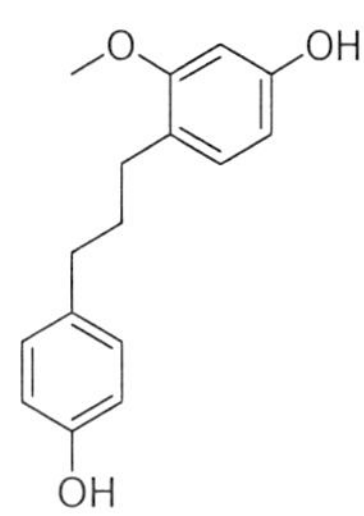

broussonin B

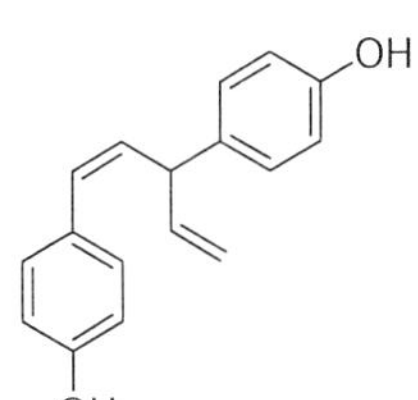

cis-hinokiresinol

Figure 3. Examples of polyphenolic compounds as proteasome inhibitors.

Curcumin is isolated from the rhizomes of Curcuma longa and displays a wide spectrum of pharmacological effects, including anti-proliferative, anti-inflammatory and anti-oxidant properties.[37] Milacic *et al.* demonstrated that curcumin inhibited proteasomal activity of purified 20S rabbit proteasome. Curcumin inhibited all three catalytic activities of the proteasome, but the strongest response was against chymotrypsin-like activity.[38] Cell-based experiments showed that curcumin could inhibit proteolytic activity and induce apoptosis in HCT-116 and SW480 cell lines. Administration of curcumin to HCT-116 to mice with colon tumors resulted in decreased tumor growth and apoptosis of the tumor tissues.[38]

In 2005, Tsukamoto *et al.* isolated four polyphenolic compounds from rhizomes of *Anemarrhena asphodeloides* for examination of their neurotrophic activity.[39] Additional *in vitro* studies revealed that three of the compounds inhibited the chymotrypsin-like activity of the proteasome.*Cis*-hinokiresinol was found to be a poor neurite outgrowth inducer, but exhibited good inhibitory activity against the proteasome with an IC_{50} value of 18 μg /ml.[39]

4. Steroid-Like Compounds

A large number of steroid-like compounds have been identified as proteasome inhibitors. Withanolides are compounds containing a steroid backbone linked to lactone or its derivatives. More than 300 withanolides have been identified, but only a few have been evaluated as proteasome inhibitors.

Withaferin A was the first withanolide isolated from *Withania smnifera*, also known as Indian Winter Cherry or India Ginseng. *Withania smnifera* has been used for centuries as a herbal medicine in India and has displayed a broad spectrum of therapeutic values, including neuroprotection,[40] anti-inflammation[41] and induction of anti-tumor effects.[42] While at least 35 natural products, including alkaloids and withanolides, have been isolated from *Withania smnifera*, withaferin A[43] has gotten significant attention due to its validated pharmacological properties. Yang *et al.* identified the proteasome as the primary target of withaferin A.[43]

Withaferin A inhibited the chymotrypsin-like activity of purified rabbit 20S proteasome with an IC_{50} value of 4.5 μM, and could suppress cellular proteasomal chymotrypsin-like activity in prostate cancer cells with an IC_{50} value of 10–20 μM. Computational modeling analysis predicted that withaferin A bound to the proteasome via covalent bonding to the threonine residue of the chymotrypsin-like subunit. Its congener withaferin Dshowed no direct inhibition on the proteasome, but was found to exert inhibitory effects against other targets in the ubiquitin-proteasome pathway.

Physalins are a class of steroidal compounds, isolated from species such as *Physalis angulate* and *Physalis lancifolia*, which were initially examined as anti-microbial and anti-parasitic agents. In 2006, physalin B and physalin C were identified as proteasome inhibitors in a high-throughput screening campaign.[44] These compounds were shown to inhibit ubiquitin-dependent protein degradation in an *in vitro* assay. Studies indicated that the proteasome may not be the primary target of these compounds, suggesting that their inhibition of proteolytic activity may be due to multiple actions exerted against the ubiquitin-proteasome pathway.

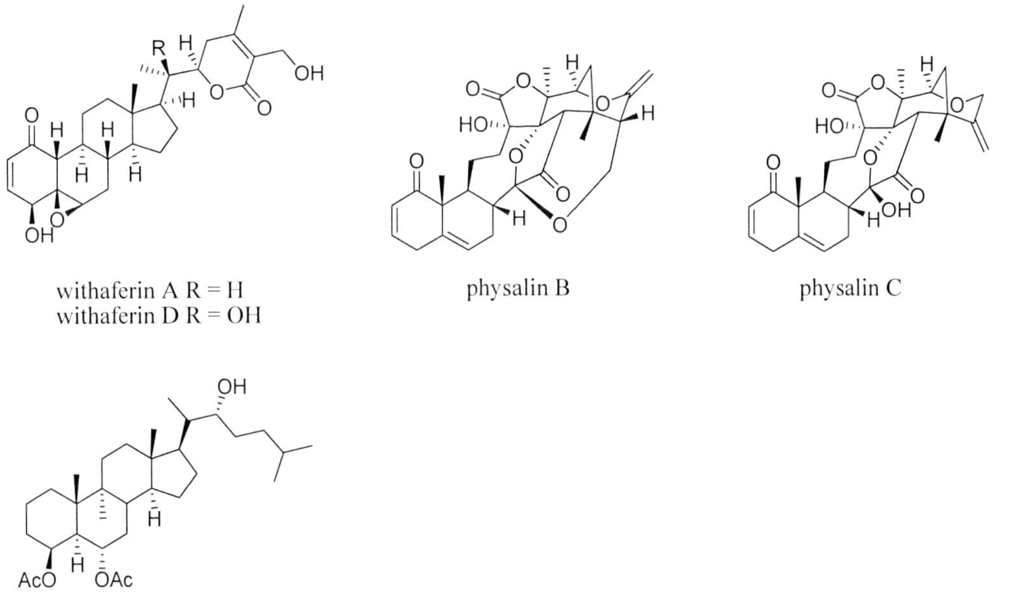

Figure 4. Steroid-like proteasome inhibitors.

Agosterols were initially isolated from the marine sponge *Spongia sp.*, and have been found to possess surprising inhibitory activity against multidrug-resistant cancer cells. In 2003, seven new agosterol derivatives were identified as proteasome inhibitors by Tsukamoto *et al.* from a high-throughput screening campaign.[45] These compounds were isolated from the marine sponge *Acanthodendrilla sp.*, and were found to exhibit moderate proteolytic inhibitory activity. Agosterol C showed the greatest inhibitory effect against the chymotrypsin-like activity of the proteasome, with an IC_{50} value of 10 μg/mL.

5. Triterpenes

Terpenes have been explored as modulating agents of the inflammatory response. A few terpenes have been identified to target the proteasome pathway. Celastrol is a quinone methide triterpene isolated from *Tripterygium regeli* or *Tripterygium wilfordii* (the Chinese plant Lei Gong Teng), and has been shown to inhibit purified 20S rabbit proteasome with an IC_{50} value of 2.5 μM.[46] Celastrol also possessed anti-proteasomal activity in a cell-based assay, with an IC_{50} value of 15 μM.[46] Celastrol has been reported to suppress NF-κB activation, inhibit IκB-α kinase phosphorylation and degradation, and down-regulate NF-κB-mediated gene expression.[46-47] Computer modeling analysis revealed celastrol interacted with the *N*-terminal threonine residue of the chymotrypsin-like catalytic domain of the proteasome.[46] Another structurally related quinone methide triterpene isolated from the *Celastraceae* and *Hippocrateaceae* families is pristimerin.[48] *In vitro* studies show that pristimerin potently inhibited the chymotrypsin-like catalytic domain of purified rabbit proteasome with an IC_{50} value of 2.2 μM. Further experiments indicate that pristimerin exhibited proteasome inhibition in androgen receptor negative PC3-prostate cancer cells with an IC_{50} value of 3 μM.

Natural Products Targeting enzymes of Ubiquitin or Ubiquitin-Like Pathways

Recognition of unwanted proteins by the proteasome is mediated through the polyubiquitination of substrate proteins. The most upstream component of the ubiquitin or ubiquitin-like pathways is the ubiquitin E1 enzyme. The E1 enzyme activates ubiquitin through the formation of an active thioester linkage in an ATP-dependent manner. Investigations suggest that ubiquitin is first adenylated at its *C*-terminal glycine residue, forming a labile and reactive center. Subsequent nucleophilic attack of ubiquitin by the E1 enzyme results in the formation of a thioester linkage.[49] The activated ubiquitin is then transferred to the E2 enzyme. Two rational approaches have been proposed to target the E1 enzyme.[50] The first approach is to block the access of ATP into the active site to inhibit the adenylation of ubiquitin. The second approach involves targeting the E1–E2 protein interface, preventing the transfer of ubiquitin to the E2 enzyme. One drawback to targeting the E1 ubiquitin enzyme, the most upstream component of the ubiquitin-proteasome system, is that multiple proteolytic pathways could be affected. This leads to indiscriminate or adverse effects on normal cellular function.

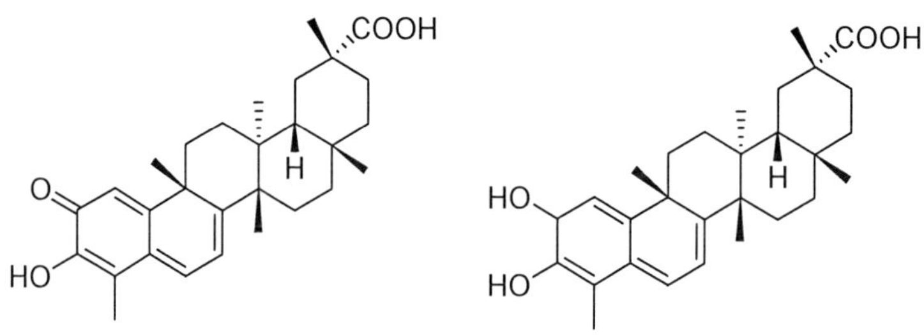

celastrol pristimerin

Fig.ure 5. Triterpene proteasome inhibitors.

Ubl conjugation pathways are very similar to the ubiquitin conjugation pathway. The Ubl conjugation signaling cascade involves the use of structurally and mechanistically similar enzymes for signal transduction.[4b] The Ubl conjugation pathways regulate distinct biological pathways important for cell proliferation and survival.[4b] One advantage of targeting the Ubl pathways is that the degradation of specific subsets of unwanted proteins can be modulated. For example, the NEDD8 E1 enzyme regulates the action of the cullin-based E3 ligases. Inhibition of the NEDD8 E1 enzyme can specifically modulate the degradation of protein substrates under the control of cullin-based E3 ligases.[51]

Another approach for inhibiting the ubiquitin or Ubl conjugation enzymatic cascades is by disrupting the interaction between the E3 enzyme and its substrate. Targeting E3 ligases may represent a selective inhibition approach as specific E3 enzymes recognize and "tag" their cognate downstream target proteins. The HDM2/MDM2 E3 enzymes recognize the p53 tumor suppressor protein, and have been the most widely studied E3 ligases.[52] HDM2/MDM2 is over-expressed in various cancer cells, leading to hyperactive ubiquitination and degradation of p53 by the proteasome. The specific targeting of HDM2/MDM2 by small molecules is a viable strategy to prevent the degradation of p53 inducing cell cycle arrest and apoptosis of cancer cells. Crystallographic analysis of the p53/HDM2 or p53/MDM2 complexes reveal that the p53 binding site in HDM2/MDM2 is large and hydrophobic.[53] The discovery of suitable inhibitors specifically targeting this large cleft may be a feasible approach to inhibit the interaction between HDM2/MDM2 and p53.

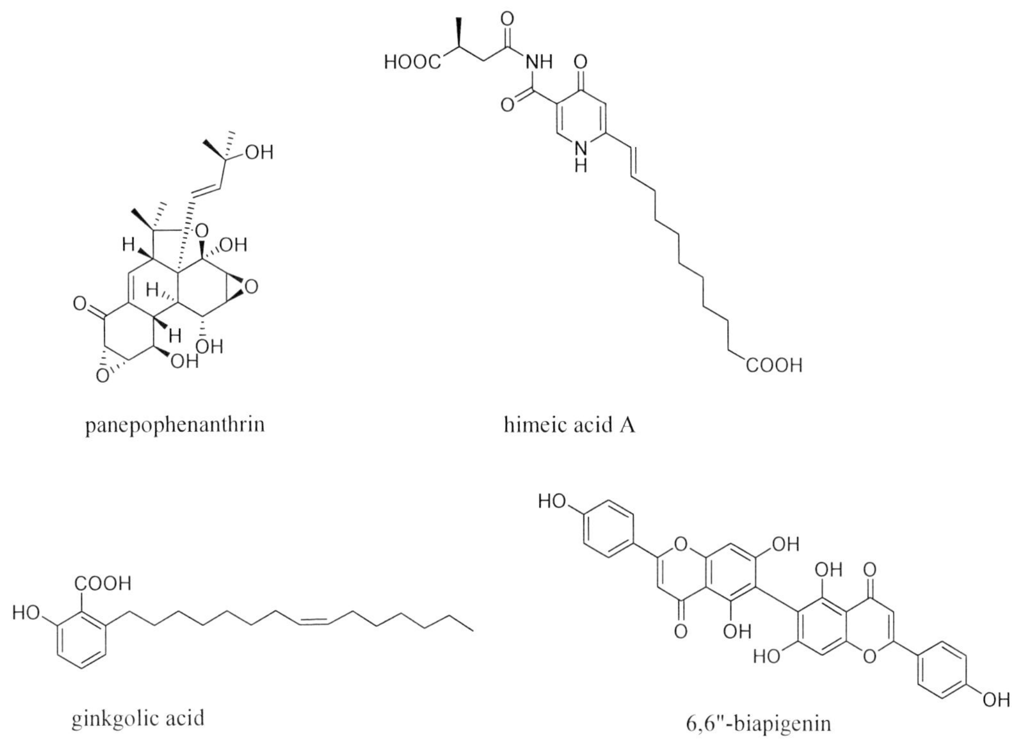

Figure 6. Natural products targeting ubiquitin or ubiquitin-like E1 enzymes.

Natural Products Targeting the E1 Ubiquitin-Activating Enzyme

The first natural product inhibitor of the E1 ubiquitin-activating enzyme (UAE) was identified by Sekizawa *et al.* from a screening campaign. Panepophenanthrin is a microbial metabolite isolated from the mushroom strain *Panus. rudis* Fr. IFO8894.[54] *In vitro* Western blot analysis showed that panepophenanthrin dose-dependently inhibited the formation of the E1-ubiquitin complex with an IC_{50} value of 17 μg/mL. Panepophenanthrin was inactive against E1 UAE in a cell-based assay at a concentration up to 50 μg/mL. Another study from the same research group identified the himeic acids as E1 UAE inhibitors. Himeic acid was isolated from the fungus strain *Aspergillus sp.* originating from the mussel *Mytilus edulis*.[55]

Western blot screening assays showed that himeic acid A dose-dependently inhibited the activity of E1 UAE, reducing ubiquitin conjugation by 65% at a concentration of 50 μM. The closely-related himeic acids B and C were inactive against E1-ubiquitin complex formation up to concentrations of 100 μM.[55] While these compounds were not as potent compared to the well-established natural product proteasome inhibitors, they represent the first two examples of E1 UAE inhibition by natural products, potentially serving valuable purpose as lead structures to generate more potent analogues targeting the E1 UAE.

Natural Products Targeting E1 Ubiquitin-Like Protein-Activating Enzymes

Post-translational modification of protein substrates by Ubls has attracted attention due to its role in controlling specific downstream degradation pathways. In 2009, Fukuda *et al.* identified ginkgolic acid and anacardic acid as inhibitors of E1 SUMO-activating enzyme (SAE).[56] Ginkgolic acid and anacardic acid are metabolites isolated from the leaves of *Ginkgo biloka*. They efficiently inhibited the formation of the SAE–SUMO complex in an *in vitro* Western blot assay with IC_{50} values of 3 μM and 2.2 μM, respectively. Both compounds were envisaged to inhibit E1 SAE through formation of thioester bonds with the cysteine residue of the enzyme active site. Interestingly, point mutagenesis experiments revealed that a cysteine-to-serine active site mutant was also efficiently bound by a fluorescent ginkgolic acid analogue, offering the possibility of an alternative SAE-inhibiting mechanism.

The pre-clinical success of MLN4924 as an inhibitor of E1 NEDD8-activating enzyme (NAE) has established NAE as a target for anti-cancer therapy.[51] MLN4924 is a synthetic adenosine sulphamate nucleotide mimic that forms an irreversible covalent adduct with NEDD8, thus inhibiting the transfer of NEDD8 to its E2 enzyme. The NEDD8 pathway is responsible for regulating cullin-RING finger ligase (CRL) activity. The specific inhibition of NAE could minimize all signal transduction pathways associated with the ubiquitin-proteasome system.

Our research group has identified 6,6≡-biapigenin as the first natural product-like inhibitor of NAE through the use of computer-aided virtual screening.[57] 6,6≡-biapigenin itself is not found in nature, but may be obtained in two steps from the natural product 6,6≡-binarigenin. 6,6≡-Biapigenin was shown to inhibit NAE activity in cell-free and cell-based assays, with IC_{50} values of 20 μM and 5 μM. Computational modeling analysis suggested that, unlike MLN4924, 6,6≡, -biapigenin does not form a covalent adduct with NEDD8. Multiple predicted hydrogen bonds with residues in the binding site suggest that the compound could act as a reversible ATP-competitive inhibitor of NAE. Blocking the access of ATP to the ATP-binding site of the enzyme would inhibit the adenylation of NEDD8, and preventsubsequent thioester bond formation with NAE.

Natural Product Targeting E3 Ubiquitin Ligases

Compared to the proteasome inhibitors, natural products targeting E3 ubiquitin ligase have received less attention. This may be due to the presence of large protein-protein interaction surfaces in E3 enzymes, usually considered to be more difficult targets. In 2006,

Duncan *et al.* conducted a high-throughput screening campaign to identify antagonists of the p53/MDM2 interaction. The group screened 53,000 compounds from a variety of microbial extracts and identified the natural product chlorofusin as an inhibitor of the p53/MDM2 interaction.[58] Chlorofusin, isolated from the fungal strain *Fusarium* sp. 22026, is a secondary metabolite produced from fungal fermentation processes. In an ELISA, chlorofusin inhibited the p53/MDM2 interaction with an IC_{50} value of 4.6 μM.

Subsequently, Tsukamoto *et al.* isolated (–)-hexylitaconic acid as an inhibitor of the p53/HDM2 interaction.[59] (–)-Hexylitaconic acid was isolated from the marine fungus *Arthrinium sp.* ELISA experiments revealed that (–)-hexylitaconic acid dose-dependently inhibited the p53/MDM2 interaction with an IC_{50} value of *ca.* 80 μg/mL. A close analogue of (–)-hexylitaconic acid, as well as two related dicarboxylic acids, did not show any inhibitory effects against the p53/HDM2 interaction. These results suggest that the presence of the terminal alkene of the dicarboxylic acid is important for inhibition of the p53/HDM2 interaction.

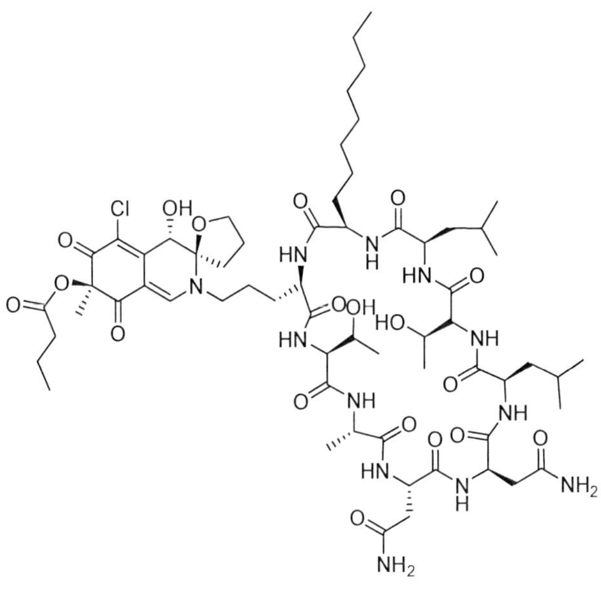

chlorofusin

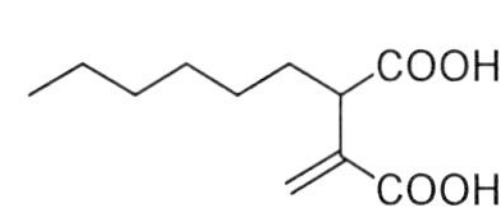

(-)-hexylitaconic acid

Figure 7. Natural products targeting E3 enzymes.

Natural Products Targeting Deubiquitinating Enzymes

Another possible therapeutic approach is to target deubiquitinating enzymes (ubiquitin isopeptidase, DUB), reversing the action of the ubiquitin conjugation pathway. DUBs are a large group of hydrolases which specifically cleave polyubiquitin isopeptide bonds and release monomeric ubiquitin for recycling. Several natural products have been reported to inhibit the action of DUB contributing to cell apoptosis. The cyclopentenone prostaglandins of the J series (PGJ) were identified by the group of Fitzpatrick[60] as the first natural product inhibitors of DUB activity *in vitro*. Δ^{12}-PGJ2, generated by the dehydration of its parent compound PGD2. Δ^{12}-PGJ2, was shown to induce apoptosis of cancer cells by suppressing the action of DUB through a p53-independent mechanism. Another class of prostaglandins showing inhibitory effects against DUB are the punagladins (PNG).[61] The punaglandins are a group of natural products isolated from the octocoral *Telesto riisei*. PNG 4 inhibited isopeptidase activity *in vitro*, and triggered cancer cell apoptosis in a p53-independent manner. These findings suggest that targeting the DUB could represent an alternative strategy for the treatment of diseases caused by dysfunctional ubiquitin-conjugation pathways.

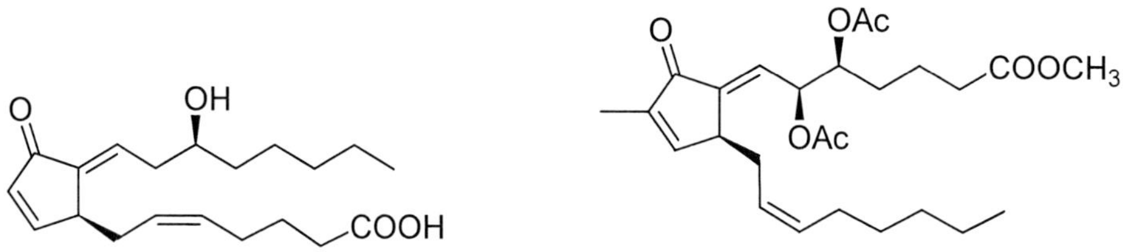

Δ^{12}-prostaglandin J_2 punaglandin 4

Figure 8. Natural products targeting deubiquitinating enzymes.

Conclusion

Since the seminal discovery of the ubiquitin-proteasome pathway, significant efforts have been put into finding chemical approaches to modulate the function of the ubiquitin-proteasome system for the treatment of various diseases. The further elucidation of the key players and mechanisms of the ubiquitin-proteasome pathway and related signaling pathways will only serve to present additional targets for the discovery of novel modulators of protein homeostasis in the future. Natural products offer numerous advantages as medical therapeutics, including bioactive substructure and diverse complex scaffolds that may be unattainable by conventional synthetic organic techniques. The unique chemical features of natural products may offer unprecedented interactions with biomolecular targets. As new products are continually isolated from various biological sources, any novel compounds exhibiting potent pharmacological effects against the ubiquitin-proteasome system can serve as a bioactive lead structure for further optimization.

We have attempted to summarize the discovery and application of natural products as inhibitors of the ubiquitin-/ubiquitin-like protein proteasome pathway, and have classified these natural product inhibitors based on their different biological targets, including the proteasome, E1 enzymes, E3 enzymes and deubiquitinating enzymes. We have also discussed their possible mechanisms of action, and envisage that this fascinating area will continue to receive critical attention in the future.This will fuel the discovery of more potent and selective natural product or natural product-like inhibitors of the ubiquitin-proteasome system.

This work is supported by Hong Kong Baptist University (FRG2/10-11/008), Environment and Conservation Fund (ECF Project 3/2010), Centre for Cancer and

Inflammation Research, School of Chinese Medicine (CCIR-SCM, HKBU), the Research Fund for the Control of Infectious Diseases (RFCID/11101212) and the Research Grants Council (HKBU/201811) and the University of Macau (Start-up Research Grant to C.-H. Leung).

REFERENCES

[1] A. Ciechanover, *Cell Death Differ.* 2005, *12*, 1178-1190;b) E. Reinstein, A. Ciechanover, *Ann. Intern. Med.* 2006, *145*, 676-684.

[2] A. Hershko, A. Ciechanover, *Annu. Rev. Biochem.* 1998, *67*, 425-479.

[3] B. A. Schulman, J. Wade Harper, *Nat. Rev. Mol. Cell. Biol.* 2009, *10*, 319-331.

[4] P. D'Arcy, S. Brnjic, M. H. Olofsson, M. Fryknas, K. Lindsten, M. De Cesare, P. Perego, B. Sadeghi, M. Hassan, R. Larsson, S. Linder, *Nat. Med.* 2011, *17*, 1636-1640;b) L. Bedford, J. Lowe, L. R. Dick, R. J. Mayer, J. E. Brownell, *Nat. Rev. Drug. Discov.* 2011, *10*, 29-46.

[5] T. A. Soucy, P. G. Smith, M. A. Milhollen, A. J. Berger, J. M. Gavin, S. Adhikari, J. E. Brownell, K. E. Burke, D. P. Cardin, S. Critchley, C. A. Cullis, A. Doucette, J. J. Garnsey, J. L. Gaulin, R. E. Gershman, A. R. Lublinsky, A. McDonald, H. Mizutani, U. Narayanan, E. J. Olhava, S. Peluso, M. Rezaei, M. D. Sintchak, T. Talreja, M. P. Thomas, T. Traore, S. Vyskocil, G. S. Weatherhead, J. Yu, J. Zhang, L. R. Dick, C. F. Claiborne, M. Rolfe, J. B. Bolen, S. P. Langston, *Nature* 2009, *458*, 732-736.

[6] D. J. Newman, G. M. Cragg, *J. Nat. Prod.* 2007, *70*, 461-477.

[7] J. Myung, K. B. Kim, C. M. Crews, *Med. Res. Rev.* 2001, *21*, 245-273.

[8] M. Muratani, W. P. Tansey, *Nat. Rev. Mol. Cell. Biol.* 2003, *4*, 192-201.

[9] D. M. Koepp, J. W. Harper, S. J. Elledge, *Cell* 1999, *97*, 431-434.

[10] C. Pickart, *Annu. Rev. Biochem.* 2001, *70*, 503-533;b) M. Verdoes, B. I. Florea, G. A. van der Marel, H. S. Overkleeft, *Eur. J. Org. Chem.* 2009, *2009*, 3301-3313.

[11] A. Julian, *Cancer Treat. Rev.* 2003, *29, Supplement 1*, 3-9.

[12] R. Rosenzweig, P. A. Osmulski, M. Gaczynska, M. H. Glickman, *Nat. Struct. Mol. Biol.* 2008, *15*, 573-580.

[13] F. Kopp, K. B. Hendil, B. Dahlmann, P. Kristensen, A. Sobek, W. Uerkvitz, *Proc. Natl. Acad. Sci. USA* 1997, *94*, 2939-2944.

[14] Y. Liu, Y. Ye, *Cell Res.* 2011, *21*, 867-883.

[15] S. Omura, K. Matsuzaki, T. Fujimoto, K. Kosuge, T. Furuya, S. Fujita, A. Nakagawa, *J. Antibiot.* 1991, *44*, 117-118.

[16] G. Fenteany, R. Standaert, W. Lane, S. Choi, E. Corey, S. Schreiber, *Science* 1995, *268*, 726-731.

[17] R. H. Feling, G. O. Buchanan, T. J. Mincer, C. A. Kauffman, P. R. Jensen, W. Fenical, *Angew. Chem. Inter. Ed.* 2003, *42*, 355-357.

[18] P. G. Williams, G. O. Buchanan, R. H. Feling, C. A. Kauffman, P. R. Jensen, W. Fenical, *J. Org. Chem.* 2005, *70*, 6196-6203.

[19] M. Stadler, J. Bitzer, A. Mayer-Bartschmid, H. Müller, J. Benet-Buchholz, F. Gantner, H.-V. Tichy, P. Reinemer, K. B. Bacon, *J. Nat. Prod.* 2007, *70*, 246-252.

[20] A. Asai, A. Hasegawa, K. Ochiai, Y. Yamashita, T. Mizukami, *J. Antibiot.* 2000, *1*, 81-83;b) M. Groll, O. V. Larionov, R. Huber, A. de Meijere, *Proc. Natl. Acad. Sci. USA* 2006, *103*, 4576-4579.

[21] S. Tsukamoto, K. Koimaru, T. Ohta, *Marine Drugs* 2005, *3*, 29-35.

[22] U. Wäspi, D. Blanc, T. Winkler, P. Rüedi, R. Dudler, *Mol. Plant Microbe In.* 1998, *11*, 727-733.

[23] M. Groll, B. Schellenberg, A. S. Bachmann, C. R. Archer, R. Huber, T. K. Powell, S. Lindow, M. Kaiser, R. Dudler, *Nature* 2008, *452*, 755-758.

[24] J. Clerc, B. I. Florea, M. Kraus, M. Groll, R. Huber, A. S. Bachmann, R. Dudler, C. Driessen, H. S. Overkleeft, M. Kaiser, *ChemBioChem* 2009, *10*, 2638-2643.

[25] Y. Koguchi, J. Kohno, M. Nishio, K. Takahashi, T. Okuda, T. Ohnuki, S. Komatsubara, *J. Antibiot.* 2000, *53*, 105-109;b) J. Kohno, Y. Koguchi, M. Nishio, K. Nakao, M. Kuroda, R. Shimizu, T. Ohnuki, S. Komatsubara, *J. Org. Chem.* 2000, *65*, 990-995.

[26] M. Groll, Y. Koguchi, R. Huber, J. Kohno, *J. Mol. Biol.* 2001, *311*, 543-548.

[27] W. Mullen, S. C. Marks, A. Crozier, *J. Agric. Food Chem.* 2007, *55*, 3148-3157.

[28] A. García-Lafuente, E. Guillamón, A. Villares, M. Rostagno, J. Martínez, *Inflamm. Res.* 2009, *58*, 537-552;b) V. Stangl, M. Lorenz, K. Stangl, *Mol. Nutr. Food Res.* 2006, *50*, 218-228.

[29] L. Bonfili, V. Cecarini, M. Amici, M. Cuccioloni, M. Angeletti, J. N. Keller, A. M. Eleuteri, *FEBS Journal* 2008, *275*, 5512-5526.

[30] S. Nam, D. M. Smith, Q. P. Dou, *J. Biol. Chem.* 2001, *276*, 13322-13330.

[31] A. Kazi, Z. Wang, N. Kumar, S. C. Falsetti, T.-H. Chan, Q. P. Dou, *Anticancer Res.* 2004, *24*, 943-954.

[32] D. M. Smith, K. G. Daniel, Z. Wang, W. C. Guida, T.-H. Chan, Q. P. Dou, *Protein. Struct. Funct. Genet.* 2004, *54*, 58-70.

[33] D. Chen, K. G. Daniel, M. S. Chen, D. J. Kuhn, K. R. Landis-Piwowar, Q. P. Dou, *Biochem. Pharmacol.* 2005, *69*, 1421-1432.

[34] Y.-X. Wu, X. Fang, *Planta Med.* 2010, *76*, 128-132.

[35] V. E. Dosenko, V. S. Nagibin, L. V. Tumanovskaia, V. Zagorii, A. A. Moibenko, *Biomed. Khim.* 2006, *52*, 138-145.

[36] T.-L. Chang, *J. Agric. Food Chem.* 2009, *57*, 9706-9715.

[37] C.-y. Jin, J.-d. Lee, C. Park, Y. h. Choi, G.-y. Kim, *Acta Pharmacol. Sin.* 2007, *28*, 1645-1651;b) A. B. Kunnumakkara, P. Anand, B. B. Aggarwal, *Cancer Lett.* 2008, *269*, 199-225.

[38] V. Milacic, S. Banerjee, K. R. Landis-Piwowar, F. H. Sarkar, A. P. N. Majumdar, Q. P. Dou, *Cancer Res.* 2008, *68*, 7283-7292.

[39] S. Tsukamoto, T. Wakana, K. Koimaru, T. Yoshida, M. Sato, T. Ohta, *Biol. Pharm. Bull.* 2005, *28*, 1798-1800.

[40] S. Kumar, C. J. Seal, M. J. R. Howes, G. C. Kite, E. J. Okello, *Phytother. Res.* 2010, *24*, 1567-1574.

[41] R. Agarwal, S. Diwanay, P. Patki, B. Patwardhan, *J. Ethnopharmacol.* 1999, *67*, 27-35.

[42] N. Sen, B. Banerjee, B. B. Das, A. Ganguly, T. Sen, S. Pramanik, S. Mukhopadhyay, H. K. Majumder, *Cell Death Differ.* 2006, *14*, 358-367.

[43] H. Yang, G. Shi, Q. P. Dou, *Mol. Pharmacol.* 2007, *71*, 426-437.

[44] F. Ausseil, A. Samson, Y. Aussagues, I. Vandenberghe, L. Creancier, I. Pouny, A. Kruczynski, G. Massiot, C. Bailly, *J. Biomol. Screen.* 2007, *12*, 106-116.

[45] S. Tsukamoto, M. Tatsuno, R. W. M. van Soest, H. Yokosawa, T. Ohta, *J. Nat. Prod.* 2003, *66*, 1181-1185.

[46] H. Yang, D. Chen, Q. C. Cui, X. Yuan, Q. P. Dou, *Cancer Res.* 2006, *66*, 4758-4765.

[47] J.-H. Lee, T. H. Koo, H. Yoon, H. S. Jung, H. Z. Jin, K. Lee, Y.-S. Hong, J. J. Lee, *Biochem. Pharmacol.* 2006, *72*, 1311-1321;b) G. Sethi, K. S. Ahn, M. K. Pandey, B. B. Aggarwal, *Blood* 2007, *109*, 2727-2735.

[48] H. Yang, K. R. Landis-Piwowar, D. Lu, P. Yuan, L. Li, G. P.-V. Reddy, X. Yuan, Q. P. Dou, *J. Cell. Biochem.* 2008, *103*, 234-244.

[49] A. Hershko, A. Ciechanover, I. Rose, *J. Biol. Chem.* 1981, *256*, 1525-1528;b) A. Ciechanover, H. Heller, R. Katz-Etzion, A. Hershko, *Proc. Natl. Acad. Sci. USA* 1981, *78*, 761-765.

[50] G. Nalepa, M. Rolfe, J. W. Harper, *Nat. Rev. Drug. Discov.* 2006, *5*, 596-613.

[51] T. A. Soucy, P. G. Smith, M. Rolfe, *Clin. Cancer Res.* 2009, *15*, 3912-3916.

[52] Y. Sun, *Cancer Biol. Ther.* 2003, *2*, 621-627.

[53] M. H. G. Kubbutat, R. L. Ludwig, M. Ashcroft, K. H. Vousden, *Mol. Cell. Biol.* 1998, *18*, 5690-5698.

[54] R. Sekizawa, S. Ikeno, H. Nakamura, H. Naganawa, S. Matsui, H. Iinuma, T. Takeuchi, *J. Nat. Prod.* 2002, *65*, 1491-1493.

[55] S. Tsukamoto, H. Hirota, M. Imachi, M. Fujimuro, H. Onuki, T. Ohta, H. Yokosawa, *Bioorg. Med. Chem. Lett.* 2005, *15*, 191-194.

[56] I. Fukuda, A. Ito, G. Hirai, S. Nishimura, H. Kawasaki, H. Saitoh, K.-i. Kimura, M. Sodeoka, M. Yoshida, *Chem. Biol.* 2009, *16*, 133-140.

[57] C.-H. Leung, D. S.-H. Chan, H. Yang, R. Abagyan, S. M.-Y. Lee, G.-Y. Zhu, W.-F. Fong, D.-L. Ma, *Chem. Commun.* 2011, *47*.

[58] S. J. Duncan, S. Grüschow, D. H. Williams, C. McNicholas, R. Purewal, M. Hajek, M. Gerlitz, S. Martin, S. K. Wrigley, M. Moore, *J. Am. Chem. Soc.* 2001, *123*, 554-560.

[59] S. Tsukamoto, T. Yoshida, H. Hosono, T. Ohta, H. Yokosawa, *Bioorg. Med. Chem. Lett.* 2006, *16*, 69-71.

[60] J. E. Mullally, P. J. Moos, K. Edes, F. A. Fitzpatrick, *J. Biol. Chem.* 2001, *276*, 30366-30373.

[61] S. M. Verbitski, J. E. Mullally, F. A. Fitzpatrick, C. M. Ireland, *J. Med. Chem.* 2004, *47*, 2062-2070.

BIOSKETCH

Dik-Lung Ma completed his PhD in 2004 at the University of Hong Kong under the supervision of Prof. C.-M. Che. He spent the years 2005–09 at the University of Hong Kong, the Hong Kong Polytechnic University, and the Scripps Research Institute with Prof. C.-M. Che, Prof. K-Y. Wong, and Prof. R. Abagyan. His research mainly focuses on luminescent sensing for biomolecules and metal ions, computer-aided drug discovery, and inorganic medicines. In 2010, he was appointed Assistant Professor at Hong Kong Baptist University.

Chung-Hang Leung completed his PhD in 2002 at City University of Hong Kong. After completing a five year post-doctoral fellowship at the Department of Pharmacology, Yale University, he was appointed Research Assistant Professor at the University of Hong Kong and then at the Hong Kong Baptist University. He is currently appointed as Assistant Professor at University of Macau. His primary research interests are in anti-cancer and anti-inflammatory drug discovery and molecular pharmacology.

In: Natural Products: Structure, Bioactivity and Applications ISBN 978-1-62081-728-5
Editors: Ramiro E. Goncalves and Marcos Cunha Pinto

Chapter 4

A REVIEW OF THE PLANT ORIGINS, COMPOSITION AND BIOLOGICAL ACTIVITY OF RED PROPOLIS

Begoña Gimenez-Cassina López
and Alexandra Christine Helena Frankland Sawaya*
Program of Bioscience and Bioactive Product Technology,
Plant Biology Department, Institute of Biology, São Paulo State University of Campinas,
UNICAMP, Campinas, São Paulo, Brazil

ABSTRACT

Propolis is a mixture of various amounts of beeswax and resins collected by bees from plants, particularly from buds and resinous exudates. The composition of propolis varies according to the kind of bee, geographic and plant origin of the samples. Propolis is important for the hive defense and has been used for its medicinal properties since ancient times. Red propolis has been found in the northeastern coast of Brazil, as well as in Cuba, Venezuela, Mexico and China. Among the most frequently cited plant sources are species of the Leguminosae and Clusiaceace families. The composition of red propolis varies both qualitatively and quantitatively, confirming different plant sources. The compounds that have been found in red propolis samples are listed herein. Red propolis has shown diverse biological activities: antimicrobial activity against different bacteria and yeast; against *Leishmania amazonensis* parasites; antioxidant, cytotoxic and potential antitumor activity; antipsoriatic, anti-inflammatory and analgesic activities; anti-obesity and hepatoprotective effects. Many new studies of red propolis are being developed due to the promising therapeutic activity; however, future studies must differentiate between red propolis from diverse geographic origins and chemical profiles.

* Corresponding author: franksawaya@terra.com.br.

INTRODUCTION

The word propolis has a Greek origin: meaning *pro*, in defense of, and *polis*, city; implying a product involved in the defense of the bee community. Propolis is used to strengthen the hive walls and coats the internal walls. It is also used to cover holes and cracks and to repair combs [1]. To prevent ants from invading the hives, bees place a ring of propolis on the entrance to make it narrow and difficult to enter. However, if an insect or small animal comes into the hive, the bees attack and kill it. If the body is too big to throw out of the hive, it is left inside and covered with propolis to prevent putrefaction [2]. As there are many bacterial and fungal species likely to grow in the hive, a layer of propolis reduces the microbial growth on the walls of the hive, also preventing wind and water entry [3].

Bees harvest propolis with their jaws and then manipulate it with the front legs and transfer it to the back legs. The salivary secretion serves as lubricant for the sticky glue texture of propolis. Returning to the hive, other bees separate each particle of the collected resin, mix with wax and the resulting propolis is pasted where it is needed inside the hive [4]. Propolis generally contains 30% beeswax, 5% pollen, 50-55% resinous substances and 15%essential oils [5].

Several species of bees produce propolis, although some have been better studied than others [6]. Different species of *Apis* live in diverse areas depending on the climatic features of each zone. *Apis mellifera* is an aggressive honey bee implicated in propolis collection [7]. This bee has a defensive behavior against intruders in their hives [8] and uses propolis as a defense mechanism. There are studies describing the different specialization levels of the bees to forage and the interaction between them [9]. Honeybee behavior has also been studied to understand better the processes involved in propolis and/or honey production [10]. In order to increase propolis production, Manrique and Egea (2002) studied why some *Apis mellifera* colonies produce more propolis than others, based on the genetic expression, showing it is possible to make a selection of bees to reach this objective [11].

The influence of the seasons on propolis production has been researched mostly in Brazil, as the mild weather permits resin collection throughout the year [12]. Environmental variables such as air temperature, humidity, precipitation, hours of solar radiation and wind speed were studied. Several different plants are the botanical source of propolis throughout the world. Since there are different plant sources for propolis depending on the geographical area of production, there are also different types of propolis. In tropical regions the botanical origin of some types of propolis is still being researched. The knowledge of plant sources of propolis is necessary to permit the chemical standardization of propolis (knowing the composition of the plant exudates, it is possible to know the quantitative composition of the propolis sample) [1].

Propolis has been used for its medicinal properties since ancient times and was used by the Egyptians to embalm cadavers because of its anti-putrefactive effect. The medicinal properties were recognized by Greek and Roman physicians like Aristoteles, Dioscorides, Phynil and Galen, and used by them as antiseptic and mouth disinfectant as well as for healing wounds. These applications of propolis were perpetuated until Middle Ages and spread to Arab physicians too. Incas used propolis as antipyretic and it was first included in the London pharmacopeia as an official drug in the seventh century. Since then, propolis has become very popular in Europe for its antibacterial activity [13]. Scientific proof of the therapeutic properties of propolis began to be studied in the 1950's, although the focus of these first studies was about bees' pollen collection [14].

Modern herbalists recommend propolis for its anti-bacterial, anti-fungal, anti-viral, hepatoprotective and anti-inflammatory properties, to increase the body's natural resistance to infections and to treat gastro-duodenal ulcers. Applied externally, propolis relieves various types of dermatitis caused by bacteria and fungi. Nowadays, propolis is a popular remedy available in the form of capsules, as an extract, as a mouthwash, in throat lozenges, creams and in powder form as well as part of cosmetic formulations and health food items [13].

Geographic and Plant Origins of Red Propolis

The physical appearance of propolis varies widely, depending on its botanical and geographical origin and on the type of bee involved in the collecting process. Propolis color may be cream, yellow, green, light, dark brown or red [15, 6]. The chemical composition of propolis is variable depending on the biodiversity and the geographical origin of this natural substance [17, 18].

There are various geographical areas where red propolis is found: the northeastern coast of Brazil, including the states of Alagoas, Bahia, Paraiba, Sergipe and Pernambuco, [19]; Cuba (Pinar del Rio), [20]; Venezuela [21]; México (Champoton) [22]; and China [23].

Studies regarding the plant sources of red propolis are very recent, dating back to 2007 and only for red propolis from Northeastern states of Brazil and from Cuba. Determining the botanical origin of propolis is a tricky task and two approaches have been applied: the comparative chemical composition analysis and the palynological analysis.

The chemical composition of red propolis samples was compared to samples of plant exudates. This approach demands detailed bee observation to be able to know which plants are visited by bees for resin collection. This observation is very important in order to determine and collect samples from possible plant sources and compare them to samples of propolis from a beehive located in the same area [17, 19].

For example, Daugsch et al. (2008) observed that bees were collecting the red exudates on surfaces of *Dalbergia ecastophyllum* to produce propolis. By comparison of propolis samples with *D. ecastophyllum,* the authors concluded that this was the botanical origin of the Brazilian red propolis. Brazilian red propolis contains isoflavonoids, and these compounds are present only in few species in the vegetal kingdom, almost exclusively in the Leguminosae family. *D. ecastophyllum* belongs to this family and contains the most important isoflavonoids that were also present in the red propolis samples tested [19]. This confirmed that this plant species is a source of resin for the production of Brazilian red propolis [17].

However, there are some compounds present in the red propolis samples that are absent in the *D. ecastophyllum* sample, suggesting that there are probably other botanical sources. It was observed that in the areas where certain propolis samples were collected, *D. ecastophyllum* was scarce, so bees also collected the exudates from other plants [17]. Piccinelli et al. (2011) detected some polyisoprenylated benzophenones in a Brazilian red propolis sample. These compounds are not present in *D. ecastophyllum*, and this result supports the theory of the contribution of other plant sources, probably from the Clusiaceace family [20].

The same study also analyzed Cuban red propolis, whose composition is similar to the Brazilian red propolis, so the botanical origin might be the same. This theory was confirmed by comparing the composition of the Cuban propolis sample to a sample of *D. ecastophyllum* [20].

Palynological analysis is another interesting approach to to identify the plant origins of red propolis. As with honey samples, the type of pollen found in propolis could be indicative of the plant species responsible for those resins. Barth and Luz (2009), carried out a study of propolis samples from the northeastern states of Brazil by the palynological method; all the data available related to this subject belongs to this single study. After analysis of the samples, 72 types of pollen were recognized and classified by family. According to this study, the red propolis samples showed great similarity with respect to the occurrence of the *Borrelia, Cocos* and *Schinus* pollen. The authors considered the presence of *Schinus* pollen to be characteristic of red propolis. In all sample localities included in this study, *Cocos nucifera* was growing along the coastal areas, which made it one of the possible plant sources. Red propolis sediments frequently contained pollen grains of Mimosa (pollen types of *M. scabrella, M. verrucosa* and *M. caesalpiniaefolia*), Arecaceae (*Cocos nucifera* and others), Cecropia (Cecropiaceae), three pollen types of Borreria (*B. densiflora, B. latifolia* and *B. verticillata*, Rubiaceae), *Symphonia globulifera* (Clusiaceae), Myrcia (Myrtaceae), Solanaceae, Tapirira and *Schinus terebinthifolius* (Anacardiaceae) [16].

The variety of plant species found indicates open land vegetation with trees and plants of hydrophilic preference. Samples from different states were found to have common pollen types, noting that some plant species are present in various areas while other are native of specific areas depending on the environmental characteristics.

In this study, in contrast with the research carried out by comparing red propolis samples and resin composition, *Dalbergia ecastophyllum* pollen was not detected, confirming that this is not the exclusive source of red propolis [16]. Furthermore, pollen of Clusiaceae family was found which correlates with the polyisoprenylated benzophenones found by Piccinelli et al. (2011) [20].

Composition of Red Propolis

Bees collect plant resins and exudates to produce propolis from different plant species, many of which are still unknown, hence the composition of propolis differs according to its botanical origin. Red propolis samples show differences in the chemical composition determined by the vegetation around the beehive. This raises a few questions that should be taken into account. Are the samples of red propolis from different geographic regions similar? How many plant sources are shared, considering there are samples from quite different geographic origins? Why are all these samples red?

At this point there are no easy answers to these questions. Some compounds have been identified to be part of propolis from some specific areas, while they are not present in red propolis from other regions. The composition not only varies qualitatively but also quantitatively. Some studies refer to a high flavonoid content, which has a direct relation with the plant source (when the resins come from *D. ecastophyllum* from the Leguminosae family), while others have higher percentage of phenolic compounds probably indicating they come from a plant belonging to the Clusiaceae family. The compounds that have been found in red propolis samples are listed in Table1.

Table 1. List of compounds identified in red propolis, divided by chemical classes

Class of compounds	Name	Reference
PHENYLPROPENE DERIVATES	trans-anethol	25
	methyl eugenol	
	trans-methyl isoeugenol	
	elemicin	
	trans-isoelemicin or asarone	
	methoxyeugenol	
	1-(30,40-dihydroxy-20-methoxyphenyl)-3-(phenyl)propane	22
	(Z)-1-(20-methoxy-40,50-dihydroxyphenyl)-2-(3-phenyl)propene	
	estragol	24
	metil-cis isoeugenol	
	1-methoxy-4-(1-propenyl)-benzene	29
	1,2,3-Trimethoxy-5-(2-propenyl)-benzene	
	scrobiculatones A and B (inseparable mixture)	21
	guttiferone E and xantochymol (inseparable mixture)	25
PHENOLIC COMPOUNDS	2,4,6-trimethylphenol	29
	methyl o-orsellinate	
	methylguaiacol	26
	homovanillic acid	
	phenolic acid, ferulic acid	
	caffeic acid-4-O-hexoside	
LIGNANS	pinoresinol dimethyl ether	28
	pinoresinol	
	syringaresinol	
TERPENS	anisylacetone	26
	methyl abietate	29
TRITERPENIC ALCOHOLS	α-amyrin	25
	β-amyrin	
	cycloartenol	
	lupeol	
	the ketone 20(29)-lupen-3-one	
	2,3-epoxy-2-(3-methyl-2-butenyl)-1,4-naphthalenedione	
MONOTERPENIC HYDROCARBONS	p-cymene	24
	limonene	
OXYGENED MONOTERPENS	1,8-cineol (eucaliptol)	24
	linalool	
SESQUITERPENIC HYDROCARBONS	α-cubebene	24
	α-copaene	
	β-gurjunene	
	β-caryophylene	
	α-bergamotene	

Table 1. (Continued)

Class of compounds	Name	Reference
	farnesene	
	δ-germacreno	
	α-selinene	
	isocariophylene	
	β-bisabolene	
	δ-cadinene	
OXYGENED SESQUITERPENS	δ-cadinol	24
FLAVONOIDS	isosativan	25
	medicarpin	
	prenylated benzophenones	
	3,4,2,3-tetrahydroxychalcone	
	volkensiflavone	
	5,7,3,4-tetramethoxyflavone	
	homopterocarpin	29
	4,7-dimethoxy-2-isoflavonol	
	chrysin	17
	pinocembrin	22
	arizonicanol A	
	melilotocarpan A	
	melilotocarpan D	
	3-hydroxy-5,6-dimethoxyflavan	
	quercetin	19
	rutin	
	Luteolin	
	formononetin	
	retupsapurpurin B	20
	retupsapurpurin A	
	(6αS,11αS)-6α-ethoxymedicarpan	28
	(6αS,11αS)-medicarpan	
	(6αS,11αS)-3,10-dihydroxy-9-methoxypterocarpan	
	(6αR,11αR)-4-methoxymedicarpin	
	(6αR,11αR)-3,8-dihydroxy-9-methoxypterocarpan	
	3-Hydroxy-8,9-dimethoxypterocarpan	17
	(6aR,11aR)-3,4-Dihydroxy-9-methoxypterocarpan	26
	vesticarpan	20
FLAVONOIDS	alnustinol	26
	(2R,3R)-3,7-dihydroxy-6-methoxyflavanone	28
	(2R,3R)-3,7-dihydroxyflavanone	
	garbanzol	
	(7S)-dalbergiphenol	
	(3R)-4′-methoxy-2′,3,7-trihydroxyisoflavanone	
	(3S)-ferreirin	26
	(3S)-violanone	
	(3S)-vestitone	28
	isoliquiritigenin	19
	2-hydroxy-4-methoxychalcone	26

Class of compounds	Name	Reference
	2′,4′-dihydroxychalcone	28
	4,4′-dihydroxy-2′-methoxychalcone	
	(αS)-α,2′,4,4′-tetrahydroxydihydrochalcone	
	mucronulatol	22
	(3S)-7-O-methylvestitol	26
	(3S)-mucronulatol	
	(3S) - vestitol	
	(3S)-isovestitol	28
	neovestitol	20
	2′-hydroxybiochanin A	28
	pratensein	
	xenognosin B	
	daidzein	19
	dalbergin	
	biochanin A	
	7,4-dihydroxyisoflavone	29
	gliricidin	26
	calycosin	
	(2S)-dihydrooroxylin A	28
	(2S)-dihydrobaicalein	
	(2S)-naringenin	
	(2S)-7-Hydroxy-6-methoxyflavanone	26
	naringenin-C-hexoside	
	Flavanone: 7-hydroxyflavanone	22
	pinobaskin	19
	pinobaskin - 3-acetate	
	liquiritigenin	
	alnusin	28
BENZOFURANS AND BENZOPYRANS	2-(2,4-dihydroxyphenyl)-3-methyl-6-methoxybenzofuran	26
	2-(20,40-dihydroxyphenyl)- 3-methyl-6-methoxybenzofuran	28
	2,6-dihydroxy-2-[(4-hydroxyphenyl)methyl]-3-benzofuranone	
	2H-1-Benzopyran-7-ol	29
ALKANES, ALCOHOLS, KETONES, ALDEHYDES AND ALIFATIC HYDROCARBONS	n-tricosane	26
	n-pentacosane	
	n-heptacosane	
	n-nonacosane	
	n-hentriacontane	
	n-tritiacontane	
	4-hydroxy-4-methyl-heptan-2-one	24
	6-methyl-5-hepten-2-one	
	octanal	
	nonanal	
	n-decanal	

Table 1. (Continued)

Class of compounds	Name	Reference
	anisaldehyde	
	n-dodecanal	
	farnesol	26
	butanedioic acid, dimethyl ester	29
	hydroxy-butanedioic acid, dimethyl ester	
	Hexadecanoic acid, methyl ester	
	10-Octadecenoic acid, methyl ester	
	tetradecanoe	24
	pentadecane	
	hexadecane	
AROMATIC HYDROCARBONS	naphthalene	24
	resorcinol	26
	benzoic acid	29
	2,2,6-Beta-trimethyl-bicyclo(4.3.0)non-9(1)-en-7α-ol	

The seasonal variability of red propolis compounds and bioactivity was studied using *Artemia salina* [24]. In this study, propolis samples were collected in the state of Pernambuco, in Brazil, at different times of the year: February, June and October. The authors observed there was a small variation in the volatile compound profile. Some sesquiterpens where only present in the sample collected in October, and some n-alkanes, a sesquiterpen and a monoterpen were only present in the sample collected in June, proving in this way that a seasonal variation in propolis composition exists. However, the main compound present in all the samples was the same (trans-anethol) and also high percentages of α-copaene and methyl-cis-isoeugenol were present.

In the study carried out by Trusheva et al. (2006), [25] the most abundant compound found was elemicin. Together with Methyl eugenol, Methyl isoeugenol and Isoelemicin, it was deemed responsible for the unusual anis-like odor of the samples they collected. β-amyrin was the main triterpenic alcohol. There was an inseparable mixture of Guttiferone E and Xanthochymol found in this sample, which coincides with the same compounds found in species of *Clusia*. This fact suggests that *Clusia* might be the botanical origin of some resins in this sample that was collected in Maceio, Alagoas State, Brazil. Righi et al., (2011) used a sample from Maceio, Alagoas State, Brazil as well. [26] However, the main compounds found in the sample were a chalcone and the isoflavans (3*S*)-7-*O*methylvestitol and (3*S*)-vestitol and the pterocarpan Medicarpin. They found that the sample was rich in phenolic compounds, corroborating previous studies carried out by Cabral et. al. in 2009 [27]. These compounds vary widely among different red propolis samples. The difference between the results obtained in both studies show that even samples collected in the same area have a different qualitative composition, which might be due to the season in which the samples were collected or the flora directly around the hives.

Daugsch et al. (2008), compared samples from the states of Bahia, Sergipe, Alagoas, Pernambuco, and Paraiba in northeastern Brazil, and the main compounds in all the samples were Isoliquitigenin, Formonetin and Pinocembrin.

The study carried out by Awale et al. (2008) analyzed samples from the South Coast of Paraiba in Brazil [28]. These authors isolated up to 41 compounds from the samples of red propolis, which presented a wide structural diversity among the flavonoid compounds. This study together with the study on red Mexican propolis [22] confirms that plants from the *Dalbergia* genus have many of the same main compounds.

Alencar et al. (2007), studied samples of red propolis collected in a mangrove area in the State of Alagoas, Brazil [29]. Isoflavones such as Homopterocarpin, Medicarpin and 4,7-dimethoxy-2-isoflavonol were the most abundant compounds.

The main compounds in the samples from Marechal Deodoro, Alagoas State, Brazil, were studied by Silva et al. (2008) [17]. The isoflavonoids Medicarpin and 3-Hydroxy-8,9-dimethoxypterocarpan represented more than 60% of the propolis extracts studied. High levels of phenolic compounds were found by Cabral et al. (2009) [27] that also analyzed red propolis from Marechal Deodoro, Alagoas in Brazil. Samples from the same area have also been studied by Oldoni et al. (2011) that reported Vestitol, Neovestitol and Isoliquiritigenin as the bioactive compounds of red propolis. [30]

Regarding Cuban red propolis composition, Cuesta-Rubio et al. (2007), studied samples from different parts of Cuba provided by La Estacion Experimental Apicola [31]. The main compounds turned out to be isoflavonoids. Seven Cuban propolis samples from Pinar del Rio, Villa Clara and Matanzas, showed some quantitative differences but similar qualitative composition. Vestitol, Formononetin and another unidentified isoflavan were always present, although their concentration varied. Once again, quantitative variations between samples collected in areas near to each other suggest that the flora, climate, age and soil can influence the composition of propolis. Isoflavans were also confirmed to be the main components in Cuban red propolis [32]. Another study on Cuban red propolis collected in Pinar del Rio, reported the presence of the isoflavonoids: Medicarpin, Vestitol, Liquiritigenin, Isoliquiritigenin, Formonetin and Biochanin A, as principal components [20]. These results obtained by Piccinelly et al. (2011) agree with other studies on red propolis from Brazil; however they evidenced the presence of polyisoprenylated benzophenones not found in samples from other geographical origins.

The composition of red Mexican Propolis from Champoton, was studied by Lotti et al. (2007) and isoflavonoids were found to be the main compounds, much like red Cuban propolis [22]. The high concentration of Pinocembrin in the sample was noteworthy as it is also found in *D. ecastophyllum* as well as Mucronulatol and Vestitol. Melilotocarpan A and D are pterocarpans that were also found in the red Mexican propolis sample, and these compounds were isolated from another species of *Dalbergia*. These data reveal that the source(s) of red Mexican propolis are from the *Dalbergia* genus.

Venezuelan propolis collected in a tropical rain forest in Trujillo State was studied by Trusheva et al. (2004) and an inseparable bioactive mixture was identified as Scrobiculatones A and B (polyisoprenylated benzophenones [21]. These compounds were also isolated in red Cuban propolis [31].

Biological Activity of Red Propolis

Red propolis has been shown to have diverse biological activities. Since differences in propolis composition exist, depending on its geographical origin and plant source, it is important to keep in mind that these activities may not be true for all samples of red propolis. Therefore, in the future, the origin of a sample should always be stated in studies of composition and activity.

The antimicrobial activity has been studied against different bacteria; *Staphylococcus aureus* [27], *Streptococcus mutans*, and *Actinomyces naeslundii* [30]. Cabral et al. (2009) suggested that the antibacterial activity against *S. aureus* was due to a synergic effect between phenolic compounds and other compounds present in the crude propolis extract [27]. However they didn't assign this activity to a specific compound. Alencar et al. (2007) also associated the antimicrobial activity mainly to the polyphenolic compounds [29].

The antimicrobial effect against *S. aureus* was also verified by Oldoni et al. (2011), who isolated and identified some of the compounds responsible for the antimicrobial activity. Vestitol and Isoliquiritigenin were the two compounds that showed strongest activity against the three species of bacteria they tested, being that Isoliquiritigenin had stronger activity [30]. They realized as well that no isolated compound showed antibacterial effect against A. *naeslundii.* Testing the possible synergetic effect of flavonoids against bacteria, they showed that Vestitol and Isoliquiritigenin were not able to potencialize each other's antibacterial activity [30].

Righi et al. (2011), confirmed the antimicrobial activity of red propolis against Gram-positive bacteria (*Bacillus subtilis*, *Enterococcus faecalis* and *Streptococcus pyogenes*) and Gram-negative bacteria (*Klebsiella pneumoniae, Pseudomonas aeruginosa, Salmonella typhimurium* and *Escherichia coli*), but the individual isolated activity of the compounds in the sample was not studied [26].

The effect of red propolis against parasites was studied on *Leishmania amazonensis* by Ayres et al. (2007) and red propolis reduced significantly the infection produced by this parasite, however it did not present direct effect on promastigotes or extracellular amastigotes. Ayres et al. (2007) didn't identify the compounds responsible for this activity [33].

The antifungal activity of propolis has been studied on yeast cells of *Saccharomyces cerevisiae* and on species of *Trychophyton* [34, 35]. Lotti et al. (2011) demonstrated that Propolis could be a source of compounds that could alleviate the multidrug resistance problem in fungi such as *S. cerevisiae*, thus it could be used for the treatment of fungal infections together with azoles. They studied the effect of some isoflavonoids of red propolis, and 7-O-methylvestitol appeared to be the most active. The study carried out by Siqueira et al. (2009) evidenced that red propolis is more efficient than green propolis for the *Trychophyton* species. The study reveals the antifungal activity of red propolis and suggests it as a possible alternative treatment for dermatophytosis caused by these species. Among all the red propolis compounds, Medicarpin, was shown to be an important compound with antimicrobial and antifungal activity [25].

Reactive Oxygen Species, (ROS) are formed in cells as a consequence of biochemical reactions and external factors. Under abnormal conditions, the excess production of these compounds causes cellular damage and thus participates in pathologies such as cancer, neurodegenerative diseases, coronary diseases and others. The antioxidant activity of red propolis was studied by Righi et al. (2011), who showed that phenolic compounds were responsible for this activity because they intercepted the free radical chain oxidation [26]. These authors suggested that chalcones and isoflavonoids might be involved in the process, as electron donors; 7-O-methylvestitol, Medicarpin and 3,4,2',3'-tetrahydrochalcone, were shown to have antioxidant activity. Alencar et al. (2007) had already verified with the Pearson test, that there was a correlation between antioxidant activity and flavonoids [29], and Cabral et al. (2009) also showed a correlation between phenolic compounds and antioxidant activity [27]. As probably all the phenolic and flavonoid compounds participate, at least partially, in the antioxidant activity, studies on isolated compounds with antioxidant activity are very few. Oldoni et al. (2011) analyzed the activity of Vestitol, Isovestitol and Neovestitol, showing that Vestitol had higher antioxidant potential, but that the others also presented this activity [30].

One of the few studies of Chinese red propolis showed that red propolis had a higher antioxidant activity than green propolis [23]. In this study, caffeic acid phenethyl ester (CAPE), which is a compound present in this red propolis sample, showed potent activity as well as other caffeic derivates.

To understand the cytotoxic activity of red propolis, it is important to take into account the physiological base of cancer. Cancer cells proliferate very rapidly and have a high demand of nutrients and oxygen. However cancer cells can acquire tolerance to nutrient starvation and become resistant in this way. Awale et al. (2008) studied the cytotoxic activity of red propolis by creating a nutrient-deprived condition for the cancer cells, and analyzing the cytotoxic activity of the individual isolated propolis compounds [28]. Of all the compounds studied, 3,8-hydroxy-9-methoxypterocarpan displayed the most potent preferential cytotoxicity against the cancer cell line studied. All the pterocarpans present in the sample showed cytotoxic activity, and their decreasing order of activity was : (6aR,11aR)-3,8-dihydroxy-9-methoxypterocarpan, (6aR,11aR)-3,4-dihydroxy-9-methoxypterocarpan, (6aS,11aS)-3,10-dihydroxy-9-methoxypterocarpan, (6aS,11aS)-medicarpan, (6aS,11aS)-6a-ethoxymedicarpan, (6aR,11aR)-4-methoxymedicarpin and (6aR,11aR)-3- hydroxy-8,9-dimethoxypterocarpan. 7-hydroxy-6-methoxyflavanone exhibited cytotoxic activity against melanoma, Lewis lung carcinoma, human lung carcinoma and human fibrosarcoma, while Mucronulatol only showed cytotoxicity against melanoma and Lewis Lung carcinoma cell lines [36]. This compound targets the control of the cell line progression, and thus it exerts cytotoxicity in cancer cells [22].

Alencar et al. (2007) assessed the cytotoxic activity of red propolis on Hela tumor cells, but didn't identify the isolated compounds responsible for it [29]. Nunes et al. (2007) worked on the bioactivity of red propolis on *Artermia salina* and suggested that a possible antitumor activity of red propolis could be due to its content of phenolic compounds [24]. Red propolis showed strong suppressive effects against vascular endothelial growth factor induced angiogenesis which is a key regulator of pathogenic angiogenesis in diseases such as cancer and diabetic retinopathy [23]. Nevertheless more studies on this topic are still needed.

Other properties of red propolis have been reported, such as its antipsoriatic, anti-inflammatory and analgesic activities [37]. Red propolis presents wound healing activity and was incorporated into collagen-based films [38]. Ledon et al. (2002) didn't find dermal and ocular toxicity on their tests on a red propolis extract, however they found a dose dependent allergic response, evidenced with the production of erythema [39]. The anti-obesity effect of propolis is another studied activity. Propolis enhances the differentiation of adipocytes by activating PPARγ, and is capable of inhibiting TNFα effects on adipocyte differentiation and adiponectin expression, thus it has been put forward as a possible diet supplement for prevention and treatment of obesity and obesity-associated disorders [40]. Red propolis appears to have a hepatoprotective effect and might have activity in the prevention of hepatitis [41].

Many new studies of red propolis are being developed due to the promising therapeutic effects shown above. However, care must be taken to differentiate between red propolis from diverse geographic origins and chemical profiles.

CONCLUSION

The samples of red propolis studied so far have come mainly from the tropical regions in South America, with the exception of one sample from China. In spite of the red color, the chemical composition has been found to vary qualitatively and quantitatively, even between samples from the same region. The plant sources of resins have been investigated by two different methods: the comparative chemical composition analysis and the palynological analysis. Many different plant sources have been suggested, being that the most frequent are plants of the Leguminosae and Clusiaceace families. Further studies are necessary to determine the main chemical markers of red propolis, or if there are different types of red propolis, and how variations in bee, season and region may affect its composition.

Red propolis has shown diverse biological activities: antimicrobial activity; against parasites; antioxidant, cytotoxic; antipsoriatic, anti-inflammatory and analgesic activities; anti-obesity and hepatoprotective effects. These results indicate that this natural product deserves to be better studied for its promising therapeutic effects, and many new studies of red propolis are indeed being developed. However, future studies must differentiate between red propolis from diverse geographic origins and chemical profiles.

REFERENCES

[1] Bankova, V. S.; DE Castro, S. L.; Marcucci, M. C. (2000). Propolis: recent advances in chemistry and plant origin. *Apidologie , 31*, 3-15.

[2] Gonnet, M. (1968). Propriétés phytoinhibitrices de la colone d'abeilles (Apis Mellifera L.). *Annales des abeilles , 11* (2), 105-116.

[3] Simone-Finstrom, M.; Spivak, M. (2010). Propolis and bee health: the natural history and significance of resin use by honey bees. *Apidologie , 41*, 295-311.

[4] Chauvin, R. (1960). Progrès récents dans la biologie de l'abeille. *Annales de l'abeille.*

[5] Derevici, A.; Popesco, A.; Popesco, N. (1964). Recherches sur certaines propriétés biologiques de la propolis. *Annales des abeilles , 7* (3), 191-200.

[6] Delaplane, K. S. (2010). Honey Bees and Beekeeping. *Learning for life*, Bulletin 1045.

[7] Seeley, T.D.; Morse R.A. (1976). The nest of the honey bee (Apis Mellifera L). *Insectes Sociaux, Paris. , 23* (4), 495-512.

[8] Atkinson, E.; ELLIS, J. (2001). Honey bee, Apis mellifera L., confinement behavior toward beetle invaders. *Insectes Sociaux*, *58*, 495–503.

[9] Drezner-LEVY, T., Smith, B. H., & Shafir, S. (2009). The effect of foraging specialization on various learning tasks in the honey bee (Apis mellifera). *Behav Ecol Sociobiol*, *64*, 135–148.

[10] Ikeno, H.; Ohtani, T. (2001). Reconstruction of honeybee behavior within the observation hive. *Neurocomputing* (38-40), 1317-1323.

[11] Manrique, A. J.; Soares, A. E. E. (2002). Inicio de um programa de seleção de abelhas africanizadas para a melhora na produção de propolis e seu efeito na produção de mel. *Interciencia*, *27* (006), 312-316.

[12] Inoue, H. T. ; DE Sousa, E.; DE Oliveira Orsi, R.; Funari, S.C.; Carelli Barreto, L.; DA Silva Dib, A. (2007). Produção de própolis por diferentes métodos de coleta. *Asociación Latinoamericana de Producción Animal, 15* (2), 65-69.

[13] Castaldo, S.; Capasso, F. (2002). Propolis, an old remedy used in modern medicine. *Fitoterapia*, *73* (Supl. 1), S1-S6.

[14] Louveaux, J. (1958). Recherches sur la récolte du pollen par les abeilles (Apis Mellifera L). *Annales des abeilles.*

[15] Salatino, A.; Teixeira, É.W.; Negri, G.; Message, D. (2005). Origin and Chemical Variation of Brazilian propolis. *eCam*, *2* (1), 33-38.

[16] Barth, O. M.; DA Luz, C.F.P. (2009). Palynological analysis of Brazillian red propolis samples. *Journal of Apicultural Research and Bee world*, *48* (3), 181-188.

[17] Silva, B.B.; Rosalen, P.L.; Cury, J.A.; Ikegaki, M.; Souza, V.C.; Esteves, A.; Alencar, S.M. (2008). Chemical Composition and Botanical Origin of Red Propolis a New Type of Brazilian Propolis. *eCAM*, *5* (3), 313–316.

[18] Sawaya, A.C.H.F.; Tomazella, D.M.; Cunha, I.B.S.; Bankova, V.S.; Marcucci, M.C.; Custodio, A.R.; Eberlin, M.N. (2004). Electrospray Ionization Mass Spectrometry Fingerprinting of Propolis. Analyst (London), v. 129, p. 739-744.

[19] Daugsch, A.; Moraes, C. S.; Fort, P.; Park, Y. K. (2008). Brazilian Red Propolis—Chemical Composition and Botanical Origin. *eCAM*, *5* (4), 435-441.

[20] Piccinelli, A. L.; Lotti, C.; Campone, L.; Cuesta-Rubio, O.; Fernandez, M.C.; Rastrelli, L. (2011). Cuban and Brazilian Red Propolis: Botanical Origin and Comparative Analysis by High-Performance Liquid Chromatography - Photodiode Array Detection/Electrospray Ionization Tandem Mass Spectrometry. *Journal of Agricultural and Food Chemistry*, *59*, 6484–6491.

[21] Trusheva, B.; Popova, M.; Naydenski, H.; Tsvetkova, I.; Rodriguez, J.G.; Bankova, V. (2004). New polyisoprenylated benzophenones from Venezuelan propolis. *Fitoterapia*, *75*, 683– 689.

[22] Lotti, C.; Fernandez, M.C.; Piccinelli, A. L.; Cuesta-Rubio, O.; HERNANDEZ, I.M. (2010). Chemical Constituents of Red Mexican Propolis. *Journal of Agricultural and Food Chemistry*, *58*, 2209–2213.

[23] Izuta, H.; Narahara, Y.; Shimazawa, M.; Mishima, S.; Kondo, S.-I.; Hara, H. (2009). 1,1-Diphenyl-2-picrylhydrazyl Radical Scavenging Activity of Bee Products and Their Constituents Determined by ESR. *Biol. Pharm. Bull.*, *32* (12), 1947—1951.

[24] Nunes, L.C.C.; Galindo, A.B.; DE Deus, A.D.O.; Rufino, D. A.; Randau, K.P.; Satiro, H., y otros. (2009). Variabilidade sazonal dos constituintes da própolis vermelha e

bioatividade em Artermia salina. *Revista Brasileira de Farmacognosia* , *19* (2B), 524-529.

[25] Trusheva, B. et al. (2006). Bioactive Constituents of Brazilian Red Propolis. *eCAM* , *3* (2), 249-254.

[26] Righi, A.A.; Alves, T.R.; Negri, G.; Marques, L.M.; Breyer, H.; Salatino, A. (2011). Brazilian red propolis: unreported substances, antioxidant and antimicrobial activities. *J Sci Food Agric* , *91*, 2363-2370.

[27] Cabral, I.R.; Oldoni, T. L.C.; Prado, A.; Bezerra, R. M.; DE Alencar, S.M. (2009). Composição fenolica, atividade antibacteriana e antioxidante da própolis vermelha brasileira. *Quimica Nova* , *32* (6), 1523-1527.

[28] Awale, S.; LI, F.; Onozuka, H.; Esumi, H.; Tezukaa, Y.; Kadota, S. (2008). Constituents of Brazilian red propolis and their preferential cytotoxic activity against human pancreatic PANC-1 cancer cell line in nutrient-deprived condition. *Bioorganic & Medicinal Chemistry* , *16*, 181–189.

[29] Alencar, S.; Oldoni, T.; Castro, M.; Cabral, I.; Costa-Neto, C.; Cury, J.; Rosalen, P.L.; Ikegaki, M. (2007). Chemical composition and biological activity of a new type of Brazilian propolis: Red propolis. *Journal of Ethnopharmacology* , *113*, 278-283.

[30] Oldoni, T.L.; Cabrala, I.S.; D'arcea, M.A.R.; Rosalen, P.L.; Ikegaki, M.; Nascimento, A.M.; Alencar, S.M. (2011). Isolation and analysis of bioactive isoflavonoids and chalcone from a new type of Brazilian propolis. *Separation and Purification Technology* , *77*, 208-213.

[31] Cuesta-Rubio, O.; Piccinelli, A.L.; Fernandez, M. C.; Hernandez, I. M.; Rosado, A.; Rastrelli, L. (2007). Chemical Characterization of Cuban Propolis by HPLC-PDA, HPLC-MS, and NMR: the Brown, Red, and Yellow Cuban Varieties of Propolis. *J. Agric. Food Chem.* , *55*, 7502-7509.

[32] Fernandez M.C. et al. (2008). GC-MS Determination of Isoflavonoids in Seven Red Cuban Propolis samples. *Journal of Agricultural and Food Chemistry*, 56, 9927-9932.

[33] Ayres, D.C.; Marcucci, M.C.; Giorgio, S. (2007). Effects of Brazilian propolis on Leishmania amazonensis. *Mem Inst Oswaldo Cruz* , *102(2)*, 215-220.

[34] Lotti, C.; DE Castro, G. M.M.; DE Sá, L. F.R.; DA Silva, B.D.S.; Tessis, A.C.; Piccinelli, A.L.; Rastrelli, L.; Ferreira-Pereira, A. (2011). Inhibition of Saccharomyces cerevisiae Pdr5p by a natural compound extracted from Brazilian Red Propolis. *Revista Brasileira de Farmacognosia/ Brazilian Journal of Pharmacognosy* , *21* (5), 901-907.

[35] Siqueira, A. et al. (2009). Trichophyton species susceptibility to green and red propolis from Brazil. *The Society for Applied Microbiology, Letters in Applied Microbiology* , *48*, 90–96.

[36] Li, F.; Awale, S.; Tezuka, Y.; Kadota, S. (2008). Cytotoxic constituents from Brazilian red propolis and their structure–activity relationship. *Bioorganic & Medicinal Chemistry* , *16*, 5434-5440.

[37] Ledon, N.; Casaco, A.; Gonzalez, R.; Merino, N.; Gonzalez, A.; TOLON, Z. (1996). Efectos antipsoriásico, antiinflamatorio y analgésico del propoleo rojo colectado en Cuba. *Revista Cubana de Farmacia* , *30* (1).

[38] De Albuquerque-Junior, R.L. et al. (2009). Effect of Bovine Type-I Collagen-Based Films Containing Red Propolis on Dermal Wound Healing in Rodent Model. *Int. J. Morphol.* , *27* (4), 1105-1110.

[39] Ledon, N.; Casacó, A.; Gonzalez, R.; Bracho, J.; Rosado, A. (2002). Assessment of potential dermal and ocular toxicity and allergic properties of an extract of red propolis. *Arch Dermatol Res*, *293*, 594-596.

[40] Iio, A.; Ohguchia, K.; Inoueb, H.; Maruyamac, H.; Araki, Y.; Nozawaa, Y.; Ito, M. (2010). Ethanolic extracts of Brazilian red propolis promote adipocyte differentiation through PPAR activation. *Phytomedicine*, *17*, 974-979.

[41] Rodriguez, S.; Ancheta, O.; Ramos, M.E.; Remirez, D.; Rojas, E.; Gonzalez, R. (1997). Effects of Cuban red propolis on galactosamine-induced hepatitis in rats. *Pharmacological Research*, 35, 1-4.

In: Natural Products: Structure, Bioactivity and Applications ISBN 978-1-62081-728-5
Editors: Ramiro E. Goncalves and Marcos Cunha Pinto

Chapter 5

NANOENCAPSULATION OF NATURAL PRODUCTS WITH MULTIFUNCTIONAL PROPERTIES IN SILICA AND HYBRID ORGANIC – SILICA HOST MATRICES

Ioana Lacatusu, Nicoleta Badea and Aurelia Meghea*
University "Politehnica" of Bucharest, Faculty of Applied Chemistry and Materials Science, Bucharest, Romania

ABSTRACT

In the large field of nanotechnology, the synthesis of organic – inorganic hybrid nanomaterials based on different polymer matrices have become a prominent area of current research and development. Sol-gel chemistry has been easily modified to the changing needs of society to produce fine-tuned sol-gel nanostructured materials for relevant applications. In this respect, there is an increasing need for natural and versatile raw materials, as well as biocompatible compounds that could be extensively used in the large field of biomedicine. The design flexibility of sol-gel technique and its ease of fabrication can create surfaces with structural and chemical features that could be compatible with biomaterials. Silicate and derivative silicate frameworks are the most abundant compounds in nature, their use in science, medicine and engineering has increased drastically in the last decade. Therefore, silica and organo-silsesquioxane matrices are the focus of this chapter because they seem to have the properties needed to encapsulate a variety of compounds with active principles. The chapter is intended to give an overview on exploitation of the sol-gel template preparation route in order to improve the main properties of some natural products (e.g. flavones, vegetal extracts and vitamins) such as fluorescence intensity and antioxidant capacity by physical entrapment in appropriate silica-derived matrices. Synthesis of some novel fluorescence nanomaterials loaded with bioactive polyphenols which are present in most plants (concentrated in seeds, fruit skin or peel) with a high spectrum of biological activity, by replacing synthetic chemicals, may open new opportunities for optical and bio-medical applications.

* Corresponding author: a.meghea@gmail.com; ioana_laca@yahoo.com; Tel/fax: +4/0213154193.

1. Fundamental Aspects of the Sol-Gel Process

One of the most changeable synthesis process in materials science field, defined by many authors as being the processing technique with the most spectacular evolution, the sol-gel method covers a developing time interval more than one and half century, and in the last decades it includes almost the entire synthesis areas of oxide and non-oxide materials, giving new developing possibilities for ceramic, glassy and other inorganic materials to go over innovative domain of nanomaterials, and by access to active biological molecules, towards smart bioactive materials [1-3].

The sol-gel process is not a novel technique, its beginning have been marked by transformation in gel of silicon tetrachloride (1850). Since 1930 the sol-gel technique has continued with aerogels discovery and after 1950 it has been used especially in phase's equilibrium studies which have opened the route for ceramic materials synthesis. In time, the sol-gel procedure became a convenient and versatile method of preparing a large variety of materials at ambient processing conditions including glass, powders, films, fibbers, monoliths etc. [4, 5] and also enables the entrapment of numerous organic, organometallic and biological molecules (proteins and enzymes) within microporous network of sol-gel derived matrix [6].

This process, considered by many researchers as an inorganic polymerization, is chemical related with organic polycondensation reactions, in which the small molecules take part at a series of chain reactions of hydrolysis and condensation, with formation of some polymeric structures (colloidal suspensions), followed by the obtaining of well defined three-dimensional networks [7]. After further processing (drying, thermal treatment), these three-dimensional networks are transformed in dense solid materials which form xerogels, aerogels etc., depending on the applied processing method [8].

The preparation of ceramic materials and glassy networks is based on the polymerisation of suitable precursors at low temperature. In general, the individual steps associated with the sol-gel process involve the preparation of inorganic matrices via four main steps: hydrolysis, condensation and polymerisation of particles, growth of particles, agglomeration and formation of networks.

The first phase in the process is the formation of the "sol". A sol is a colloidal suspension of solid particles in a liquid. Colloids are solid particles with diameters of 1-100 nm. During the sol-gel transformation, the viscosity of the solution gradually increases, which means that the colloidal particles and condensed silica species are linking to each other to form a "gel". A gel is an interconnected and rigid network with pores of sub micrometer sizes and polymeric chains whose average lengths are grater than one micrometer.

The components of the sol-gel process are: the precursor, water, a catalyst and a solvent such as an alcohol. The precursors or the starting materials used in the preparation of the sol are usual inorganic metal salts or metalloid surrounded by different ligands [9, 10]. For example, the usually inorganic precursor for silica is Na_4SiO_4 and for aluminium oxide are inorganic salts such as $Al(NO_3)_3$ and organic compounds such as $Al(OC_4H_9)_3$ [10]. However, the most suitable precursors for oxides are the molecules that already have metal – oxygen bonds, named metallic alcoxides $M(OR)_n$ or oxoalcoxides $MO(OR)_n$ (R = unsaturated organic groups, alkyl, aryl), β-diketonates $M(\beta\text{-dik})_n$ (β-dik = RCOCHCOR') and metallic carboxilates, $M(O_2CR)_n$. Metallic alcoxides [(M(OR)n] which belong to organo-metallic compounds family, sometimes known as hybrid organic-inorganic monomers, are among the most used precursors in sol-gel researches. Metal alkoxides have an organic ligand (R is typically an alkyl group) attached at a metal or metalloid, where M represents a network-forming element such as Si, Ti, Zr, Al, B, etc. [6]. They are the most usual precursors due to the fact that they are advantageous at high purity and rapidly react with water, being very reactive reagents against nucleophylic species. The commonly used precursors in sol-gel processes are tetramethylorthosilicate (TMOS, $Si(OCH_3)_4$) and tetraethylorthosilicate (TEOS, $Si(OC_2H_5)_4$).

The basic sol-gel reaction begins with the hydrolysis reaction of metal or semi-metal alkoxide, in presence of a solvent and acid or base catalyst, that leads to a hydroxylated product (e.g. silanol groups, ≡Si–OH) and the corresponding alcohol. The schematic representations of sol-gel reactions are shown in figure 1. Generally, both the hydrolysis and condensation reactions occur simultaneously once the hydrolysis reaction has been initiated. In the second step, condensation reactions between an alkoxide group and a silanol group or two silanol groups produce siloxane bonds (≡Si–O–Si≡) resulting into production of alcohol and water as by-products.

Condensation or polymerization reaction is a more complex process that may take place by a series of olation and oxolation equilibria. These equilibria may be represented by reactions from figure 1. In general, olation is a more rapid process than oxolation; as a result the olation is the prevalent way for condensation. These monomers and oligomers are architectural base pattern from which the polymeric particulates (colloids) are obtained by further aggregation.

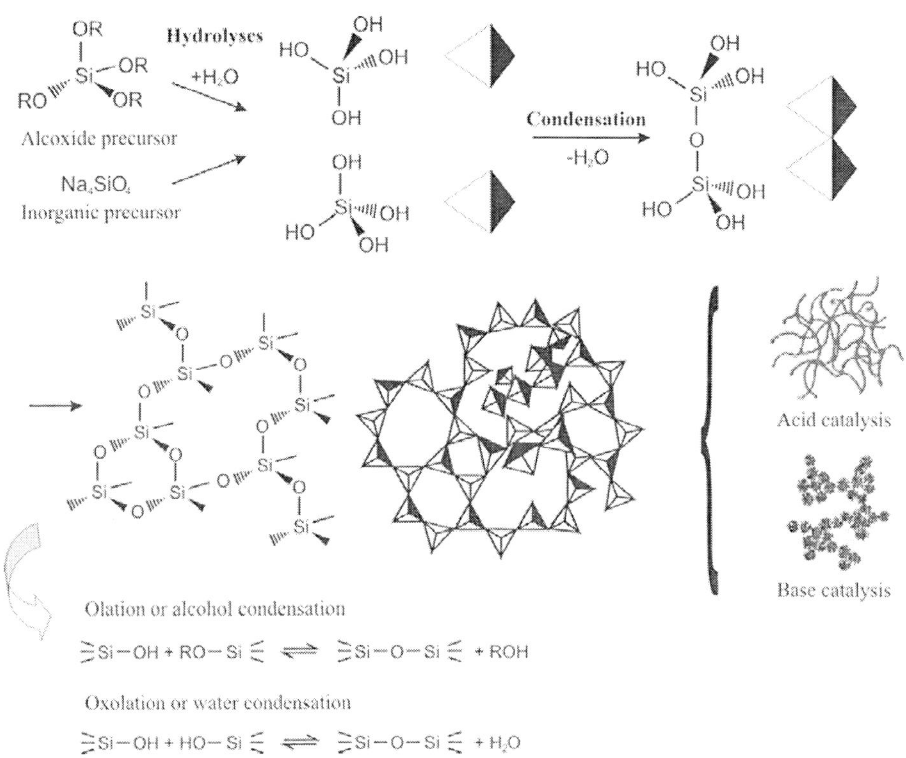

Figure 1. Processing steps involved in preparing sol-gel derived silica materials.

These chemical reactions occurring during sol-gel process strongly influence the properties and composition of the final material. Further processing of the sol, while it is still in the solution phase, enables to obtain sol-gel materials in different configurations. Thus, the sol can be cast into thin films, fibbers, or monoliths. Thin films can be produced on a piece of substrate by various coating techniques like dip, spin and spray coating [6, 11, 12].

The polycondensation between the sol particles results in a nano-porous, glass-like, three-dimensional network that is optically transparent. After the sol-gel transition, by further drying and heat-treatment, the gel can be converted by grinding into crushed fine powder. During this step, aging and drying of the sol-gel composite occurs. Ageing or pre-polymerization of the sol causes aggregation due to hydrolysis and condensation reactions, and consequently it results into an increase in viscosity. During this step, the sol is allowed to stand either at room temperature or at a higher temperature for a period of time during which hydrolysis and condensation reactions cause aggregation and cross-linking. This process is accompanied by the loss of the liquid phase (solvent which can be water or alcohol) in the process called "syneresis", which increases the strength and decreases the porosity of the sol-gel processed material.

The method used to dry the gel can have a dramatic influence on the final material properties. When drying process is performed under ambient conditions and sometimes at temperatures not exceeding 100 °C (e.g. evaporation), substantial shrinkage occurs and the resulting composite is known as "dry" gel, called "xerogel" [13]. A xerogel is a relatively sturdy, typically transparent but porous material. The pore size depends on such factors as time and temperature of the hydrolysis and the type of catalyst used. The diameter of the pores is directly related to the shrinkage of the "wet" gels [14]. Heat treatment of a xerogel at elevated temperature produces viscous sintering (shrinkage of the xerogel) and effectively transforms the porous gel into a dense glass [6]. As the viscosity of a sol is adjusted into a proper viscosity range, ceramic fibbers can be drawn from the sol. Ultra-fine and uniform ceramic powders are formed by precipitation, spray pyrolysis, or emulsion techniques [6]. When the solvent is removed under supercritical conditions, the network does not shrink and a highly porous, low-density material known as an "aerogel" is produced.

The "gel point" is defined as the point at which the entire solid mass becomes interconnected. The physical characteristics of the gel network depend upon the size of particles and extent of cross-linking prior to gelation.

The physical-chemical properties of sol-gel derived materials mainly depend on the composition of sol and processing conditions. There are a number of variables and factors that influence the hydrolysis and condensation rates and hence the detailed microstructure of the sol-gel glasses. Among them, there are few that are considered to have a greater impact: pH, nature and concentration of catalysts, precursor type, nature of solvent, water/sol-gel precursor molar ratio, time and temperature of ageing and drying processes [10].

One of the most important parameters that plays a major role, not only regarding the reaction mechanisms, but also the influence of microstructure of final material is the pH (figure 1) [15]:

- acid catalysis leads to a more polymeric form of gel with linear chains (with small rate of cross-linking) as intermediates [16]. Acid conditions (slow hydrolysis) result in fine gel features that dry to a high bulk density (low porosity). The molecular tails interpenetrate to each other, with apparition of some supplementary branches as result of gelation process. In other words, by using acid catalysed reactions, in a first stage an opened network structure is formed which further lead to the condensation of small clusters [17];

- base catalysis yields to colloidal gels where gelation occurs by cross-linking of the colloidal particles. This catalysis develops coarser gels that dry to a low bulk density (higher porosity) and may lead to variation in the homogeneity of final materials. In base conditions (rapid hydrolysis) more branched clusters are formed which are not interpenetrated before drying process. They behave as discrete species and the gelation occurs by clusters binding [16].

1.1. Why Sol-Gel Process for Nanomaterial Synthesis?

The sol-gel process was used to produce pure, dense, stoichiometric, and monodisperse materials. Besides the fact that the sol-gel procedure could be used for the synthesis of multicomponent materials, of materials with binary and ternary composition, using double and mixture of alcoxides [31, 32], the sol-gel process became an extremely attractive process to obtain nanostructures with a wide range of structures such as nanoporous membranes, thin film coatings, ceramic fibres and so on. In addition to the advantages of conventional preparing process, the sol-gel derived materials are attractive materials due to their physical-chemical properties that can be tuned by choice of precursor(s) and the processing protocol. For example, sol-gel derived silica materials have several clear advantages [6, 18-21]:

- these "host" sol-gel materials may be prepared by low-temperature preparation procedure with a high purity, homogeneity and porosity, a wide and controllable range of surface areas, low pore dimensions and narrow pore size distributions, case in which they are called "monodispersed", good thermal stability, well beyond most dopant molecules;
- they can be miniaturized to micron- or nanosized scale and can also allow the control of electrical conductivity through the choice of the metal or metal alkoxide;
- they manifest compatibility with many organic and inorganic molecules;
- chemically, physically, thermally and photochemically stable without being hazardous to humans or the environment, as compared to organic polymers;
- exhibit higher mechanical strength and negligible swelling in organic solvents compared to most organic polymers;
- a broad optical window, being optically transparent glasses they are suitable for various spectroscopic based analytical measurements (e.g. allow the use of modern spectroscopic tools to study dopants from a xerogel) and also it is possible to couple optics and bioactivity to make photonic devices and biosensors;
- can be cast into desired shapes, as monoliths, coated as thin films on glass slides and fibber and can be ground into powders;

Moreover, the broadening of applicability area of products obtained by this technique, clearly demonstrates that porous sol-gel materials possess many promising features that may be exploit in various applications. The main utilisation of sol-gel process in the last years was correlated with the synthesis of hybrid nanoporous materials, by encapsulation of some biologic or organic molecules. Even the sol-gel procedure is known by 150 years, only in the last fifteen years it has been explored for preparation of hybrid organic – inorganic materials, where the phases coexist at nanometric scale. The occurrence of the reactions of hydrolysis and polycondensation of alkoxy compounds, which underline the processes of the sol-gel synthesis, makes it possible to introduce a wide variety of organic and inorganic dopants into oxide and hybrid matrices in order to confer necessary functional properties to the synthesized materials [22, 23]. One of the most important advantages remains the preparation of doped sol-gel materials which have the ability to preserve the physico-chemical properties of different dopants and most important – sometimes to enhance them. Since 1984, the sol-gel method has allowed the physical trapping or the covalent bonding of many types of large organic and inorganic species inside inorganic networks, this leading to a new generation of advanced materials with important properties [24].

Hybrid organic-inorganic materials synthesized by the sol-gel method have attracted the particular attention of researchers owing to the possibility of synthesizing nanocomposites (coatings, films, bulk materials) with technically valuable properties [25]. From a chemical and material standpoint, sol-gel derived materials have a combination of properties which can hardly be achieved by other materials. The fact that the process works at room temperature and close to neutral pH, enables its use in bio encapsulation-related applications. These features make these materials as exclusive hosts for a series of biological molecules that may be used in bio-medical applications. The sol-gel technology advantages allow the construction of biomedical sensors, materials for laser, drug delivery and so on [6]. Applications exploiting porous materials to encapsulate sensor molecules, enzymes and many other compounds are developing rapidly. The last researches in bio-medical domain reported that sol-gels with entrapped various molecules may be used in construction of implants and coatings with bioactive properties [26]. Potential applications with emphasis on biomedical and environmental aspects are reviewed by Podbielska and co-workers [14].

Therefore, the importance of sol-gel technology in the last years was much amplified, being an intensive and attractive research field, mainly due to the exceptional increase of applications in all science domains: from preparation of ceramic materials, glasses and composites with highly homogeneity, medical nanocapsules and Nan engine in medicine field, quantum dots and metallic nanoparticles, as catalysts, Nan manufacture of electronic equipments, towards at nanobiosenzors and decontamination agents for bio-security. Thus, the broad range of possible applications of sol-gel derived materials and biomaterials marks this technology as one of the most promising fields of contemporary material sciences.

1.2. Silica and Organo-Silica Matrices Characteristics

In the large field of nanotechnology, the different polymeric silica matrices based Nan composites have generated a significant amount of attention in the recent literature, it becoming a prominent area of current research and development. Since the discovery of ordered mesoporous silica materials in 1990s by the researches from Mobil Corporation, synthesis and applications of these mesoporous solids have received intensive attention. These aspects are based on the fact that silica networks have utility as porous encapsulation matrices owing to their highly ordered structures (e.g. they possess well defined internal networks, cages, cavities and channels), high surface area and also they have mechanical strength and are compatible with many types of photochemical experiments. Due to the mild sol-gel processing conditions, significant concentrations of many types of biologically active agents can be incorporated in a liquid sol. The active compounds are embedded in the matrix of the gel, which after condensation and drying becomes a porous, glassy solid.

Silicon and siliceous compounds such as silica have been understood to play an active role in organism health and development. Silicate and derivative silicate frameworks are the most abundant compounds in nature. Their use in science, medicine and engineering has increased significantly in the last decade [27]. Silica materials obtained by the sol-gel process are amorphous and porous, and have found applications in biomedicine as a coating on implants and medical products [28], biocatalysts [29], biosensors [30] and matrix for release of different drugs [31-33].

Room temperature processed silica based sol-gel materials have been studied and used for biomedical applications that include tissue, cell and enzyme encapsulation and controlled release of drugs, as can be seen later (chapter 1.3.1), due to their multiple advantages such as [34-41]:

- biocompatibility; are resorbable materials with a good compatibility and a favourable tissue response;
- bioactivity (e.g. materials function like the tissue in which they are implanted);
- excellent photonic media as a result of their low optical losses;
- good mechanical properties, chemical, photochemical and thermal stability;
- the low ecological impact of silicate chemistry, they are not a food source for micro-organisms;
- the ability to modify the surface hydroxyls on silica surface with amines, thiols, methacrylate and other coupling agents.

Other characteristic features of the sol-gel silica matrices are the optical transparency of the resulting insoluble support, high porosity, low thermal conductivity, low dielectric constant and high laser damage threshold [42]. Beside these important features, the principal advantage of utilized silica materials in the biomedicine area is that the sol-gel method is non-toxic, uncomplicated, inexpensive, takes place at room temperature (important for thermo sensitive drugs), and it does not require the use of pharmaceutically unacceptable solvents. These materials are biocompatible *in vivo*, cause no adverse tissue reactions and degrade in the body to silicon acid, i.e., $Si(OH)_4$, which is eliminated through the kidneys [43].

Silicon-containing materials represented by silica and polysilsesquioxane are important components to produce high performance organic-inorganic hybrid nanomaterials. Silica and organosilsesquioxane matrices are the focus of next chapter (chapter 2) because these matrices have the appropriate properties needed to encapsulate compounds with bio-active principles and the sol-gel technique is suitable for vegetal extract entrapment since they require maintaining of their specific properties such as fluorescence intensity, by physical adsorption.

The two materials – inorganic and organic – from composition of a hybrid nanomaterial, are neatly different from each other, both in their properties and as referring to their advantages and drawbacks. Due to these novel, special properties, not found at macro scale, preparation, characterization and application of hybrid organic – inorganic materials became an increasingly expanded research field in the science of material chemistry. Not only the property of organic component, but also the property of inorganic silicon based component is crucial to realize high performance materials with well-balanced properties. Since interface between organic and inorganic materials determines the miscibility of the components, modification and control of the interface structure in these systems is essential to obtain high performance materials with long stability [44]. Generally, organic materials possess a good flexibility, hardness, hidrophobicity, an excellent shaping capacity and novel optical and electronic properties [45]. Organic polymers, has also some drawbacks like heat instability and natural degradability by ageing. In contrast, inorganic materials such as glasses and ceramics of silica show high thermal stability and mechanical strength, but sometimes they are more fragile and brittle. Therefore, if a homogeneous combination could be obtained between inorganic and organic tails (that means a monophase material), it may offer unique possibilities that combine the advantages of both organic-inorganic materials and allow the design of optical, electronic and mechanical properties needed for numerous applications.

Since silica manifest a low miscibility with organic components, modification of silica surface by silane coupling agents has been mainly accomplished to improve the silica inorganic properties. The polysilsesquioxane compounds can be considered as molecularly surface-modified silica materials by organic substituents, and have attracted a lot of interest as components for synthesis of high performance hybrid nanomaterials, being used to improve the properties of various polymeric systems without need of further chemical modification [44].

Silasesquioxane or silsesquioxane is the general IUPAC name for a family of polycyclic compounds that contain oxygen and silicon atoms in the $SiO_{1.5}$ ratio. The name sil(a)sesquioxane is derived from *sil-oxane* (compounds of silicon and oxygen) and *sesqui* (from Latin, meaning one and half) and the general name reflects the ratio of silicon and oxygen in the completely condensed silsesquioxane [46].

Polyhedral oligomeric silsesquioxanes (POSS) and related polyhedral oligomeric silicates (POS), also known as spherosilicates, have received considerable interest among the academic community and industries and show excellent properties derived from their unique structures. The polyhedral oligomeric silsesquioxanes are produced as materials with various structures, such as random, ladder, incompletely and completely condensed cage structures, as illustrated in figure 2 [44, 47].

Silsesquioxane compounds are distinguished by their unique polyhedral frameworks with varying degrees of symmetry, with silicon atoms at corners and oxygen atoms interspersed between them in a tetrahedral configuration. The tetrahedral coordination forms a three-dimensional structure by a series of Si–O–Si bonds, creating an inner inorganic framework (a silica cage) made up of silicone and oxygen, as shown in figure 2 [27]. The "silsesquioxane" term is generally associated with all structures with the empirical formula $[RSiO_{1.5}]_n$, in which each silicon atom is connected in a theoretical representation at 1.5 oxygen atoms and at a hydrocarbon tail (R). When n = 4, 6, 8, 10, 12, 14, or 16 (n > 4), the resulted compounds are called "polyhedral oligomeric silsesquioxanes" (POSS); if *n* is an undefined number, this compound is simply named "polysilsesquioxane" [48-50]. The most available and widely used compounds are the octahedral species (n = 8), in which the spherical size of the cubic silsesquioxane cage is 0.54 nm [51].

The organic groups (R) are attached on the surface of silica matrix through the non-hydrolysable Si – C bonds and function as the network modifier [52]. The R constituent group may be any number of organic groups (e.g. alkyl, cicloalkyl, alkylen, aryl, arylene), any organofunctional derivates from these organic or reactive/polimerizable groups such as acryl, α-olephine, styrene, epoxy, carboxylic acid, isocyanides, amine, alcohol and silan (e.g. stiryl-POSS, methacrylate-POSS, norbornyl-POSS, vinyl-POSS, epoxy-POSS, siloxan-POSS etc.) although most applications of silsesquioxanes incorporate methyl, halogen, vinyl or phenyl chains [53, 54]. R groups placed at surface are chemically accessible and present a great reactivity. The number and type of the functional corner groups on POSS cages can be readily varied, which makes POSS molecules to be excellent nanodimensional building blocks for preparing various organic-inorganic hybrid nanomaterials [55, 56].

The silsesquioxane derivatives are building blocks that combines a truly inorganic – organic architecture, known as hybrid precursors because they are the smallest possible silica particles with a cage shape (with typical size of 1-3 nm) that can be easily functionalized with organic moieties [45, 57]. Their potential to mimic silica surfaces arise from the similarity of silsesquioxanes to siliceous clusters that possess the structural and electronic features of hydroxylated silica surface [58]. The low hydrophilicity of silsesquioxanes as comparing with those of silica is easily to be explained by the presence of organic hydrophobic groups at solid surface, instead of hydrophilic OH of silica.

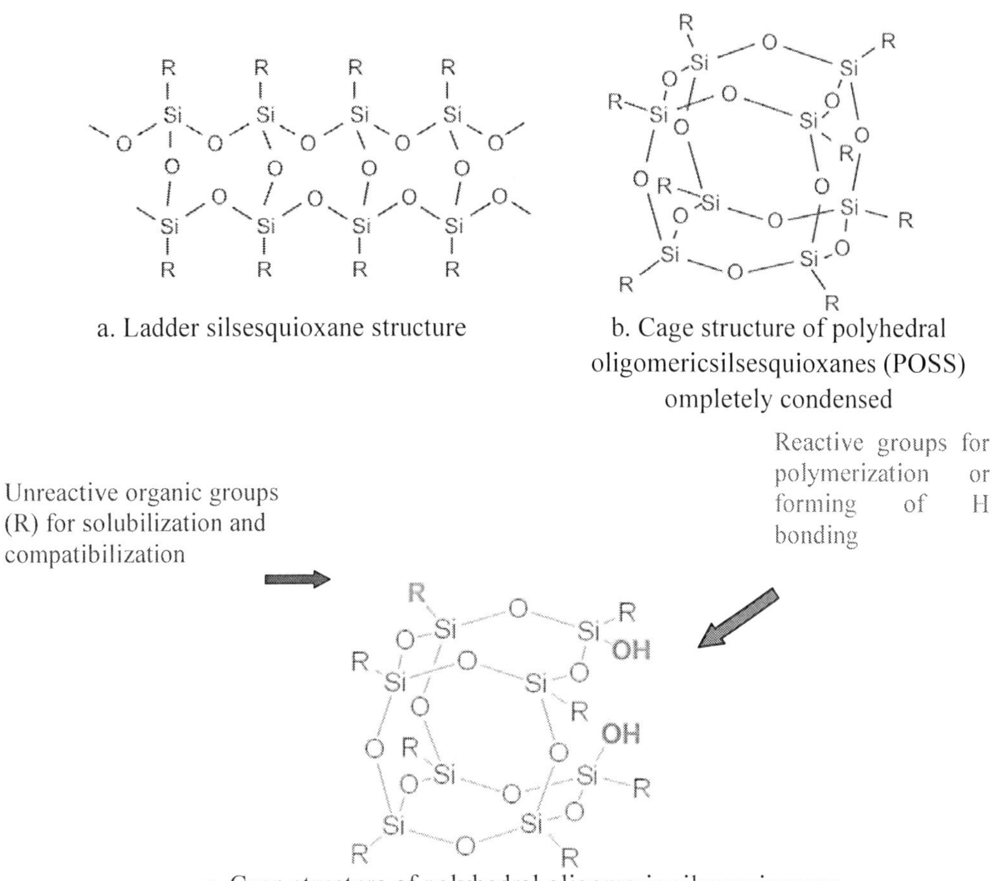

a. Ladder silsesquioxane structure

b. Cage structure of polyhedral oligomericsilsesquioxanes (POSS) ompletely condensed

c. Cage structure of polyhedral oligomeric silsesquioxanes (POSS) incompletely condensed

Figure 2. Structures of silsesquioxanes.

Another characteristic of this compound is its chemical composition, being a hybrid intermediate between silica (SiO_2) and silicon (R_2SiO), and this leading to a thermally and chemically robust matrix [47]. The molecule of such a silsesquioxane derivative presents reactive functionalities like OH groups, adequate for polymerization, formation of hydrogen bonds or polymeric graphing towards branching polymeric chains and nonreactive organic functionalities, R for solubilization and compatibilization [59]. Thus, the entire silsesquioxane molecule confers a precise three-dimensional structure, the nanostructured units may be easily incorporated into the common polymeric materials via copolymerization, grafting or blending [60]. POSS compounds are distinguished by their capability to polymerize yielding elastomers, thermoplastics, and thermally cured systems with an ordered structure and improved properties [61-63].

At present there is a variety of POSS compounds that contain different combinations of unreactive substituents and/or reactive functionalities. The incorporation of POSS derivatives into polymeric materials can lead to significant improvements in polymer properties which include increases in use temperature, oxidation resistance, surface hardening and improved mechanical properties [52]. Moreover, the utilization of POSS monomers does not need significant modifications at processing. The monomers are simply mixed and copolymerized. As long as the POSS monomer is soluble in the monomer blend, it is incorporated in a molecular dispersion of the final copolymer. As a result, no phase separation occurs, even some aggregation of POSS units connected with the polymer may appear.

These POSS molecules, present as monomers and polymers, are emerging as a new chemical technology for preparing novel organic-inorganic hybrid nanomaterials. The organic-inorganic hybrid materials with covalently bonded phases can be prepared based on POSS without a substantial change in the conventional polymer synthesis procedures [46]. As a result, a rapidly increasing number of POSS monomers are being synthesized and novel applications are continuously being proposed. Of all the synthesized and modelled silicate compounds, silsesquioxanes have emerged in the forefront for their numerous applications and potential uses. Application of silsesquioxane compounds covered area from catalysis [49, 64], metal complexes [48, 65], barrier materials [66], mimicked-silica surfaces [67], nanostructure polymers [68], film materials [59], to synthesis of new porous materials [57, 61], encapsulation material for communications, optoelectronics [69], until the uses in biotechnology and medicine [58, 64], as carriers for drug delivery process [59]. Most of the researches on these compounds have been conducted towards the development of polymer-inorganic nanocomposites with properties focusing on materials science. For instance, Zeng and co-workers [68] has recently reported a series of POSS-containing hybrid materials, which exhibit superior thermal properties and good solubility [70, 71]. Hence, POSS is incorporated into organic NLO chromophores to produce POSS-based organic-inorganic hybrid functional materials and expected to endow the hybrid materials with novel optical properties and enhanced thermal stability.

Possible uses in engineering and materials science have received only a partial emphasis as compared to applications in biotechnology and medicine. POSS derivatives have been shown to present a variety of thermal and chemical conditions that may be turned for some *in vivo* stability studies. For example, some cationic polyhedral oligomeric silsesquioxane units exhibit a variety of attributes (e.g. nanoscale size, three-dimensional functionality, efficient cellular uptake, low toxicity and high solubility) that make them attractive as biocompatible drug delivery systems. These advantages were used by McCusker and co-workers to use some quaternary ammonium functionalized polyhedral oligomeric silsesquioxanes as new carriers for drug delivery processes [72]. The potential developmental toxicity of two silsesquioxane derivatives (quaternary silsesquioxane and phenylsilsesquioxane) were evaluated by toxicity studies realised on rats and rabbits [73, 74]. The obtained results demonstrated that oral administration of quaternary-silsesquioxane as high as 1000 mg/kg/day did not produce teratogeneity or other indications of developmental toxicity in the rat concepts. In the case of phenylsilsesquioxane used widely in the personal care industry, being a common component of skin and oral care products, no treatment-related clinical signs of toxicity were observed and no marked effects upon maternal food consumption, body weight gain, or uterus or liver weight were detected. The group leaded by Mousavi has developed a novel nanocomposite based on polyhedral oligomeric silsesquioxane-poly(carbonate-urea) for use as tissue implants [75]. These nanocomposites have enhanced interfacial biocompatibility, better biological stability and stronger mechanical properties as compared with conventional silicone biomaterials, thus making them safer as tissue implants [75].

Therefore, these three-dimensional organosiliceous compounds, due to their octahedral structure, nanometer size, as well as good optical properties, represent potentially nanoconstruction sites very useful for obtaining of hybrid matrices that may be successfully used in an encapsulation process. In this respect, it is interesting to evaluate the suitability of new template silica-silsesquioxane hybrid networks – as host matrices used for some sensitive vegetal extracts immobilization in order to investigate the role of silsesquioxane compound as building blocks in sol-gel encapsulation process and how these matrices may preserve and even enhance some specific optical properties of natural extracts such as fluorescence properties.

1.3. Advances in the Sol-Gel Encapsulation Process

1.3.1. General Aspects

The field of nanotechnology is one of the most popular areas for current research and development in basically all technical disciplines. Materials science represents the domain in which the most ideas and revolutionary concepts from other science like biology, cybernetics, informatics, have been transferred in the last century. This integrator principle is sustained firstly by the possibilities apparently unlimited of domain that may offer in any moment an alternative synthesis way which may be developed towards the proposed goal.

The optical characteristics, the great specific surface and the relatively easy synthesis of hybrids sol-gel materials in various shapes and sizes are only several properties that have attracted a considerable interest for the study of sol-gel materials for different biomedical applications. The hybrid nanomaterials prepared with the help of sol-gel chemistry have been applied to different research fields for years. In the last twenty years these studies bring out the developing of smart materials, called „living ceramics", due to the immobilization of biological compounds in sol-gel matrices. As a result, the fastest growing research directions in the field of sol-gel science and technology have been oriented towards the developing of new alternative to realize the combination of sensitive molecules with rigid support materials, by means the immobilization of various biomolecules in sol-gel matrices with retaining of their bioactivity.

From its beginning, the sol-gel nanoencapsulation have opened a new and challenge way in immobilisation processes of biological materials. Even the sol-gel science represents an old method, the bioencapsulation process by using this approach has not been realized until 1990. Sol-gel encapsulation of biomolecules in silica matrices has been firstly reported by Braun and co-workers in 1990 [76]. Twelve years later, Jin *and colab.* published a key paper describing the entrapment of proteins into alkoxysilane-derived silicate glasses via the sol-gel method [77]. This group demonstrated that a series of enzymes, including aspartase and alkaline phosphatase, could be entrapped into glasses derived from tetraethyl orthosilicate (TEOS) with retention of enzymatic activity. The biomolecules preserve their bioactivity after encapsulation and remain accessible for external factors by spreading within porous silica. Since than, the researches in this domain have been continuously developed and a great number of bioencapsulation examples by sol-gel technique have been presented in specialty literature [78-82].

Similar with conventional sol-gel process, the starting point for sol-gel encapsulation process is the precursor hydrolysis (e.g. alcoxysilane, alcoximetalate, alkyl silicate, or a mixture of them [48-53]), in acid or base catalysis, with formation of hydroxy derivatives by liquid prehydrolizable oligomer type (silicic acids, hydroxometalates, hydroxysilans etc.). Subsequent reactions of condensations lead to colloidal soluble polymers and futher to highy cross linking network polymers with phase separation (polysilicates, hydrated metal oxides, polysiloxanes, etc). These water-insoluble oligomers known as sol-gel intermediates are used for encapsulation or can be converted into more biocompatible water-soluble derivatives by transesterification with glycerol [83]. Encapsulation of active molecules takes place by mixing of sol (at pH $\geq$7) with a buffer solution or a suspension of biomolecules. At this point the gelification process is initiated.

By hydrolysis and condensation reactions of precursors an inorganic/hybrid porous matrix are formed that is growing around the biomolecules in a three dimensional network. The polycondensation process leads to the formation of soluble macromers and then colloidal solutions, which by coalescence increase the solution viscosity towards sol-gel transition, when gelifiation occurs. This process leads further to a hydrogel that contains nano- and microparticles with biological compounds embedded or interstitial captured, together with an interstitial liquid phase. By ageing process, the hydrogels is usually accompanied by shrinkage and syneresis phenomena (removing from volume the liquid phase), with transformation in an ageing hydrogel. A further controlled drying under atmospheric conditions, leads to the cracking of pores and structural consolidation, thus resulting a xerogel with encapsulated biomolecules.

Typical methods to immobilise biomolecules onto inorganic, organic or polymeric surfaces have been based on physical adsorption, covalent binding to surfaces, entrapment in semipermeable membranes and microencapsulation into polymer microspheres and hydrogels [77]. Thus, the biomolecules can be added during the sol-gel manufacturing process or introduced into gel by adsorption of an active molecule/biomolecule onto the surface of the gel. In the first case, during aggregation of the formed colloidal particles, the biomolecule is incorporated in the lattice of silica gel polymer. In the second case, a gel of desired porosity is immersed in solution of a biomolecule to be encapsulated. This approach is important for labile agents (e.g. biomolecules) to avoid their decomposition under the conditions used for gel production.

The good performance of sol-gel encapsulated biomolecules is assured by some unique features [2, 3, 84-87]:

- the sol-gel matrices used for encapsulation are highly porous materials. For example, the doped silica-matrix pores meet two requirements: firstly, the pores are large enough to allow unrestricted transport of molecules including buffer ions, substrates and products of the reaction and analytes and secondly, pores are small enough to prevent leakage of encapsulated macromolecules. In addition, their porosity may be controlled to a significant degree by the judicious selection of precursors, polymerization conditions and so on;
- encapsulated biologicals in sol-gels are highly active and essentially permanently entrapped. Typically, they show improved resistance to thermal and chemical denaturation, biological degradation and to proteolytic attack with a direct result in increasing of storage and operational stability.
- sol-gels can be fabricated as self-supporting structures or applied to a variety of inorganic and organic supports.
- bio-doped sol-gels function in gaseous, liquid (aqueous, aqueous-organic, low-water organic), solid-liquid (suspension, eutectic) and subcritical and supercritical environments.
- sol-gels have unparalleled optical properties, thus the number of potential applications of organic doped sol-gel glasses is very large.

Presently, the sol-gel chemistry that represents an inherent low temperature and biocompatible process, offers new and interesting possibilities for successfully immobilizing of heat-sensitive and fragile biomolecules inside silica and/or organic-inorganic materials. Following the specialty literature, a wide variety of biomolecules ranging over enzyme, protein, antibodies, ADN, ARN, antigens, bacteria, fungi and whole cells of plant, animal and microbes, have been embedded in sol-gel matrices (silica, metallic oxides, organosiloxanes and sol-gel hybrid polymers) [7, 88-92].

The last decade has seen a revolution in the area of sol-gel derived materials since it has been demonstrated that these materials can be used to encapsulate biological species in a functional state. The main advantage of these "living ceramics" is the fact that they may provide applications in various domains. The processes of immobilization and preservation of biomolecule activity are very complicated and they are determined by the complex of chemical and structural characteristics of the matrices, as well as biological experimental conditions. The interactions between the biomolecule and the inorganic, organic or hybrid material determines the degree to which the biomolecule retains its native properties, and such interactions can be tuned to provide optimised biomaterials that are suitable for a variety of applications [77]. Typical applications of sol-gel derived biomaterials include selective coatings for optical and electrochemical biosensors, controlled release agents, diagnostic devices and even bioartificial organs, catalysts, stationary phases for affinity chromatography, immunoadsorbent and solid-phase extraction materials, solid-phase biosynthesis and unique matrices for biophysical studies.

With a rapid advance, techniques such as polymers nanoengineering, encapsulation protocols and preparation methods of new multifunctional nanomaterials have opened new sight that seems to revolutionize science and technology at the frontier of knowledge. Thus, in the following chapter are discussed the advances in sol-gel encapsulation process, by highlighting the evolution in the few years and last applications of hybrid nanomaterials obtained by different approaches of synthesis.

1.3.2. Recent Developments of Sol-Gel Nanomaterials

The preparation of organic – inorganic hybrid materials is one of the most attractive fields of sol-gel chemistry. The sol-gel technique has been proved to be a suitable approach not only for bioencapsulation of biomolecules (e.g. enzymes, proteins, cells, antibodies and so on) into ceramic matrices, but also it can be easy by extended for encapsulation of some functional organic materials, by using as host matrices, inorganic silica or organic-inorganic hybrid sol-gels, accessible for advanced optics, coatings, microprocessing and so on.

The generic "sol-gel" processes and their different kind of physical or chemical packing of functional groups from various active molecules have demonstrated that it is possible the physical/chemical entrapment of organic molecules inside silica derived-matrices. By using both kinds of mineral or organic precursors of silicic acid and their further processing by competitive hydrolyses reactions combined with polymerisation reactions on controlled directions, the silica materials thus obtained have formed a new class of materials, named "hybrid silica materials" [45]. The preservation of specific properties of encapsulated compounds and sometimes their improving together with the great flexibility in finalising the forms designed, make that these hybrid materials an attractive solution for application in biotechnology, medicine, environmental technology, sensors, photonic media, etc.

Conceptual, the "hybrid organic-inorganic" term represents the result of features interpenetration of two important chemical compounds, both of them with significant contributions in developing of materials science, each of them with characteristic properties, advantages and drawbacks at specific domain:

- "brittleness" of organic molecules with thermal stability on a limited temperature range (< 250°C) and special reaction conditions, has represented along the time the starting point for fundamental research of materials chemistry;
- "rigidity/stiffness" of inorganic materials, often used for fine synthesis, with target applications towards bio- and nanodomains.

However, the most evident advantage of organic-inorganic hybrid materials remains the fact that they may favourable combine different properties of organic and inorganic components in a single material. Due to the greater combining possibilities of components, this domain is very creative; it offers the opportunity to invent almost unlimited sets of hybrid materials with a wide spectrum of known or unknown properties.

The term of "hybrid material" is used for many systems that include a large domain of different materials such as highly crystalline coordination polymers, amorphous sol – gel compounds, materials with and without interactions between organic and inorganic units, etc. Starting from classical theorizing approaches regarding the structure of hybrid organic-inorganic materials [93-95] and having in view the affiliation of both organic and inorganic compounds, there is a detailed classification that makes the distinction (depending on structural differences) between the possible interactions that take place between organic and inorganic species:

- *hybrid organic-inorganic materials of class I* include all the hybrid materials resulted by assembling of organic and inorganic components by weak interactions (van der Waals, hydrogen bonds, hydrophobic-hydrophilic or weak electrostatic interactions). In these materials the organic molecules are physically trapped into the inorganic silica matrix, and only electrostatic interactions can be established between the organic molecule and the silica cage [96]. Synthesis of hybrid materials of class I is usually performed by formation of inorganic base network in the presence of a tailored organic phase [45].
- *hybrid organic-inorganic materials of class II* are those materials that present strong chemical interactions between components. In this approach, inorganic precursors have to carry functional groups that may react with organic phase during sol-gel process. These materials could be synthesized where the organic molecules are attached to the silica matrix by covalent or/and ionic-covalent bonds. However, these materials were not deeply investigated due to the difficulty to bind the organic molecule to the silica matrix by means of a covalent bond [45, 96].

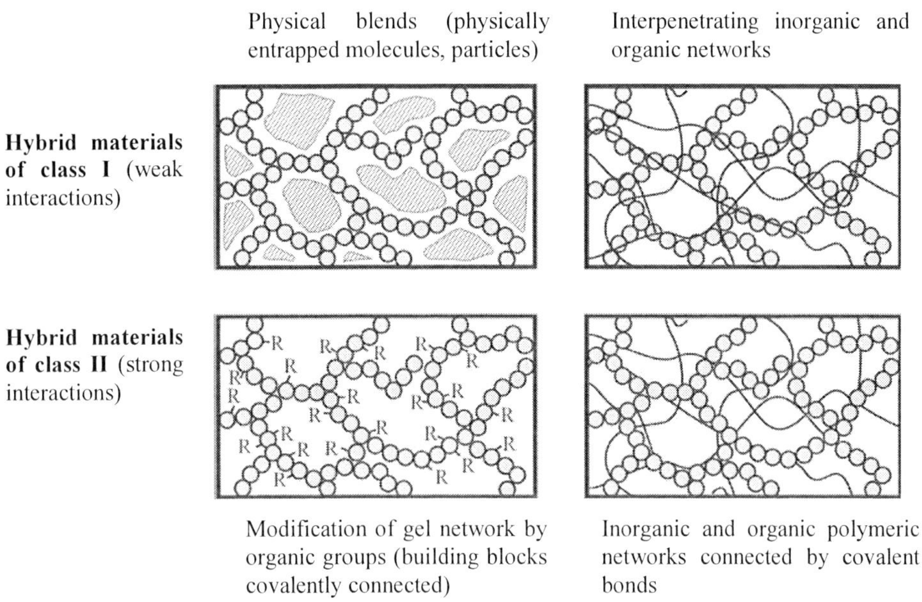

Figure 3. Classification of hybrid materials.

Compounds of "blends" type, belonging to class of hybrid materials are formed when there is no any chemical strong interaction between the organic and inorganic architectural patterns [45]. An example for this kind of material is the combination of some particles or inorganic clusters with organic polymers, without implication of a chemical interaction between components (figure 3a). In this case, the material formed consists, for instance, in an organic polymer with entrapped discrete inorganic chains in which, depending on the components functionalities, for example weak cross linking occurs by the entrapped inorganic units through physical interactions or the inorganic components are entrapped in a cross linked polymer matrix.

If an organic and an inorganic network interpenetrate each other without strong chemical interactions, "interpenetrating networks – IPN" that belong to hybrid materials of class I are obtained (figure 3b). An example in this case is a sol-gel material formed in the presence of an organic polymer or vice versa. The hybrid materials of class II are obtained when the discrete inorganic building blocks (e.g. clusters) are covalently bonded to the organic polymers (figure 3c) or organic and inorganic polymers are covalently bonded to each other (figure 3d) [45].

The organic-inorganic hybrid materials may be prepared in various ways. The simplest method relies on dissolution/incorporation of organic molecules in a liquid sol-gel during the sol-gel process, thus the organic molecules get trapped within the oxide gel network during the hydrolysis and polycondensation reactions [84]. This method, called as pre-doping method, is largely preferred since it allows eliminating or reducing the leaching drawbacks typical of impregnation. In this case, sol-gel synthesis is mandatory since it allows operating near room temperature, so that the organic molecules can be easily trapped without thermal degradation. Another advantage of this process is that the components are mixed at a molecular level, and a large amount of dye can be incorporated before fluorescence quenching takes place because of dye clustering [97-103].

The other way is the impregnation of a porous gel (e.g. xerogel) with organic molecules [104]. The impregnation or post-doping method [105, 106] requires the sol-gel synthesis of a porous silica glass compatible with the size of molecule that have to be encapsulated. In this case the open porosity can be filled by immersing the bulk xerogel in a solution of organic molecules or biomolecules upon capillarity phenomenon; the carrier fluid has to be removed by repeating the drying step. Impregnation is essentially a physical adsorption process that can be further complicated by several post-doping treatments, such as dehydration or partial densification.

Organic – inorganic hybrids may be applied in many areas of materials chemistry owing to their simple processing and easily to change in order to realize an appropriate design at molecular scale. There are four major directions regarding the synthesis and applications of organic – inorganic hybrid materials:

- molecular engineering;
- their organization at nanometric level;
- their combination with bioactive compounds;
- their processing from functional hybrid towards multifunctional hybrid.

Through careful selection of precursors and additives, these materials can be designed for specific applications, and can produce useful and robust devices with good analytical parameters. For example, microencapsulation of biomolecules and living cells in calcium alginate gel beads coated with a semipermeable membrane has been widely investigated for industrial, pharmaceutical and medical applications. In a study elaborated by Chernev and co-workers are exposed the experimental results about the formation, properties and structure of sol – gel silica based biocomposite containing calcium alginate as an organic compound [107]. Inorganic-organic silica hybrid transparent materials with calcium alginate are obtained via sol-gel method at room temperature. Two types of silicon precursors have been used in the synthesis: tetramethylortosilicate (TMOS) and ethyltrimethoxysilane (ETMS). All studied hybrid nanocomposite have an amorphous structure. Cell immobilization experiments showed that the synthesized hybrid nanomaterials were successfully applied in maintaining cell viability and enzyme stability for 8 reaction cycles of 4 – cyanopyridine degradation where on the 8*th* cycle, 41 % of enzyme activity was maintained [107].

The emerging synthetic techniques, such as soft and hard organic templates and molecular recognitions between hybrid interfaces, have impelled the research progress on multi-functional hybrid nanomaterials with hierarchical structures and new properties. The silica surface consists in two types of functional groups, siloxane (Si–O–Si) and silanol (Si–OH). Thus, silica gel modification can occurs *via* the reaction of a particular molecule with either the siloxane (nucleophylic substitution at the Si) or silanol (direct reaction with the hydroxyl group) functions, although it is generally accepted that it is the reaction with the silanol function that constitutes the main modification pathway [108].

There are three main methods in which functional groups are attached to the silica surface [109]:

- by reaction between organosilanes or organic molecules and silica surface functions;
- through chlorination of the silica surface followed by reaction of the Si – Cl with an appropriate functional molecule;
- incorporation of functional groups *via* sol-gel methodology followed by post-modification, wherever necessary.

Thus, the combination of inorganic and organic components into a complex material can promote and optimize the functionalities and properties of the composites, such as optical, electrical, magnetic, mechanic, catalytic, and sensing properties. In a publication from 2008, an overview on the recent advances in the self-assembly of the inorganic-organic hybrids in terms of the synthesized strategies, principles and emerging techniques, have been given [110]. The synthetic methods discussed for assembling functional hybrids include *in-situ* self-assembly, template-induced self-assembly, evaporation-induced self-assembly, and layer-by-layer assembly.

Silica gels can be chemically modified using organic precursors producing organically modified silica (ORMOSIL), a class of novel materials for hosting varieties of organic and inorganic substrates. Modern chemical research using ORMOSIL matrix has tremendous utility in chemical, medical, optical, electrochemical and many more areas of interest. ORMOSIL matrix materials show an enhanced activity during catalysis, photochemical activities like absorption and emission, electrochemical sensitivities, sensing of gases, solvents, pH of solution and biomolecules, etc [111]. ORMOSIL-based materials can be used as efficient protective coatings and can be utilized in designing wave guides that can carry out excellent photonic transmission of information. The main applications of these ORMOSIL materials are largely discussed in a paper published by Dash *et al.* about molecular entrapment, gas sensor, solvent and pH sensors, biomolecular sensors, protective coatings, catalytic and scavenger activity and photochemical applications [108].

Having in view the fact that the porous silicon films are currently under intense investigation for optical, thermal and electronic applications, a recent study elaborated by the

Liu and co-workers developed a sol-gel synthesis of SiO_2 film with two opposite structures (nanoporous and protuberant), prepared by an acid one-step and acid-base-two steps catalytic sol-gel process, under the action of CTAB template [42]. The main conclusion is related to the catalyst acidity that can remarkably adjust the structures of the silica films: the films deposited from acid catalyzed sols have a porous structure, while the SiO_2 film fabricated from the acid-base two-step catalyzed sols under the action of CTAB have a protuberant surface. The introduction of organic groups into an inorganic network improves mechanical properties, leading to easier processing of thick films [42]. In particular, organic-inorganic hybrid waveguides in thin film configuration become more important, as they can be obtained at a low heat treatment temperature, which would allow direct integration onto the same chip with the active device with the pump source and other optoelectronic components [112].

An effective and reproducible method of preparing highly monodisperse spherical colloids of organic-inorganic hybrid silica was studied by Deng and colab. consisting in a sol-gel reaction in aqueous solution, by using different silica precursors (vinyltriethoxysilane phenyl-triethoxysilane, tetraetylortosilicate and (3-mercaptopropyl) trimethoxysilane) [113]. This method has also been extended to design and prepare other organic-inorganic hybrid materials especially in monodisperse surface-modified silica spheres.

In an interesting research, a novel non-surfactant method was described to synthesize mesoporous silica using a kind of organic dye with special large plane structure, named basic fuchsine (BF) as template [114]. The typical synthesis of dye-template mesoporous silica followed a three stepped reaction, in which aminopropyltriethoxysilane, m-xylenediisocyanate, BF, and tetraethoxysiloxane were used as reactants and dichloromethane as solvent. This method avoids the use of surfactant templates and hence may be less expensive than the conventional surfactant-directing methods [114].

Recently, the cell immobilization technology to produce biocatalysts has attracted much attention. The application of different biocatalysts reduces the energy expenses and their employment could decrease the distribution of various pollutants in the environment. It has been found by Samuneva et al. that there is a strong correlation between the structural features of the hybrids and the activity of the biocatalysts [115]. The aim of this study was to investigate the influence of the silica precursor and quantity of organic component on the structure, properties and possibility for biological application of the silica hybrids with j-carrageenan. Hybrid nanocompsites containing different precursors (TMOS, ETMS) and j-carrageenan were synthesized via sol-gel method at room temperature. The results from the biological experiments were compared for different matrices about their residual activity and operational stability depending on the hybrid. During the performed investigations of synthesized silica hybrids it was established that these hybrid materials are appropriate as carriers for obtaining active biocatalysts. The possibility of application of the synthesized biocatalysts in an enzyme degradation process of the toxic, carcinogenic and mutagenic substances benzonitrile, fumaronitrile, o-, m-, and p-tolunitriles was investigated at batch experiments. The biodegradation process performed in the two steps bioreactor proved no washout, allowing high and stable process efficiency even at high temperature [115].

By combining the hardness and optical transparency of silica with the optical properties of different organic dyes it is provided a large area of optical properties, such as fluorescence, laser emission, photocromism, nonlinear optics or photochemical hole burning. The mean size of organic and inorganic phases are mostly of the order of few nanometres, therefore they are transparent and can be used for optical applications. Moreover, due to their improved mechanical properties, hybrid sol-gel matrices can be polished down to one nanometre in surface roughness. A great variety of work has been developed in which the structural and optical properties were studied [116-118]. The requirements for achieving new optical gel-glasses are related to the control and processing of sol-gel glasses, by means of a precise control of compositional and a pore surface variation in cage trapped molecules [117]. The modification of an inorganic network structure with organic groups gives a larger space for the isomerization of organic photoactive molecules as compared to inorganic glasses, thus the organic-inorganic hybrids are anticipated as desirable materials for photonic applications, which can trap organic molecules [118].

Recently, optically, transparent and homogeneously dispersed organic-inorganic hybrids materials containing organic components have been widely synthesized by the hydrolytic sol-gel process through hydrolysis-condensation reactions of alkoxide precursors [112, 119]. For example, the coupling of low-temperature processing and gel-glass doped with organic photoactive or electroactive molecules has opened new opportunities for optical and electro-optical applications [120]. Beside the hydrolytic route, the exploitation of the non-hydrolytic sol-gel method is also useful in the synthesis of hybrid organic/inorganic materials for optical applications, since the residual hydroxyl groups can be reduced or eliminated by forcing the condensation reaction [121]. Dire *et al.* uses a non-hydrolytic sol-gel process for the condensation reaction of methacryloxypropyl trimethoxysilane and diphenylsilanediol, in order to synthesize nanostructured molecular units for the preparation of hybrid organic-inorganic coatings and to overcome some drawbacks of the hydrolytic sol-gel process [120]. Nanostructured molecular units of different molecular complexity were synthesized starting from methacryloxypropyl-trimethoxysilane and diphenylsilanediol using the non-hydrolytic condensation reaction with suitable promoters. The obtained samples were used to prepare hybrid organic-inorganic matrices by polymerization of the organic functions.

The hybrid composites that contain various luminescent organic molecules, complexes of transition metals and rare-earth ions, and metallurgical and semiconducting nanoparticles have been widely studied in view of the possible use of them as optically active components of laser systems, elements of nonlinear and integral optics, luminescent screens, and photorefractive materials [112, 123]. Glasses, also including the hybrid materials obtained by the sol-gel method, do not represent an optically inert medium and manifest natural luminescence in different spectral regions, which depends on the technique according to which the materials were prepared. At the present time, the nature of this luminescence has not been completely studied [124], but there is no doubt that the luminescence centres formed by the structure of the matrix exert a substantial effect on the processes of energy relaxation of the active centre included in it.

Due to their structural flexibility, the hybrid materials are being increasingly used, in particular, in the field of integrated optics [125, 126]. Wenxiu Que and colab. proposed a strategy to prepare TiO_2/γ-glycidoxypropyltrimethoxysilane and methyl-trimethoxysilane hybrid organic-inorganic material, which contains azobenzene groups and is doped with neodymium ions, by a low temperature sol-gel technique [112]. This hybrid material manifest valuable optical multifunctional properties, which include optical data storage, optical switching and upconversion luminescence for photonic applications based on the doping of azobenzene and neodymium ions. These results indicate that the new prepared hybrid materials with multifunctional photonic properties are promising candidates for integrated optics and photonic applications.

The sol-gel glasses may be poured in desired shapes and are optical transparent, this last aspect being useful in coupling of optical properties with bioactivity, in order to build the photonic devices and biosensors. Sensor development has been driven, in large part, by the need for new devices that are less expensive and simpler to construct and operate, provide adequate detection limits and selectivity, and are accurate and reliable [127]. In a generic biosensor, an immobilized biorecognition element (e.g., an antibody, cell, DNA oligonucleotide, enzyme, lectin, protein,) serves to selectively recognize the target analyte and the binding or conversion (if the analyte is a substrate) event is used to produce an optical, mass, thermal, or electrochemical signal that is related to the analyte concentration in the sample [128].

Molecular imprinting is an exciting and promising technique that is being increasingly adopted as a platform for creating responsive materials like chemical sensors. Molecularly imprinted materials appear to offer an inexpensive, robust, and reusable alternative to expensive and labile biorecognition elements. An overview of recent progress in molecular imprinting in sol-gel matrices, the potential analytical applications of these tailor-made materials and their limitations are published by Diaz-Garcia [19]. Molecular imprinting involves arranging polymerizable functional monomers around a template followed by polymerization and template removal. Arrangement is typically achieved by non-covalent interactions (e.g. H-bonds, ion pairing) or reversible covalent interactions. A properly designed molecularly imprinted polymer (MIP) can then bind the template or structurally similar analytes [129].

The recent growth of interest in organic-inorganic hybrid materials prepared by sol-gel chemistry along with the growing interest in molecular imprinting in inorganic matrices may stimulate the design of unique materials with controllable pore size, structural rigidity, thermal stability and enhanced recognition properties. A review realised by Holthoff *and colab.* summarizes recent research efforts on the development of molecularly templated (sometimes called molecularly imprinted) organic and inorganic polymers as possible replacements for expensive/labile biorecognition elements [127]. The review starts with a briefing on biosensing and the appropriate issues and limitations. The main objective of study has been focused on the molecular templating within organic and inorganic polymers in order to create new materials with analyte binding characteristics related to a biorecognition element. Several recent developments are discussed, in which the analyte recognition and an analyte-dependent transduction methodology are simultaneously incorporated directly within the templated materials.

Probably one of the most researched area with a growing interest in the last years are based on the combination of silicon chemistry with life sciences, that makes possible the application of hybrid bionanomaterials as carriers for different bioactive compounds. This silica material, originally developed for engineering applications, is currently also being studied as a polymer for the entrapment and controlled release of drugs [130]. For a controlled release drug delivery system biodegradability and biocompatibility are the fundamental requirements. As stated above, the specific characteristics of silica (amorphous, porous, resorbable materials, good compatibility etc.) allow its exploitation as a matrix for entrapping bioactive compounds, with several applications in biotechnology and biomedical sciences.

Controlled drug delivery systems can achieve precisely spatial and temporal delivery of therapeutic agents to the target site. Generally, the controlled drug delivery systems can maintain the concentration of drugs in the precise sites of the body within the optimum range and under the toxicity threshold, which improve the therapeutic efficacy and reduce toxicity.

There has been a rapid growth in the area of drug delivery, in searching of new drug delivery systems. In recent years, mesoporous materials, which have unique pore size, higher surface area and pore volume, have been widely employed as carriers for controlled drug delivery. Many molecules including drugs, proteins and growth factors can be released from sol-gels and the quantity and duration of the release can vary widely. Even the release mechanisms of drugs from sol-gel derived silica materials have not been extensively studied, some data from literature [131-133] showing that controlled release of antibiotics and proteins is possible from these porous materials and also demonstrating that the release is dependent on synthesis parameters such as type of precursors, molar ratio between silica precursor and water and the concentrations of bioactive drugs. Other literature reports indicate that the drug release is controlled by simple diffusion through solvent-filled capillary channels while other papers concluded that the release of drug occurs according to a combined process of the diffusion and the erosion process of the matrix [32, 33, 134]. Recently, the silica sol-gel micro spheres are reported by Radin *and colab.* as excellent controlled release materials, being ideally biodegradable materials with generally good biocompatibility [26]. In this study, the authors were focused on the synthesis of sol-gel porous microspheres by using a novel process with two steps. Controlled release sol-gel silica microspheres containing an antibiotic – vancommycin or an analgesic – bupivacaine were synthesised using a new acid-base catalysed sol-gel process followed by emulsification. The novelty is related to selection of an appropriate catalysis, in order to shorten the time to gelation of the sol. The drugs incorporated in the microspheres showed a slow and long-term release and longer *in vitro* dissolution durations than those of the granules.

Teoli *and colab.* accomplished a completely study in which the wet sol-gel derived silica gels are highly promising materials for parenteral administration and sustained release of protein compounds [135]. Two gel formulations, with different levels of polymer concentration, were loaded with model proteins (e.g. avidin, bovine serum albumin – BSA and ribonuclease-A) and protein release was measured upon immersion in physiological buffer. Preliminary *in vivo* experiments were carried out in mice to evaluate the bioerodibility and local toxicity of the implanted material [135]. Results indicate that sustained release and quantitative protein recovery can be achieved within a time-frame that can be modulated by formulation parameters. Embedded proteins maintain their original conformation and are stable to both thermal denaturation and protease degradation. The gels formulations are highly erodible, as demonstrated by both *in vitro* and *in vivo* experiments and no local or systemic toxicity was observed after subcutaneous administration in mice [135].

SBA-15 is another important mesoporous SiO_2 material, synthesized by Lopez and co-workers, using tetraethylorthosilicate, TEOS, as precursor and Pluronic P123 as the organic structure-directing agent appropriate for synthesizing an ordered porous oxide matrix with a control pore size, shape and surface area [136]. The specific aim of this study was to determine the feasibility of loading valproic acid and sodic phenytoin molecules inside the pores of ordered mesoporous SiO_2 and study the biocompatibility of such reservoirs with brain tissue [136]. The use of these drug-containing reservoirs based on nanostructured materials represents an alternative to deliver the drug without causing secondary effects, for temporal lobe epilepsy which is one of the most frequent types of human neurological diseases.

In a complementary study published in 2009, Wang *et al.* presented an investigation on progress of this novel application of mesoporous materials as carriers for various drug delivery, comparing the different behaviour of drug and mesoporous-solid systems [137]. This paper contains some important aspects treating the textural and structural properties of several mesoporous materials such as M41S, SBA, MSU, and HMS regarding to drug loading and release profile, the influence of functionalisation of mesoporous materials by organics and the bioactivity of various mesoporous solids, the important factors influencing the drug loading and release mode as well as the release kinetics, the biocompatibility of the drug systems and their emerging application for tissue engineering.

In the meantime, Prokopowicz *et al.* also reported a study in which organically modified silica xerogels were prepared by a two-step acid/base catalyzed sol-gel process that provides a slow release of an anticancer drug – doxorubicin hydrochloride (DOX) [138]. The aim of this work was to prepare silica/polydimethylsiloxane xerogels with encapsulated doxorubicin hydrochloride that would provide a slower drug release compared to non-modified silica. This research represents a supplement of a previous work of authors in which doxorubicin hydrochloride-loaded non-modified silica xerogel was prepared by the sol-gel method and examined for its stability and drug release [139].

The influence of different amounts of silanol-terminated poly(dimethyl-siloxane) (PDMS) added on the properties of xerogels intended for the release of the drug and the dissolution of xerogels was investigated. The rate of release of the drug under the *in vitro* conditions used is affected by PDMS content, which depends on both the microstructure and chemical properties of the materials. An increase in PDMS content results in an increase in the hydrophobicity and the decrease in porosity of materials, leading to the decrease in drug release. These silica/polydimethylsiloxane xerogels loaded with doxorubicin may be a promising carrier material used as an implantable drug delivery system for long-term disease control (e.g. bone disease) [138].

2. Templating of Sol-Gel Microenvironment for Synthesis of Novel Fluorescent Nanomaterials

This chapter provides an overview on the recent advances in the self-assembly approach in terms of the synthesized strategies that allow a facile access to designed synthesis of novel hybrid organic-inorganic materials, principles and functional properties. In this chapter we will give a brief account of our work and also currently researches in progress on the synthesis of novel fluorescent nanomaterials and its related applications in nanotechnology, especially in biochemistry. The fluorescent hybrid nanomaterials are synthesised using a self-assembly approach in which the surfactants are assembled with inorganic and hybrid silica matrices into sophisticated nanostructures through favourable molecular interactions. More extensive literature reviews and experimental details are avoided here, but they can be found in the publications cited in the references section.

The purpose of this chapter is to summarize the recent attempts to investigate the influence of the silica and hybrid silica-silsesquioxane matrices on the synthesis and properties of hybrid nanomaterials containing active principles and to improve the fluorescence properties of conventional dyes by extending the molecular encapsulation towards entrapment of some active principles from flavonoids and mixture of flavonoids from several natural extracts (e.g. Begonia extract, orange peel extract, ornamental bush extract and green tea extract). To accomplish this objective, a series of systematic aspects of the self-organization process and the evolution of colloidal templated sol in sol-gel encapsulation should be covered.

2.1. Self-Organisation Approach for the Sol-Gel Microenvironment with Template Agents

Self-assembly process is one of the novel chemical strategies, which is able to direct the assembly of the various structurally well defined nanomaterials into the complex architectures with hierarchical structures. The organization of growing the inorganic or hybrid networks has been templated by organic structure-directing agents, especially surfactants [140-142]. The aqueous behaviour of surfactants derive directly from their molecular structure, i.e. from the fact they have an amphiphilic structure [5, 143] with two parts with opposing solubilities in water: the hydrophilic or solvophilic part ("polar hydrophyl head") is usually a short polar group that may be ionic or uncharged. The hydrophobic part ("hydrophob unpolar tail") is a long chain, usually of hydrocarbon nature or aromatic-based composition. A vast class of chemical compounds meet these requirements, thus the surfactants can be classified by the presence of formally charged groups in its head, from synthetic anionic and cationic surfactants to natural zwitterionic lipids and block-copolymers [144, 145]. Some commonly encountered surfactants of each type include:

- *ionic:* anionic (e.g. sodium dodecyl sulfate (SDS), ammonium lauryl sulphate, and other alkyl sulphate salts; lauryl ether sulphate (SLES); alkyl benzene sulfonate; soaps or fatty acid salts); cationic (e.g. cetyl trimethylammonium bromide (CTAB) and other alkyltrimethylammonium salts; cetylpyridinium chloride (CPC); polyethoxylated amine (POEA), etc.); zwitterionic (e.g. dodecyl betaine; dodecyl dimethylamine oxide, etc.)
- *nonionic* (e.g. alkyl poly(ethylene oxide), copolymers of poly(ethylene oxide) and poly(propylene oxide) commercially called Poloxamers or Poloxamines, alkyl polyglucosides including octyl glycoside and decyl maltoside, fatty alcohols such as cetyl alcohol, oleyl alcohol etc.

The amphiphilic or dual solubility character of a surfactant implies that, firstly it adsorbs at any polar-polar interface, reducing the interfacial tension, and secondly it self-organizes in bulk water into aggregates containing many molecules and having well defined average shape and size, a process usually called surfactant self-assembly [145]. Self-assembling processes, also called self-organization process, is a bottom-up approach that basically consists on designing atoms and molecules that undergo a physical, chemical or biological process that ends up with the atoms and molecules being at the right place forming the right structure. The bottom-up approach relies on self-organization of molecules or nanometer sized compounds, being for many scientists the more elegant possibility to form large complex hierarchical structures [45].

The self-assembling process is based on the principle that one or more of a variety of forces drive atoms and/or molecules to self-organise into structures (which will be those that represent lowest energy configurations within the particular constraints of the environment). The main thermodynamic driving force for both adsorption and self-assembling processes is the hydrophobic effect that means the large free-energy gain associated with the isolation of the surfactant chains from contact with water. The hydrophobic effect is characterized by a subtle interplay between a large positive entropic term and an enthalpy term whose sign is temperature-dependent [145].

Self-assembling of surfactant in solution gives rise to a large variety of structures: micelles, vesicles, sponge phases, liquid crystal lyotropic phases (hexagonal, cubic, and lamellar). Micelles and lamellar phases were the most extensively studied systems, but the variety of structures observed with surfactants also attracted considerable interest, and recent progresses allowed rationalizing the physical chemistry principles of their molecular organization [146]. These dynamic properties are important in synthesis of mesoporous materials (with the size of $2\ nm < d < 50\ nm$).

Thus, it can be said that the production of nanoporous structures with controlled size of pores, morphology and size distribution usually involves the use of tailored templates that can fill space and/or direct the formation of specific structures. Moreover, the use of templates provides the conditions for self-assembling of raw materials in the desired way and is absolutely required for long-range arrangement of the nanoporous materials.

Templates have been technologically used in materials processing for a long time. The applications of surfactants have been developed from the classical field of cleaning or solubilization, to be used as powerful tools in formation of nano- and mesostructured porous materials. The accuracy and reproducibility of self-assembled processes were observed by obtaining a high control of material architecture at nanometric scale. This approach of multifunctional nanomaterial preparation with controlled properties has started a remarkable interest with connectivity between different domains such as surface and materials chemistry, physics, biology and so on. In most of researches some ionic and neutral surfactants have been used as template that may direct the mesophase formation based on electrostatic and hydrogen bonds interactions, respectively. For example, the ionic surfactants may form ordered hexagonal porous structure. In concordance with the technology developed by the researches from Mobil Corporation, the quaternary ammonium salts with long tail manifest a minimum of its energy value in solution, by assembling in micelle structure. Their ability to form micelles like rods and hexagonal arrangements on a large domain (with diameters in mesoporous range of 2-4 nm) has been known for long time ago. The formation of this organized structure is strongly dependent by the surfactant concentration, the length of alkyl chain and temperature of solution [147].

The proposed mechanism for pores formation, by using ionic surfactants as template is called „liquid-crystal templating mechanism – LCT". The LCT mechanism has been introduced by Beck [148] in order to explain the formation of mesoporous materials such as MCM-41, in which the supramolecular assemblies of alkyltrimethylammonium surfactants serves as efficiently structural component for directional formation of silica mesophase.

The neutral surfactants do not follow the LCT mechanism due to the absence of strong electrostatic interaction. Instead, on this way of templating some aggregates are formed due to the physical agglomeration or hydrogen bonds [149]. The hydrogen interactions between hydrophilic surfaces of flexible micelles and hydrolysis intermediate products have proved to be the driving force in mesophase formation. The size of pores may be established by varying the length and structure of surfactants. Mesoporous materials prepared by neutral surfactants sometimes present channel arrangement ordered on an extensive domain. These kinds of materials have a textural/fibber mesoporosity and a larger thickness of walls, and as a result they present a better thermal stability. The most used neutral surfactant is that based on poly(ethylene oxide) derivatives, and recently alkyl polyglucosides, which represent a convenient alternative, being neutral, non-toxic and biodegradable. For example, some authors have been used different sugar-based surfactant molecules such as commercially available octyl-β-D-glucoside, dodecyl-β-D-maltoside or tailor-made glucoside alkylpolyethers and the corresponding acetylated derivatives as templates in the synthesis of mesostructured thin silica films produced by sol-gel processing. For example, octyl-β-D-glucoside surfactants formed preferentially lamellar mesophases, dodecyl-β-D-maltoside surfactants a worm-like silica structure and the glucoside alkylpolyethers a well-ordered hexagonal silica mesostructure. Their self-assembling capability in combination with a prehydrolyzed silicate solution was investigated by Štangar *and colab.* [150].

The recent literature developed a new class of template namely "non-surfactant template". Nonsurfactant molecules such as glucose, fructose, maltose, dibenzoyl-L-tartric acid, citric acid, urea, glycerol, cyclodextrins, hydroxyethyl methacrylate, oligopeptides etc., can be used as templates to direct the mesostructure formation during the sol-gel reactions. This non-surfactant approach has unique advantages to mesoporous materials referring to biocompatibility. In order to elucidate the mechanism for the mesophase formation the group leaded by Wei has investigated cca 100 compounds [151]. They demonstrated that only those compounds that contain highly polar functional groups can work as template which allows the obtaining of narrow distributed mesopores. Their presumption is based on the strong interactions and hydrogen bonding between the nonsurfactant molecules or its aggregates and the inorganic species (e.g. silicate intermediates) that may play an important role in directing the mesophase formation.

The non-surfactant-templated materials have no discernable packing or orientation of the mesopores or channels. For instance, D-glucose, D-fructose and other related compounds which represent some atypical templates, may fulfil the role of a conventional directing agent by virtue of its highly polar functional groups which permit the formation of some collectivity (known as aggregates or glucidic assemblies) that may physically retain the organic molecules. The organic molecules are bound to the sites of the sugar layer by non-covalent interactions, such as hydrogen bonds and Van der Waals interactions. Moreover, this glucidic collectivity may participate to strong polar interactions and H bonds with OH groups of inorganic species (intermediate silicate species) and thus can favour or even catalyze the sol-gel reaction. One of the possible explanations is based on the fact that the strongly polar interactions and the hydrogen bonds between glucidic aggregates and inorganic silica species play an important role to direct the formation of the interconnected channels. These glucidic templates do not form an ordered structure (with uniform pores) as usual templates, but this disordered structure of glucidic assemblies is sometimes advantageous due to the fact that it allows the spreading of properties along all directions.

2.2. The Template Effect on Evolution of Colloidal Sol in Sol-Gel Encapsulation

The ability of surfactants to self-assembly in well defined structures has the advantage of synthesis and design of new inorganic/organic materials with nanometric size. Template – directing synthesis represents a convenient and versatile method for generating hybrid nanostructures. There are two types of templates, usually encountered as "soft template" and "hard templates" [110]. Surfactants, polymer and other organic small molecules can be used as soft templates to induce self-assembly of different kinds of inorganic-organic hybrid nanomaterials. The commercial powders and prepared nanoparticles with different morphologies can play the hard templating role in directing the formation of the functional nanocomposites. The method based on hard template can be used to synthesize and organize the metal-based hybrid materials with good quality and unique structures. Soft templates play key rolls in the fabrication and self-assembly of the organic-inorganic hybrid materials with novel structure and uniform shapes, thus the applied area of this method is more universal and effective.

One of the most popular techniques intensively used for the preparation of new materials with specific applications in the top scientific domains is the synthesis in sol-gel porous matrices assisted by template. Different types of surfactants may be used as templating agents because features such as structure, composition, pore diameter, pore volume, and surface area can be tailored by the inorganic source material, the molar composition, the template type, and the condensation/hydrolysis processes.

The porous sol-gel matrices are formed when inorganic oxides (e.g. silica) polymerize in the presence of surfactants with „template" role, which serve as structure-directing agents for the oxide framework and assure the presence of some cavities with pre-established shape and size [45]. The shape and size of self-assembled template aggregate is the result of a fine balance between different free-energy contributions [152, 153]: a transfer term related to the water removal of the chains into the liquid-like apolar core of the aggregate; headgroup repulsion terms, associated with electrostatic or steric repulsions between the hydrated head groups; the packing constraints of the chains in the core; the minimization of oil/water interfacial contact in the aggregate surface; the contact of the head groups with the aqueous medium and the repulsions between them, that ensure the formation of aggregates and not a macroscopic phase separation.

These porous materials formed in presence of a template are of scientific and technological importance because of the presence of voids of controllable dimensions at the atomic, molecular, and nanometer scales, enable them to discriminate and interact with molecules and clusters. Nanoporous materials can have open (interconnected) pores or closed pores and can have amorphous, semi-crystalline or crystalline (e.g. lamellar, cubic, hexagonal) frameworks. These two characteristics influence very much the applications of a specific nanoporous material and for what is suitable. Nanoporous materials combine the advantages of porous materials with the physico-chemical-biological functionality of the material itself. Normally, properties of materials are enhanced or inhibited by the nanometer-sized porous structure, but still depend on the material chemical composition.

There are two main approaches to produce sol-gel porous materials [19]:

1) A non-covalent approach which is based on the self-assembling process. In this case the complex formation is the result of non-covalent or metal ion coordination interactions (e.g. van der Waals, π-stacking, electrostatic), between the template and the sol-gel network. The template is directly added to a sol-gel solution prior to acid-catalysed hydrolysis and condensation. As the template molecules are removed (by solvent extraction, calcination or other appropriate methods), the porosity of materials increases resulting in a mesoporous inorganic matrix (e.g. silica) with interconnected channels of regular diameters which may play host role to a variety of organic/inorganic molecules.
2) A covalent or pre-organized approach that employs the formation of some reversible covalent bonds. This approach usually involves a prior chemical synthesis step to link the precursors to the template or to a structurally similar molecule to create a "spacer". This conjugate is then polymerised using an excess of the metal oxide precursor. Once the sol-gel is formed, the spacer or the template is chemically removed, leaving a pocket that has the ability to bind molecules of the appropriate size and shape.

For instance, the feasibility to produce some colloidal sols with a controllable pore size, by using the first approach, has been investigated in a recent study [154] starting from triethoxymethylsilane (TEOS) as silica precursor together with cationic surfactants (monododecyltrimethyl-ammonium bromide = C12-1, didodecyldimethyl-ammonium bromide = C12-2 and trioctadecylmetil-ammonium bromide = C18-3) as template agents for the silicon oxide framework. The templated sols obtained by using these types of templates were further used to encapsulate of an active compound from flavonoid class (quercitin) *via* physical associations between the silanol groups and polar head groups of quercitin.

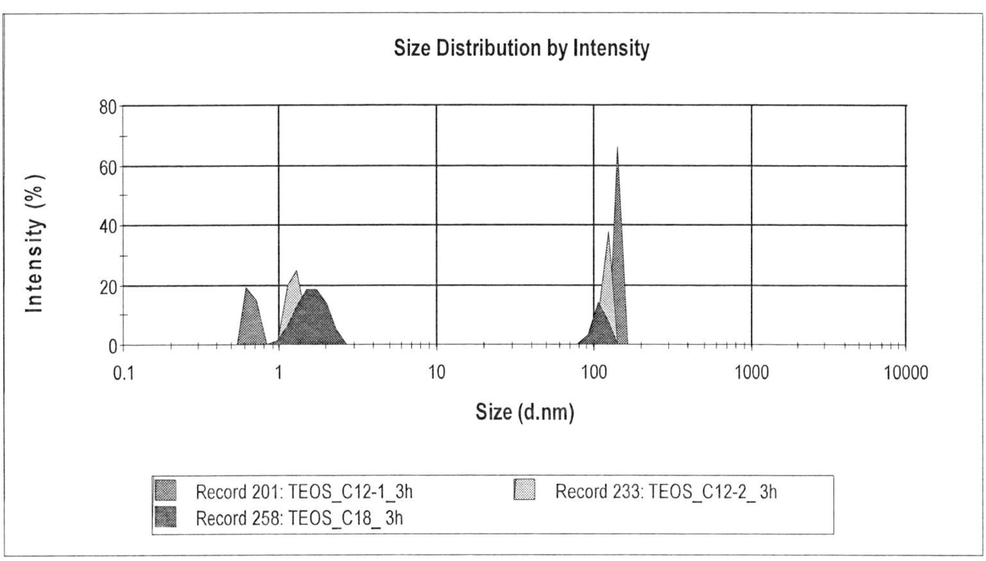

Figure 4. The evolution of colloidal templated sols made with C12-1, C12-2 and C18-3 surfactants by DLS [154].

In the optimized conditions (TEOS/H_2O/Et-OH/surfactant molar ratio of 1:2:3.8: 0.0005 w/w), which permits a slow hydrolysis degree and a desired porosity of templated sols, the TEOS precursor is partially hydrolysed, to produce various templated sols. The evolution of templated sols prepared with the three surfactants having different lengths of alkyl chains and, as a consequence, with different hydrophobicity degrees, was observed by dynamic light scattering (DLS) technique (figure 4).

From figure 4 it can be observed that after one hour of hydrolysis, the estimated size of colloidal suspensions can be included in two size orders (between 0.6 – 1.6 nm and 109-142 nm), as a result of competitive processes of hydrolysis and beginning of polycondensation reactions. As can be seen from figure 4, in case of C12-1 templated sol the hydrolysis degree is more advanced (0.66 nm, 33.8% and 142 nm, 66.2%), comparative to C18-3 templated sol, when after one hour of hydrolysis only a percent of 25% of colloidal size of 109 nm is obtained. The increase of hydrolysis degree in case of C12-1 surfactant may lead to a larger size of final products, as will be shown later by the results from fluorescence analysis (figure 5). This slowing of hydrolysis processes in the C18-3 sol could be due to the three octadodecyl chains of surfactant. This aspect is also sustained by the hydrolysis degree of C12-2 templated sol which is between those of C12-1 and C18-3 templated sols (figure 4). The evolution in time of these competitive hydrolysis and condensation processes is exemplified for C12-2 templated sol (table 1). It can be observed that after 60 minutes of hydrolysis, the sizes of C12-2 colloidal sol are 1.25 nm (52.9%) and 119 nm (47.1%). After an interval of 30 minutes, the sizes of colloidal sol are modified, these being 1.53 nm (for the size I) and 421 nm (for the size II) [154].

Table 1. The evolution in time of size distribution for colloidal sol templated with C12-2 surfactant

Time [min]	Size I		Size II	
	nm	%	nm	%
60	1.25	52.9	119	47.1
75	1.36	47.1	207	50.8
90	1.53	43.8	421	56.2

This increase in time of the second type of aggregate size (hundreds nanometers), accompanied by simultaneous decrease of percent for the first type of aggregates less than 2 nm, suggests the advance of polycondensation processes. Polycondensation processes lead to the formation of soluble macromers and then colloids, which coalesce and raise the solution viscosity to the sol-gel transition, when the bulk gelation occurs. Thus, the chemical species resulted from hydrolysis steps react during polycondensation step, to form Si – O – Si bonds leading to the silica network formation. Since silanol condensation is highly accelerated by bases, this is achieved by simply ensuring that the $pH > 7$. At this stage, the templated silica sols may be used for encapsulation of quercitin molecule, by mixing of sol with a given amount of quercitin and initiating the gelation step. Organic biomolecules added to the sol became physically entrapped into the cavities of the network formed upon gelation [154]. This results in a hydrogel carrying embedded or interstitially captured quercitin. The hydrogel is aged in an oven and further polycondensation leads to cross-linking and pore-network development (3D network) which surrounds the organized structure of surfactant and is typically accompanied by shrinkage and syneresis (bulk extrusion of the liquid phase) to produce the aged hydrogel doped with quercitin.

A clear indication of quercitin encapsulation inside silica matrix has been provided from fluorescence measurements (figure 5). By comparing the pure quercitin fluorescence spectrum with those of resultant encapsulated samples it can be observed that in all the encapsulated samples the fluorescence intensity was enhanced due to the conformational arrangement of quercitin inside the silica network and also due to the excellent optical properties of silica matrix.

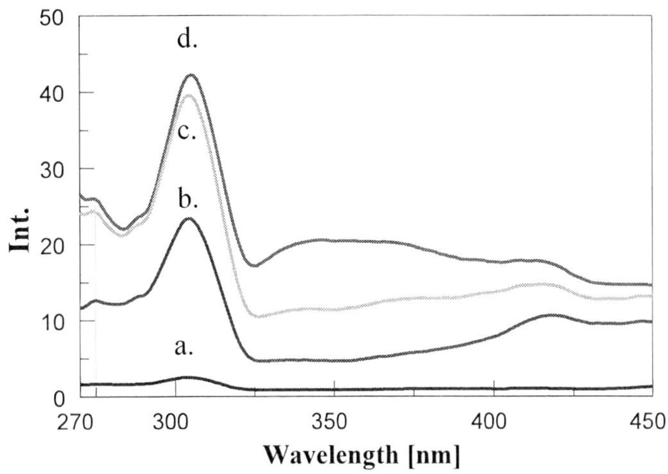

Figure 5. Fluorescence spectra of pure quercitin (a) and encapsulated quercitin samples b = with C12-1 template; c = with C12-2 template and d = with C18-3 template intensity [154].

As it can be seen in the figure 5 there is a significant difference in the fluorescence of the samples prepared with the three cationic surfactants, especially in the sample prepared with C12-1 surfactant compared to the C12-2 and C18-3 samples [154]. Moreover, there is an important enhance of fluorescence intensity in both encapsulated quercitin samples prepared with C12-2 and C18-3 surfactants (16 times) as compared to the native quercitin.

2.3. Synthesis and Fluorescence Properties of Hybrid Nanomaterials with Active Principles

The sol-gel method has allowed the physical trapping or the covalent bonding of many types of organic species inside inorganic networks that lead to a new generation of advanced materials with important properties. Sol-gel hybrid organic-inorganic materials are largely exploited as novel materials in the top research areas, due to the control of the structure at a nanometric scale and often conferring superior properties of silica-derived matrices. Sol-gel optics is undoubtedly the most active field in the use of hybrid materials. The advantages may derive from optical properties of organic dyes combined with the hardness and optical transparency of silica. Among the various inorganic hosts/inorganic oxide glasses, silica is preferred being indeed the best host candidate for its superior mechanical, thermal and optical properties, as well as for its good chemical stability. In fact, they guarantee low optical losses, resistance against dry processes, high flexibility in the final material property design (e.g. refractive index tunability), and the possibility of embedding large amount of organic chromophores.

Modern research on organic dyes includes investigations for conjugated polymers, hydrogen bonded assemblies, chromogenic sensors, molecular shuttles, solar energy cells, photonics and various approaches to photodynamic therapy. There are some common shortcomings of organic dyes, especially those with long-wavelength absorption bands, regarding to their susceptibility to chemical and photochemical degradation. The reason for the enhanced reactivity is the inherently small HUMO-LUMO energy gap, which means that the dyes are potentially reactive with both nucleophyles and electrophyles [155]. Another potential drawback of organic dyes is their tendency to aggregate, which induces multichromophoric interactions that alter the colour quality and quench the photoluminescence. In principle, these problems can be attenuated by encapsulation strategies that isolate the individual dye molecules and prevent self-aggregation or similar interactions with the chemical environment [155]. From an ideal point of view, the hosted organic molecules entrapped inside host inorganic or hybrid silica matrices should satisfy the following conditions:

- to be homogeneously distributed;
- to be trapped in a cage, so that collisional deactivation of the main properties (e.g. antioxidant, fluorescence) is avoided;
- no dimmers or aggregates that inhibit the fluorescence activity should be formed; besides, the samples should exhibit some photostability.

The main limitation of the organic dye materials, more than the thermal stability, is related to their photo stability which is determined by the host nature and by its physical-chemical interactions with the dye. In the literature, many hosts have been employed to entrap organic dye-molecules [103, 105]: (a) high surface area powdered materials in slurries or as colloids; (b) high surface area porous glasses; (c) polymeric materials in which the dye is not simply adsorbed but homogeneously distributed; (d) monolayer of low surface area supports; (e) inorganic oxide glasses obtained by sol-gel, where the dye is embedded. Among them, the host matrices obtained by sol-gel synthesis is mandatory since it allows operating near room temperature, so that the organic molecules can be easily trapped without thermal degradation. An advantage of organic molecules incorporation in a liquid sol during the sol-gel process is that the components are mixed at a molecular level, and a large amount of dye can be incorporated before fluorescence quenching takes place because of dye clustering [101, 102]. In the sol-gel process, since chemically stable guest molecules are already present in the precursor solution for the synthesis and the reaction conditions are relatively mild, encapsulation without extensive modification or other negative reactions would be the possible process occurring and no new covalent bonds are formed during the synthesis [156]. By these means, some organic chromophores with good optical functions can be perfectly incorporated into the inorganic or hybrid silica based materials.

The mechanisms of organic molecules retention during adsorption on silica surfaces occur by physical and/or chemical adsorption. Of special interest is the physical adsorption because it allows the maintaining of specific properties. In a recent research elaborated by

Garcia-Sanchez *and colab.* it is underlined by a comparative fluorescence study that the silica xerogels physically grafted with some free and metalled tetraphenylporphyrins derivatives that contain amino, hydroxy, carboxy substituted groups assured a preservation of fluorescence, the best results in terms of fluorescence preservation being obtained by using a tetraphenylporphyrin substituted with – NH_2 group in the *orto* position [24]. On the other hand, a covalent interaction of the tetraphenylporphyrins species with silanol groups of the silica surface inhibits the fluorescence.

The formation of hydrogen bonds between the electronegative atoms of the adsorbed molecules and the hydrogen atoms of the silanol groups from silica are primarily responsible for organic compound adsorption. The organic compounds are optically or electronically active due to their structure and composition. Compounds with electron donor oxygen, such as acids, ethers, alcohols and esters [157], can form hydrogen bonds with surface hydroxyl groups. Carbon-carbon double or triple bonds lead to a strong delocalization of π electrons, which can be evidenced by dark colours or nonlinear optical phenomena, fluorescence or photoluminescence [158]. Moreover, the specific properties of molecules in solution, aggregate molecular crystals, polymers or sol-gel glasses can enrich the set of molecular states where electron delocalisation leads to optical properties [119].

The possibility of entrapment of organic molecules with dyes characteristics (optically active molecules) into porous materials produced by the sol-gel templated method offers several advantage for the optical applications, with respect to use of these ideal materials for various applications in optics (for example in the fabrication of optical sensors, waveguides, filters, etc.) [25]. A large number of organic dyes have therefore been entrapped within sol-gel silica matrices. Some examples include the encapsulation of organic dye fluorescein into MCM-41 mesoporous molecular sieve by a sol-gel method [156]. The organic dye fluorescein encapsulated into MCM-41 mesoporous material proved to be present as an additive within the surfactant micelle of the mesophase, which possessed a hexagonal mesostructure with short-range order and a uniform nanosize. Furthermore, the lack of aggregation at high concentration was discussed in terms of the effect of the host-guest interaction on these properties. Spectral changes, longer luminescence lifetime and less aggregation at high concentration were observed to the dye-functionalized mesophase in contrast with that dissolved in water [156].

More recently, Khamova *et al.* prepared a series of hybrid organic-inorganic materials with optical properties in the form of gels and thin-layer coatings by the sol-gel method [25]. The work was devoted to the investigation of the fluorescence of materials prepared from two types of sols that contain Nile Red dye: (1) a mixture of the TEOS and 3-glycidoxypropyl-trimethoxysilane (GPTMS) precursors and (2) TEOS and a mixture of ED-20 epoxydiane resin and DEG-1 aliphatic epoxy resin as precursors. The fluorescence parameters of these products were investigated as a function of the Nile Red concentration and the composition of precursors forming hybrid sol-gel matrices (tetraethoxysilane, modified alkoxysilane, epoxy compounds). The experiments performed have demonstrated that both hybrid sol-gel matrices can be used for immobilizing the Nile Red dye in the design of bulk nanocomposite materials and nanocomposite coatings that exhibit fluorescence properties and which hold much promise as materials for the use in the fabrication of optical gas sensors.

Other investigations have been conducted on encapsulation of some hydrophilic active dyes (e.g. Lithol rubine B, U1-Red) in silica microspheres matrices by sol-gel process combined with a water-in-oil microemulsion [60, 159]. Lithol rubine B dye doped silica microspheres were prepared by combining the two techniques formed from sodium silicate and dye aqueous solution in cyclohexane medium [60]. An important objective achieved by authors was the stabilization to leaching of water soluble dye against water. The doping of GPTS (3-glycidoxypropyltrimethoxysilane) in sodium silicate and dye mixture solution greatly enhanced the stability against leaching of the dye. It was ascribed that GPTS serves simultaneously as an intermediate for the chemical bonding between the dye and silica, and as an agent for the formation of hybrid sol responsible for the shrinkage of pore size. The average particle size of these microspheres decreased as the homogenizing speed for formation of W/O emulsion and the weight ratios of water to oil increased, while the concentration of sodium silicate solution increases. It was deduced that GPTS might serve either as physical entrapment with the decrease of pore size or as covalent entrapment in silica matrices [60].

The use of photocromic fluorescent compounds in biology and other fields may be limited by their low solubility in aqueous media, cytotoxicity, pH dependence of the fluorescence and so on. A promising solution to avoid these problems was presented in a study reported by Folling [160] who realised the encapsulation of some fluorophore (diarylethene and rhodamine fluorescent dye) inside silica nanoparticles. The result of these experiments was successful synthesis of a new and efficient fluorescent photochromic compound, functionalized with an amino reactive group, that can be used for tagging other molecules or biomolecules of interest and for imaging and switching in optical microscopy. Moreover, the biocompatibility of silica makes them promising for *in vivo* imaging applications, by applying standard protocols for biomolecular coupling such as DNA and antibodies, thus opening up a large range of applications.

Other researches have treated the encapsulation of rhodamine dye recognised as one of powerful dyes used extensively in biotechnology applications such as fluorescence microscopy, flow cytometry, fluorescence correlation spectroscopy and ELISA [96, 161, 162]. Rhodamine 6G (Rh6G), can be immobilised in SiO_2 both physically (materials of class I) and by covalent bonds (class II materials). One of the studies describes a novel synthesis of mesoporous silica particles with encapsulated rhodamine 6G perchlorate dye, in which the dyes are physically entrapped inside the nanosize channels/tubes of micrometer size silica particles, through a one-step self-assembling process [161]. The authors underline a significant increasing of fluorescence brightening, as comparing with the maximum obtained from the same dye dissolved in aqueous solution, mainly due to the higher concentrations of the dye molecules inside the pores without dye dimerization which may quench fluorescence. The high concentrations of the dye without dimerization are explained by the presence of surfactant molecules inside the channels which can act as dispersant and also by silica walls that do not let the dye molecules to dimerize perpendicular to the channels; potentially, interaction with the silica surface could decrease the fluorescence. The interaction of the dye molecules with the silica walls is diminished by the fact that the silica wall is coated with the surfactant head groups. The size of several micrometers of the synthesised particles allows them to be used as ultrabright fluorescent markers in a broad variety of applications, without the use of special fluorescence microscopes [161]. In contrast, Grandi *and colab.* synthesized for the first time a rhodamine-SiO_2 class II hybrid in which rhodamine is chemically entrapped inside the silica matrix [96]. The research was motivated by the many efforts that have been performed in the last few years to embed dye molecules in various solid matrices with the aim to obtain solid-state dye laser devices that could replace the liquid ones. For the hybrid materials synthesis, a suitable sol-gel precursor was used and the formation of the bonds between 3-isocyanatepropyltriethoxy-silane and rhodamine was checked by FT-IR analysis. The photoluminescence spectra of the final hybrids confirm that they are promising candidates for applications in solid state dye lasers. The optical spectroscopy measurements of class II samples show that they are better materials for dye laser application with respect to class I samples.

As a results of all these recent developments published in the last five years, of particular interest is the encapsulation of some natural extracts originated from vegetable plants that contain polyphenolic compounds with pharmacologic properties used in medicine and related fields, discussed largely in the next chapter. One of the first studies that relates the entrapment of a vegetal extract into a silica xerogel matrix was published with recently [119]. The structural evolution of barley leaves extracts embedded in silica xerogel, in terms of the interaction between pigments embedded within the inorganic matrix, as well as the chlorophyll fluorescence of the doped glass were evaluated under a heat treatment and compared with the pure barley leaves. The results obtained in this study offered information about the formation parameters of chlorophyll-doped organically modified silica fluorescence glass, respectively the thermal treatment conditions necessary for obtaining a glass fluorescent suitable for optical applications [119].

2.3.1. Natural Products with Multifunctional Properties

The use of vegetable extracts for pharmaceutical and cosmetical purposes already has a long tradition and still retains its relevance today having in view the fact that they are typical components in the natural and physiological cosmetic and drug formulations. Frequently, singular substances with a definite effect can successfully be isolated from an extract; however very often this specific effect is closely related to the synergistic effects of different substances present as a mixture – usually with multifunctional properties – in contrast to chemical individual, homogenous substances. In spite of researches conducted in the last years that generated a growing interest in the potentially important role of flavonoids in maintaining human health and of numerous studies performed on polyphenol systems and factors influencing stability, solubility and other characteristics of polyphenol complexes [163, 164], their application for construction of hybrid nanomaterials remains very narrow. In this context, our group was focused on a research in the area of entrapment into different silica and silica-silsesquioxane polymers of several natural extracts such as flavonoid mixtures from Begonia plant, orange peel extract, ornamental bush extract and green tea extract.

Phenolic compounds are regarded as such group that are synthesised by plants during development and that response to some damaging conditions such as infection, wounding, UV radiation, etc. [165]. Approximately 8000 naturally occurring compounds belong to the category of "phenolics". A straightforward classification attempts to divide the broad category of phenolics into simple phenols and polyphenols, based exclusively on the number of phenol subunits present. The term "plant phenolics" encompasses simple phenols, phenolic acids, coumarins, flavonoids, stilbenes, up to hydrolysable and condensed tannins, lignans, and lignins. The polyphenols, to which the flavonoids belong, possess at least two phenol subunits; compounds possessing three or more phenol subunits are referred to as tannins (hydrolysable and non-hydrolysable). The basic flavonoid structure is the flavan nucleus, which consists of 15 carbon atoms arranged in three rings (C6–C3–C6), which are labelled A, B, and C (figure 6).

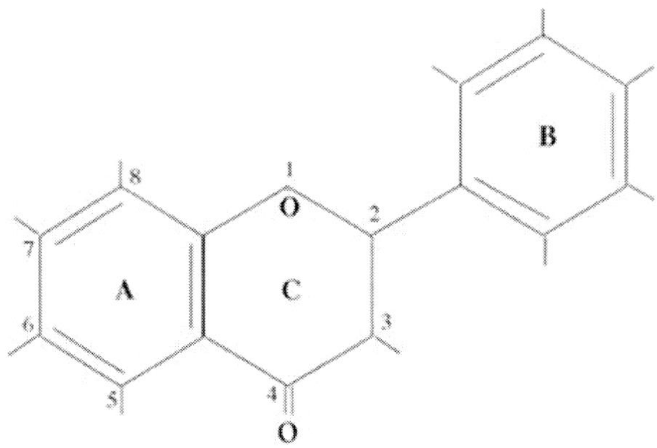

Figure 6. General structure of bioflavonoids.

Flavonoids or bioflavonoids represent the single, most widely occurring group of phenolic phytochemicals compounds with recognized biological activity, widely present in plant kingdom, vegetables, concentrated in the seeds, fruit skin or peel, bark and flowers, and grains [166]. They are the most common pigments together with the chlorophyll and carotenoids. Apart from their physiological roles in plants, flavonoids are considered as important components in the human diet [163], providing various health-promoting benefits. They are of current interest due to their important biological and pharmacological properties, especially for the anti-oxidative and anti-inflammatory properties, antimutagenic and tumour growth reduction activities, protective effect for liver, etc. [164, 167]. Flavonoids have powerful antioxidant activities *in vitro*, by virtue of their ability to scavenge a wide range of free radicals such as reactive oxygen and nitrogen species, chlorine species (hydroxyl radical, OH^-; superoxide, $O_2^{\bullet-}$; peroxyl radicals, $RO_2^{\bullet}$; alcoxy radicals, RO•; hypochlorous acid, HOCl; peroxynitrous acid, ONOOH and, in some cases chelating transition metal ions in a structure dependent manner [168]. Flavonoid compounds exhibit a high affinity against iron ions, which catalyze many processes responsible for free radical formation, their antiperoxidic activity being related to the chelating capacity of iron [169].

Antioxidants are organic molecules which can prevent or delay the progress of lipid oxidation, and their ability to do this is based mainly on their phenol-derived structure. There are two types of antioxidants: endogenous non-enzymatic (e.g., uric acid, glutathione, bilirubin, thiols, albumin, and nutritional factors, including vitamins and phenols) and enzymatic (e.g., the superoxide dismutases, the glutathione peroxidases, and catalase) [170]. The most important source of antioxidants is provided by nutrition, many of them belonging to the phenol family. Oxigen reactive species initiate within living organisms some peroxidative processes of lipid membranes, while very reactive by products are obtained able to further with biological substrate. Some relatively recent investigations have correlated the properties of various plants with their antioxidative capacity to protect the organism against oxidative stress generator of numerous diseases of digestive, cardiovascular and nervous systems [171]. Growing number of researches on the role of antioxidants suggest that there is strong association between high intake of antioxidants and low incidence of diseases linked with free radicals.

Synthesis of some novel fluorescence nanomaterials loaded with bioactive polyphenols which are present in most plants (concentrated in seeds, fruit skin or peel) with a high spectrum of biological activity, by replacing synthetic chemicals, may open new opportunities for optical and bio-medical applications. Frequently, singular substances with a definite effect can successfully be isolated from an extract; however very often this specific effect is closely related to the synergistic effects of different substances present as a mixture – usually with multifunctional properties – in contrast to chemical individual, homogenous substances.

In this respect we selected for hybrid nanostructured synthesis natural extracts from different vegetal plants originated from *Begoniaceae Specia* (called as Begonia extract), *Sambucus ebulus* (named as ornamental bush extract), fruit peel and leaves (orange peel extract and green tea extract), which contain flavonoids and other polyphenolic compounds with well known recognized functions. Among the natural extracts with benefic activities that may be valorised in the bio-medical field, orange peel extract (OPE) is a rich source of flavonoids, being a mixture of polymethoxyflavones highly bio-active (as major constituents), compounds associated with antioxidant, anti-inflammatory, antitumor activities and with potential chemopreventive properties [172, 173]. Unlike many phytochemicals, orange peel extract is a soluble lipid, a property which is desirable in many drug products since facilitate passage across biological membranes, and thus the bioavailability is enhanced. Orange peel and its extracts have been used in a variety of herbal drug products in combination with many different plant components and extracts [174]. These naturally bio-active compounds present in plants such as flavonoids, phenolic compounds, terpenes and many others active principles are believed to exhibit disease preventive properties. Diets containing some of these substances have been shown to be protective against diseases such as colon and breast cancer in animals [175]. For example, a recent invention relates the beneficial effects of orange peel extract, by providing of some compositions and methods of inhibiting tumour cell growth and treating and preventing cancer based on administration of an orange peel extract either alone or in combination with other phytochemicals [176]. A similar study presents a method for preparation of some plant extract powder and oral compositions containing plant extract porous powder carrier having activities of prevention and treatment for periodontal diseases or tooth decay preventing effect [177].

Green tea extract is a rich source of bioflavonoid associated with several health benefits, these including potential cancer-fighting properties and a strong antioxidant effect that protects the body from against damaging effect of free radicals [178]. Green tea has long been used by the Chinese and Indian people as medicine to treat headaches, body ache, and poor digestion and to improve well-being and life expectancy. Green tea extract is derived from leaves of Camellia sinensis, the plant from which green and black teas are made. The active ingredients in green tea extract are high level of polyphenols in the form of flavinoids like catechins and epigallocatechin gallate (EGCG). Polyphenols, flavinoids, catechins and EGCG are powerful antioxidants that appear to interfere with and reduce the spread of certain types of cancer cells. The antioxidant activity of EGCG in green tea extract is up to 100 times more powerful than that of vitamin C or E, in counteracting free radicals and pro-oxidant action [179].

The active principles of these flavonoid mixtures can be incorporated into silica materials either in inorganic silica or modified silica network with polyhedral oligomeric silsesquioxanes, before the gelifiation. By combining different three-dimensional silica frameworks and hybrid silica-silsesquioxane building blocks in various ratios with appropriate vegetal extract compositions, new hybrid materials with specific optical properties can be obtained, by controlling their mutual arrangement at the nanoscopic scale. As a consequence, in the following chapter the main discussion is focused on the evaluation of the behaviour of the main properties of four natural extracts at immobilization in a templated silica matrix and in a hybrid silica-silsesquioxane network, using as templates different surfactants: conventional surfactants such as ionic (e.g. cationic ammonium salts), and nonionic, neutral, non-toxic and biodegradable surfactants from poly(ethyleneglycol) class (PEG), alkenyl succinic class (ASA) and a high biocompatible non-surfactant from glucidic class (D-Glucose). The topics of this research are motivated by:

- the importance of natural extracts which present as main components flavonoids with recognized activity, being a family of synergistic active phytochemicals with a variety of biological effects;
- the advantage of using a silica matrix and silica-silsesquioxane network (e.g. optical transparency, biocompatibility, chemical inertness, thermal stability and so on);
- the uses of high biocompatible template from the glucidic class;
- obtaining of some new hybrid materials which contain a natural extract with enhanced fluorescence properties, by entrapment in adequate host silica-silsesquioxane networks suitable for bio-optical applications.

2.3.2. Sol-Gel Entrapment of Flavonoids and Mixtures of Flavonoids with Different Kinds of Templates

The sol-gel technique offers a wide range of processing approaches that can produce three-dimensional matrices in different configurations such as bulk structures, porous materials, thin films and so on. The sol-gel reaction of metal alkoxide is one of the most effective methods for the preparation of hybrid materials. Specific modifications in the physics and chemistry of the sol-gel processing allow the preparation of new sol-gel nanomaterials such as fluorescence glass. Here we briefly report the experimental works conducted in order to evaluate the structural evolution of polyphenolic compounds from vegetal extracts embedded in two silica and hybrid silica-silsesquioxane networks, as well as the interaction between entrapped extract with the silica-derived matrix, followed by the connection with optical properties of doped polymeric glasses in correlation with fluorescence of native extracts. The optimization of the synthetic process is necessary in order to obtain reproducible results, but mainly to preserve the specific properties and activities of encapsulated compound.

Silica has no crystalline properties, but constitutes a framework of tetrahedral coordinated Si–O–Si bonds (a polymeric oxo-bridged SiO_2 network) in which many vertices are ended with OH groups. Colloidal silica surfaces have been shown to provide strong interactions among functional groups. These terminal hydroxyl groups involved in hydrogen bonding lead to formation of a pseudo-crystalline cluster of silica. Furthermore, by including within silsesquioxane monomers with rigid inorganic cores, some flexible organic substituents and

OH groups useful not only as polymerizable groups for copolymerization reactions, but also for involving in hydrogen bonding, an appropriate matrix with good performance for physical interactions with the functional groups from natural extract is obtained.

The entrapment of natural extract by a templated sol-gel procedure involved two distinct stages, which pass though an intermediate gel state that can be dried and sintered into a solid glass. Hydrolysis and condensation occur at localized regions where silica polymerizes in stages to nuclei of silica, which then can lead to the formation of a homogeneous gel structure or a particulate structure during basic condensation. We have chosen the acid hydrolysis, firstly, due to the slow rate of hydrolysis reaction that permits to obtain fine gels that can be dried to a high bulk density (low porosity), while fast hydrolysis develops coarser gels that dry to a low bulk density (higher porosity) materials and secondly, the slow condensation rate at acid pH assures enough time to evaporate the large amount of ethanol produced in TEOS hydrolysis, before adding the flavonoid. The first stage is represented by pre-forming of a homogeneous templated silica sol using the acid-catalyzed hydrolysis and condensation of tetraethylorthosilicate (TEOS) in the presence of different templates: D-glucose, thetraethylenglicol monohexadecylether, ASA derivative. This templated sol is usually a nanosuspension with diameter of tenths nanometers. The preparation of homogeneous hybrid matrix have been realized by incorporation of an incompletely condensed octameric silsesquioxane monomer (octaisobutyltetracyclo [7.3.3.1^{5,11}] octasiloxane-*endo*-3.7-diol = Sq) in a templated pre-hydrolyzed sol derived from TEOS sol-gel processing, followed by copolymerization reactions with functional groups of organoalkoxysilane. The further stage involves the combination of hydrolyzed precursor sol with an aqueous vegetable extract solution. In order to realize the entrapment of extract inside the hybrid silica-silsesquioxane network, the extract is added together with silsesquioxane alcoholic solution in the liquid templated sol, before increasing of viscosity (gel transformation), and prior the sol becomes interconnected to form a rigid network during ageing period. As polycondensation continues, the degree of cross-linking among the nuclei increases and the flavonoid molecules can participate in these reactions with the alcoxide precursors resulting in physical entrapment or/and covalently bonded active molecule within the xerogel structure, which can evolutes under heat treatment, and their optical properties are related to the local environment of the fluorescence species. During subsequent ageing process, the formation of new Si–O–Si bonds by further polycondensation reactions results in gel shrinkage expelling solvent from the pores and encapsulating the flavonoid structure inside the silica network. In our research the mechanism of flavonoid encapsulation in silica network occurs by physical adsorption, that means the flavonoid molecules are embedded into the creating network structure due to the formation of hydrogen bonds between the oxygen electronegative atoms of carbonyl and hydroxyl groups of the flavonoid adsorbed and the hydrogen atoms of the residual silanol groups from surface of silica polymeric network. These hydrogen bonds are mainly responsible for flavonoid adsorption, but sometimes stronger donor-acceptor interactions could be also involved.

The natural extract immobilization in such polymeric networks assures an optimal conformational arrangement. The steric constraints inhibit the mobility leading to the immobilization of organic molecules from natural extract in a conformational arrangement favorable for transmission and amplification of fluorescence properties by achievement of some weak interactions (hydrogen bond, π electrons extended conjugation donor-acceptor interactions). In this mode the weak extract – matrix interactions which appear between free

OH groups of polymeric matrix and carbonyl and hydroxyl groups of flavonoidic and polyhydroxilic compounds from natural extract, allow their adsorption on microporous surface of matrix, thus enhancing the thermal stability of immobilized extracts (as is exemplified for a native flavonoid – rutin and Begonia extract, figure 11), without affecting the specific properties as occurs in the case of some strong chemical bonds. To our knowledge, if some chemical interactions between active organic molecules with polymeric matrices could exist, these may lead to the inactivation of flavonoid molecules, losing their active properties. Therefore, the fluorescence signal would be minimized or fluorescence quenching effect might appear facts that have not been encountered in this study.

By using optimum experimental conditions the encapsulated samples synthesized were homogeneous and transparent to visible light, without precipitation, observations that indicated the absence of a possible phase separation or the template expulsion. The homogeneity of obtained glass hybrid materials may be due to perfect integration of silsesquioxane oligomers into the pre-formed TEOS network, having in view the structure of silsesquioxane with two OH groups for polymerization or forming of H bonds and some unreactive organic groups (R) for solubilization and compatibilization and also by the low amount of silsesquioxane that was used for the encapsulation process.

The effective conversion of templated silica sol in hybrid silica materials containing orange peel extract (OPE) or ornamental bush extract (OBE) entrapped inside the polymeric networks and other information regarding the weak interactions by hydrogen bonds type between silica and silica-silsesquioxane network and organic molecules from vegetal extracts (mixture of flavonoid and polyhydroxylic compounds), have been confirmed by FT-IR and UV-VIS spectroscopy. The spectral investigations on final products of all extracts encapsulated in polymeric silica and hybrid silica-silsesquioxane matrices showed the changes of silica network after encapsulation process (FT-IR) and some small shifts of wavelengths of chromphore groups from native extract (UV-VIS-NIR), as a result of weak interactions of silanol groups with functional groups of flavonoid compounds [180, 181].

Some proofs of entrapped process of vegetable extract (exemplified for OBE and OPE extracts), by adsorption on microporous surfaces of selected matrices (figure 7, 8) were as follow:

1) a first observation that suggests the modification of silica polymeric network as a result of extract immobilization into network pores is provided by the broadening of the IR domain from 1070-1200 cm^{-1} which includes the characteristic bands of asymmetric Si-O-Si and C-OH vibrations. This aspect is also sustained by the shift of specific bands for O-Si-O bond from 440-580 cm^{-1}; for example, in case of orange peel extract, from the shift is 580 cm^{-1} and 446 cm^{-1} (for immobilized extract with PEG template), 565 cm^{-1} and 448 cm^{-1} respectively (for immobilized extract with D-glucose template), as comparing to its positions in templated silica matrices (574 and 450 cm^{-1} for PEG templated silica matrix; 560 and 458 cm^{-1} respectively, for D-glucose templated silica matrix) [180, 181].
2) the implication of Si – OH bonds in forming some weak interactions (hydrogen bonds) with other functional groups from flavones structure is evidenced by the broad band from ~ 3400 cm^{-1} and by the slight shifting of Si – OH band: (i) for orange peel extract, from 940 cm^{-1} (in templated matrices) towards higher values in extract immobilized samples (948 cm^{-1}); (ii) for ornamental bush extract, from 949

$cm^{-1}$and 942 cm^{-1} respectively – in templated matrices spectra, towards higher values 952 cm^{-1} and 948 cm^{-1} in the entrapped samples with ornamental bush extract.

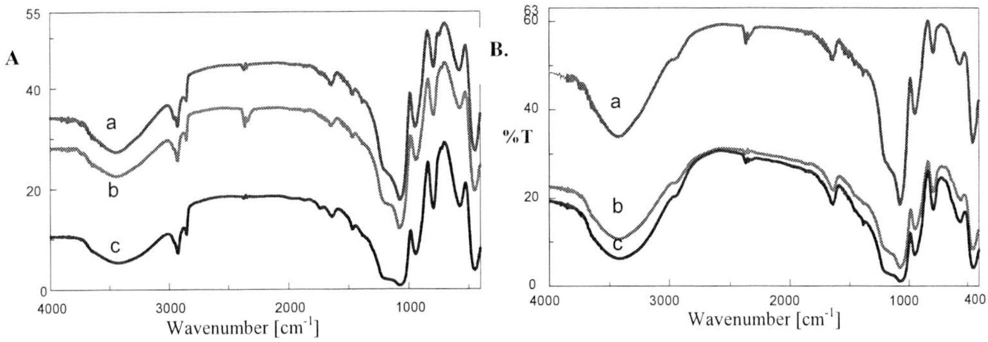

Figure 7. FT-IR spectra of OBE samples: A. PEG templated and B. D-Gl templated samples. (a) silica templated matrix; (b) ornamental bush extract entrapped in inorganic silica matrix and (c) hybrid silica-silsesquioxane matrix [180].

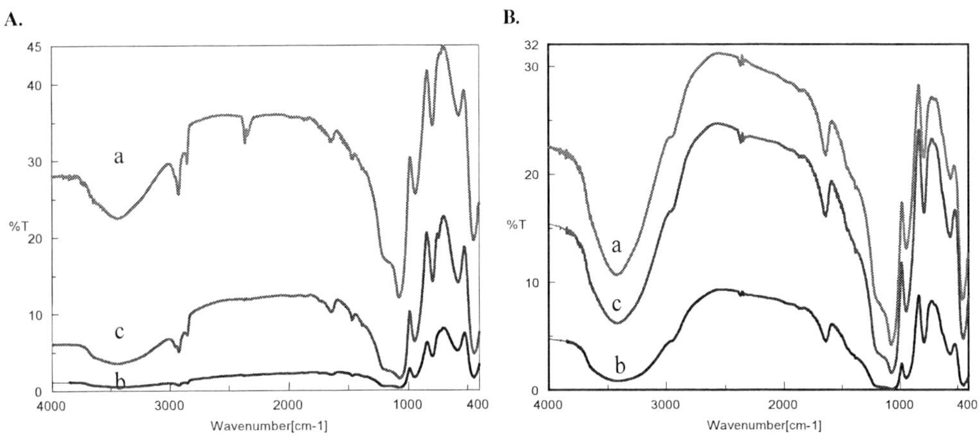

Figure 8. FT-IR spectra of OPE samples: A. PEG templated and B. D-Glucose template: (a) silica matrix, (b) OPE imobillized in inorganic silica matrix, (c) OPE imobillized in hybrid silica-silsesquioxan network [181].

As pigments responsible for the colour of leaves and petals, flavonoids strongly absorb in visible domain. Therefore an adequate tool for structural analysis of flavonoids mixture encapsulated in silica-derived matrices was UV–Vis spectroscopy. By comparing the maximum absorption bands characteristic for flavonoidic compounds which appear in native extracts spectra, with those of encapsulated extract samples in inorganic silica and hybrid silica-silsesquioxane networks, the following observations can be made (figure 9, 10) [180, 181]:

- both absorption bands present in ornamental bush extract at 304 and 342 nm, and also 256 nm and 316 nm in orange peel extract, characteristic for n → π* transitions of auxochrom and chromophore groups of COOH and C = O type from aromatic flavonoidic structure, are shifted towards higher wavelengths (284 nm/280 nm and

332 nm respectively, bathocrom effect), in the electronic spectra of orange peel extract immobilized extract samples with PEG and D-glucose templates. In contrary, these bands are also evident in the ornamental bush extract entrapped samples spectra, but shifted towards lower wavelengths, 290 nm and 325 nm (hipsocromic effect).

- the broad band of native ornamental bush extract from 558 nm assessed to aromatic compounds with chromophores of some OH multiple bonds originated from polyphenols, was modified in all the four entrapped samples. This band is recognized in entrapped samples as a much flattened band at 540 nm in case of using a PEG template or as a more sharpened band centred at 524 nm in case of D-glucose. These last observations suggest the implication of these groups in forming of weak bonds with OH groups of the matrix [180, 181].

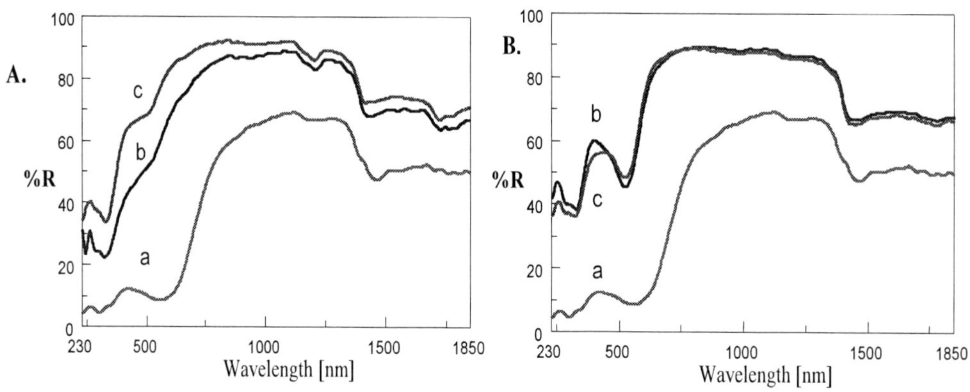

Figure 9. UV-VIS-NIR electronic spectra for OBE samples: A. PEG templated and B. D- Gl templated samples. (a) native ornamental bush extract; (b) entrapped in inorganic silica matrix; (c) hybrid silica-silsesquioxane matrix [180].

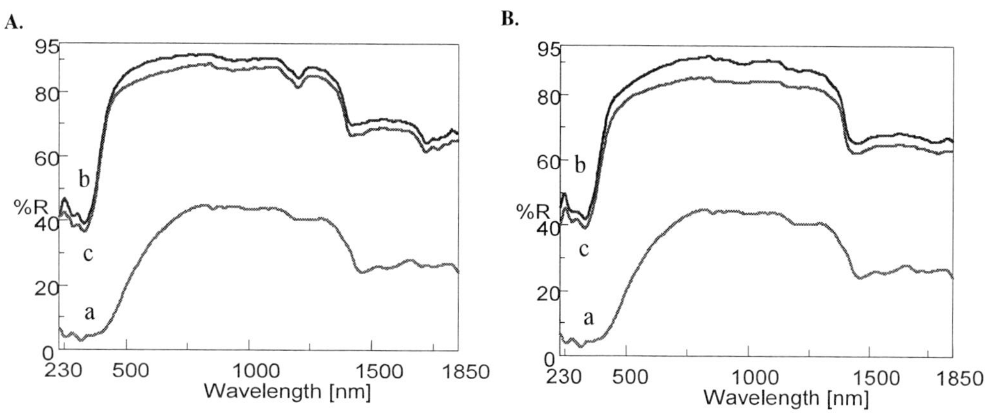

Figure 10. UV-VIS-NIR electronic spectra of OPE samples obtained with: A. tetraethylenglicol monohexadecilether template, B. D-Glucose template: (a) native orange peel extract-OPE, (b) OPE immobilized in inorganic silica matrix, (c) OPE immobilized in hybrid silica-silsesquioxane matrix [181]

In NIR domain, both extracts present two distinct regions at 1460-1560 nm and 1700-1780 nm, characteristic for free and associated OH groups. The first absorption region appears significantly shifted towards lower wavelengths in immobilized extract samples, for example as a broad shoulder in sample spectra with OBE extract entrapped and as a broad band in case of OPE, between 1420-1460 nm for samples obtained with tetraethylenglicol monohexadecylether and 1420-1510 nm for those templated with D-glucose. In contrast, the bands from 1704 and 1758 nm – for OBE and 1700 and 1750 nm – for OPE, were shifted towards higher values and recognized: (i) for OBE, as a broad band (at 1780 nm) in case of samples templated with D-glucose, and as two distinct bands (1720 nm and 1775 nm), in case of samples templated with PEG; (ii) for OPE, as two distinct bands (1720 nm and 1764 nm) for PEG templated matrices, while they almost disappear in D-glucose templated matrices.

Therefore, the presence of vegetal extract in the silica based matrices has determined some changes in its characteristic vibrations (towards lower or higher values) that show a more disordered structure (figures 7 and 8). This aspect is the result of presence of vegetal extract that hinder the ordering of the network. As a consequence, the network surrounding the organic part resulted to a more disordered network as comparing to those of silica templated reference. Regarding to electronic spectra, as have been seen, the NIR domain exhibits similar vibrations with those encountered in native extract with minor shifts (figures 9 and 10). In this way it was proved that polyphenolic compounds from vegetal extract do not interact with the network by chemical reaction, but only by physical interactions [180, 181]. We consider that such hydrogen bonding interactions were strong enough to permanently integrate the natural extract as an integral component of inorganic silica and hybrid silica modified with silsesquioxane networks. Moreover, a clear evidence of vegetal extract immobilization by physical interactions in both matrices was shown by fluorescence measurements (figure 13).

Biomolecules entrapped in sol-gel matrices typically exhibit improved resistance to thermal and chemical denaturation, as well as increased storage and operational stability. For this reason, a thermal analysis study was realized in order to observe the behaviour at encapsulation of vegetable extract with flavonoid structure and on its performance to remain encapsulated in silica sol-gel matrix while keeping at the same time or even enhancing its fluorescence properties. In order to understand the main characteristic properties of vegetable extract, rutin was chosen as a model flavonoid substance from the same class. Rutin is one of the typical glycosides from the flavonoids series widely distributed in natural plants; pharmacologically it is a potent inhibitor of lens aldolase reductase (potentially useful for prevention of diabetic cataracts), has a direct constrictor action on the capillary bed and decreases the permeability and fragility of the vessels [14].

Thermal decomposition studies of the encapsulated rutin into nano-silica templated matrix revealed a high stabilization for the flavonoid complex formed at the inner surface of nano-silica network [182]. The TG curves show that the decomposition of the encapsulated rutin has a maximum of mass loss associated with the total oxidation of rutin complex stabilised at inner surface of silica network (figure 11 A). There is clear evidence on the existence of a rutin-surface interaction due to the very sharp signal associated with rutin-silica complex decomposition process, for a temperature range around 363°C on DTG curve. Phase transitions are accompanied by free energy changes and are due to either an alteration in enthalpy (ΔH) or entropy (ΔS) of the systems. Enthalpy changes result in either endothermic or exothermic signals, depending on whether the transition is due to consumption of energy, e.g. recrystallization of an isotropic melt. In this case, a strong exothermal reaction occurred, with a maximum at 372 °C which is the most probable associated to the total oxidation of this surface-rutin complex (DTA curve). Maximum of decomposition effect is located at 363°C when the differential curve has its maximum change (DTG). Some structural changes in silica morphology occurred near 430 - 440 °C.

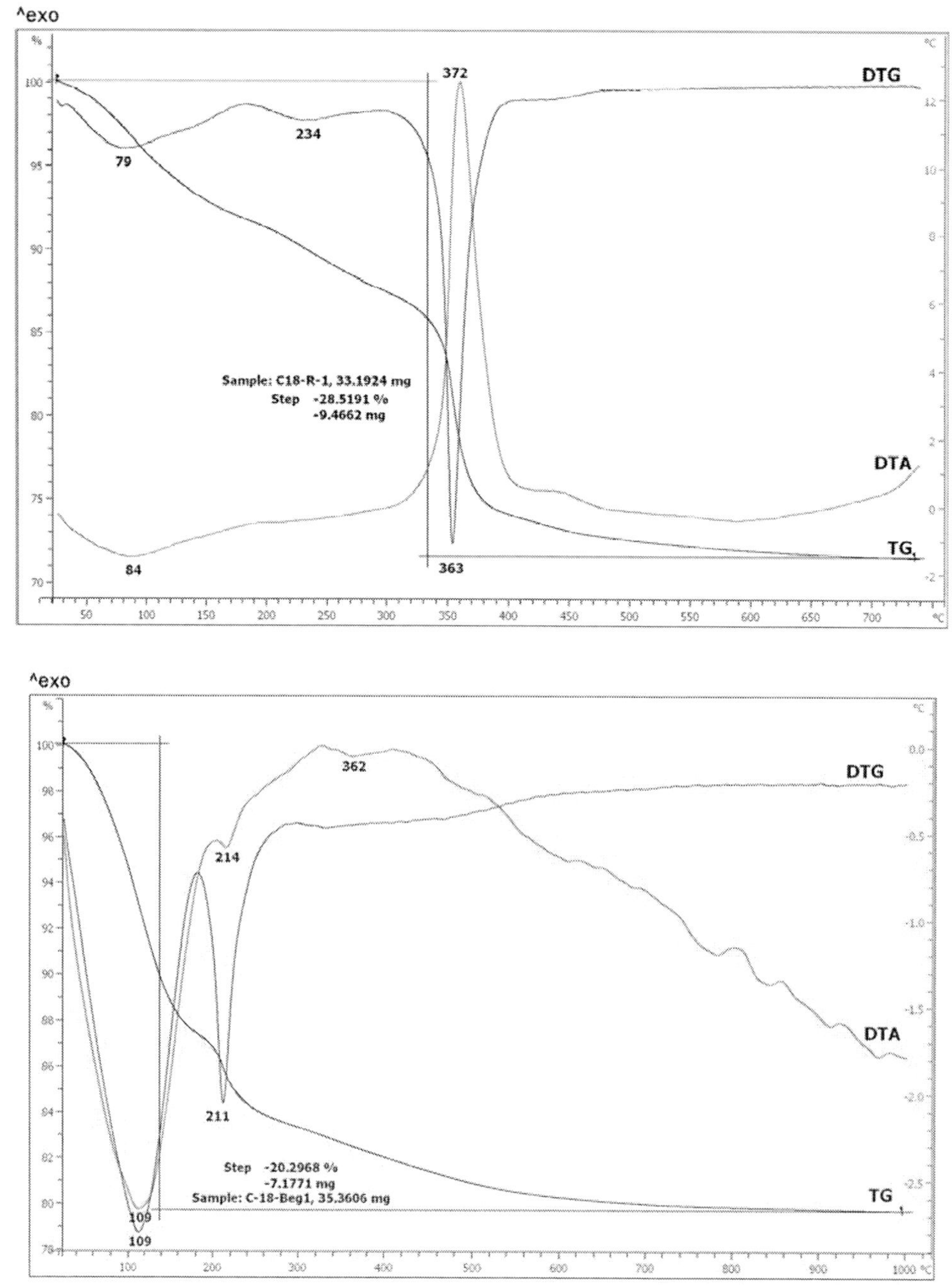

Figure 11. Thermal analysis of: A. rutin encapsulated sample and B. Begonia extract [182].

According to the above thermal analysis, the presence of rutin in the matrix is obvious even at temperatures higher than its melting point (195°C), which confirms a physical structural complex interaction between silanol groups of matrix surface and rutin. This fact is explained by the increasing of thermal stability with cca. 180°C of rutin encapsulated compounds during controlled thermal decomposition [182].

The behaviour of thermal decomposition of encapsulated Begonia extract is quite different as expected (figure 11 B), due to the structural complexity of this sample. The oxidation process of the vegetal extract seems to be initially completely covered by solvent/water evaporation process, with a main mass loss near 109°C and a characteristic high endothermic effect (DTA). A residual decomposition occurred at 211°C accompanied by a secondary structural water evaporation process. Due to the complexity composition of vegetal extract, the decomposition occurs on a broad temperature domain around to the value of rutin encapsulated, while active principles persist for a longer time into the silica matrix [182].

As it was observed above by FT-IR and UV-VIS spectroscopy and thermal analysis, it is obviously the presence of the flavonoid compounds in the final encapsulated silica materials. This presence was also confirmed by EDX/SEM analysis which is presented in figure 12 A and B. According to SEM/EDX analysis, the encapsulated materials contain mainly silicon and oxygen atoms which form the highly branched silica network. Beside these main peaks, another smaller peak observed is attributed to carbon atoms, as a result of organic flavonoid encapsulation in silica network.

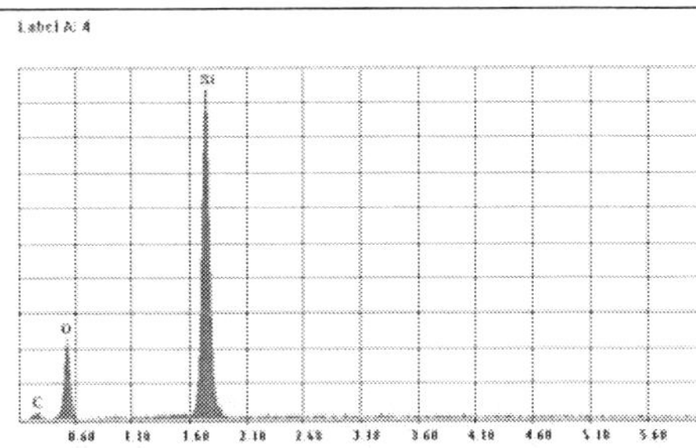

A.

Element	Wt %	At %
C	17.89	26.45
O	45.28	50.26
Si	36.84	23.29
Total	100.00	100.00

X radiation emission spectrum - qualitative compositional analysis

Chemical quantitative composition of encapsulated rutin

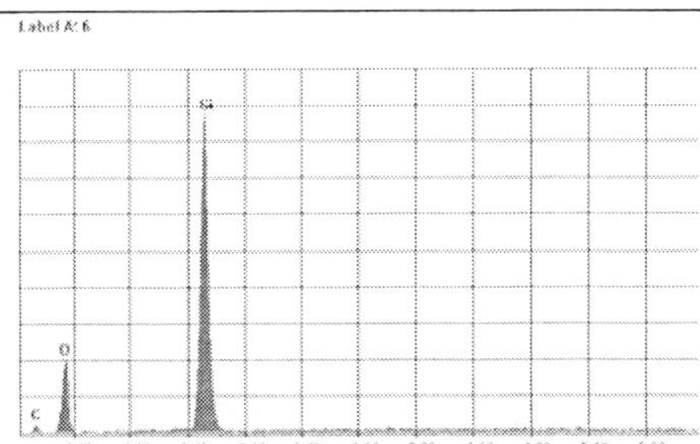

B.

Element	Wt %	At %
C	19.99	29.00
O	45.57	49.63
Si	34.45	21.37
Total	100.00	100.00

X radiation emission spectrum - qualitative compositional analysis

Chemical quantitative composition of encapsulated Begonia extract

Figure 12. SEM/EDX diagram of flavonoid encapsulated in silica network showing elemental distribution. A. encapsulated rutin with C18 surfactant; B. encapsulated Begonia extract with C18 surfactant [182].

To sustain the conclusion obtained by FT-IR and UV-VIS-NIR spectroscopy, the synthesized samples were also studied by means of fluorescence effect. These observations were in agreement with the comparative results observed in the fluorescence spectra. The weak interactions between organic molecules and different sol-gel matrices appear to play an important role in the optical properties of organic dyes and also of natural extracts. Some of these effects were observed by using of templated inorganic silica and templated organic-inorganic hybrid matrices with different kinds of templates for encapsulation of an individual falvonoid (rutin) and mixtures of flavonoids from vegetal extracts [182].

Fluorescence activity of the encapsulated flavonoids was determined by measurement of the intensity of fluorescence emission. The synthesized hybrid materials were excited at λ = 260 nm, obtaining a characteristic emission of carbonyl groups from flavonoid compounds at $\lambda_{Em} \approx 305$ nm and a maximum excitation located between 420 – 480 nm, assigned to multiple OH groups (more evident in case of samples with orange peel extract and ornamental bush extract).

In order to eventually study the structural evolution of extracted components from plants or leaves embedded in silica derived-matrices, as well as the interaction between the pigments embedded with the matrix and then correlates them with optical properties, we compared the spectral characteristics of the vegetal extract doped glass with the spectral fluorescence of native natural extracts [180-183]. These results provide us information about the formation parameters of extract-doped modified silica fluorescence nanomaterials. By comparing absorption and emission spectra in glassy matrices with vegetal extract entrapped, much more significant differences than in UV-VIS absorption spectra can be noticed. As can be seen in figure 13, by comparing the native extracts fluorescence spectra with those of immobilized extract spectra in inorganic silica and hybrid silica-silsesquioxane networks, a clear evidence of immobilization process can be observed (figure 13). As a result of vegetal extract entrapment, all synthesized hybrid materials manifest a significant increase of fluorescent signal (except the case of green tea extract).

Moreover, some notable changes have also occurred to the emission spectra. On one hand, the near resemblance in characteristic emission bands of extract (~ 305 nm) in the both matrices indicated that the dye was encapsulated into silica-derived matrices without chemical changing y under the relatively mild reaction conditions of the sol-gel process. On the other hand, compared to the native extracts, the emission band ranging from 420 to 480 nm has considerably broadened. Also, the differences that exist between the native and encapsulated extract may represent a clear indication of the lack of dimerization process of dyes molecules from extract. There might exist a strong host-guest interaction, concerning the local environment of the guest, the position present in the mesophase (e.g. confinement and isolation effects of the host matrices on the guest molecules of extract) that hinder the formation of dye aggregates. Generally, the tendency of dimerization and aggregation at moderate concentrations in aqueous solution significantly reduces the fluorescence quantum yield of these organic dyes and thus hinders their use. The simplest explanation of the decrease of fluorescence is a gradual denaturation of organic dye during the sol-gel process or the formation of quencing centers or non-fluorescing aggregates. Therefore, it is necessary to avoid the undesirable dimerization and aggregation during the encapsulation process, even though the concentration of the doped dye must be significantly reduced. If the fluorescence decreases when the natural extracts are embedded in glass, then the denaturation is an evident process [120]. This situation can be due to the photoprotective mechanism in which the nonphotochemical quencing such as the high energy state quenching (qE) is activated in the formation of materials; this may be promoted during the interaction between inorganic matrix and vegetal extract. Thus, this mechanism may represents an important step to form efficient fluorescence glass.

In the case of ornamental bush extract embedded in silica-derived glass, it has been observed that a more significant enhancement of optical signal was obtained when the extract was entrapped in hybrid silica – silsesquioxane network (14 times higher for sample templated with PEG and 9 times for samples templated with D-glucose), than in case of a simple wrapping in an oxide network of silica (6 times for PEG template/5 times for D-glucose template) (figure 13 A) [180]. These results may be explained, on one side, by the fact that the incorporation of organic groups from silsesquioxane molecule into an inorganic silica matrix decreases the volume and pore size and, on the other side, a significant part of vegetal extract was entrapped in the cage structure of silsesquioxane compound. From figure 13 A and B it can be observed that much more evident is the extended conjugation of hydroxyl groups from polyphenolic compounds present in OBE, and specially in OPE, which appears in emission fluorescence spectra as broad bands ranging between 430 – 480 nm, than emission bands of carbonyl groups from flavonoid structure (from 305 nm) [180, 181].

Similar results were obtained for orange peel extract entrapment, as in all the immobilized samples the fluorescence intensity was more than seven times amplified (for carbonyl region) and fourteen times (for emission bands of OH phenolic, in case of encapsulated samples in hybrid silica network) [181]. A better behaviour was observed in case of extract entrapment inside silica matrix templated with PEG, when the amplification of the second region was almost twenty times (figure 13 B).

Regarding the green tea extract, the fluorescence experiments showed that the optical signal was amplified only three/four times when the tea extract was entrapped in both silica/silica – silsesquioxane networks by using two types of templates (PEG and ASA) [183]. This smaller intensification of green tea extract fluorescence may be explained by the complex composition of active ingredients from tea extract with high content in polyphenols like catechins. The chemical instability of tea polyphenols are well known, they are rapidly oxidized when are exposed on many different factors such light, heat and oxidants. Moreover, the catechins have the peculiar property of forming polymers with themselves.

The allure of emission spectra of samples prepared with ASA and PEG templates is different [183]. An unusual behaviour is observed for PEG samples where the presence of multiple OH bonds from PEG completed with those of GTE is evident (figure 13 C and D), probably due to existence of a double bonds conjugation and multiple OH groups. As it can be observed, much more evident is the electronic effect of hydroxyl groups from sample prepared with PEG template which appears in emission fluorescence spectra as a broad band located between 300 – 425 nm, as compared to the sharp emission band of carbonylic groups from flavonoid structure of GTE (305 nm).

The best results, in terms of fluorescence enhancing, are obtained in case of Begonia extract, by using classical ionic surfactants (cationic ammonium salts with two and three alkyl chains – C12-2 and C18-3) [182]. As it can be seen in the Fig. 8 there is not a significant difference in the fluorescence of the samples prepared with the two cationic surfactants, even the samples prepared with the C18 surfactant generally exhibited a better behaviour than C12-2 surfactant. Nevertheless, there is an important increase of fluorescence intensity in the encapsulated rutin (10 times, figure 13 E) and especially in begonia extract (40 times, figure 13 F) as compared to the native flavonoid. Moreover, by comparing the individual flavonoid fluorescence spectrum (rutin) with those of Begonia vegetable extract, the emission bands in the encapsulated extract with both template agents were considerably increased due to the existence of some mixture of flavonoid components with synergistic effect in enhancing the fluorescence properties. B.

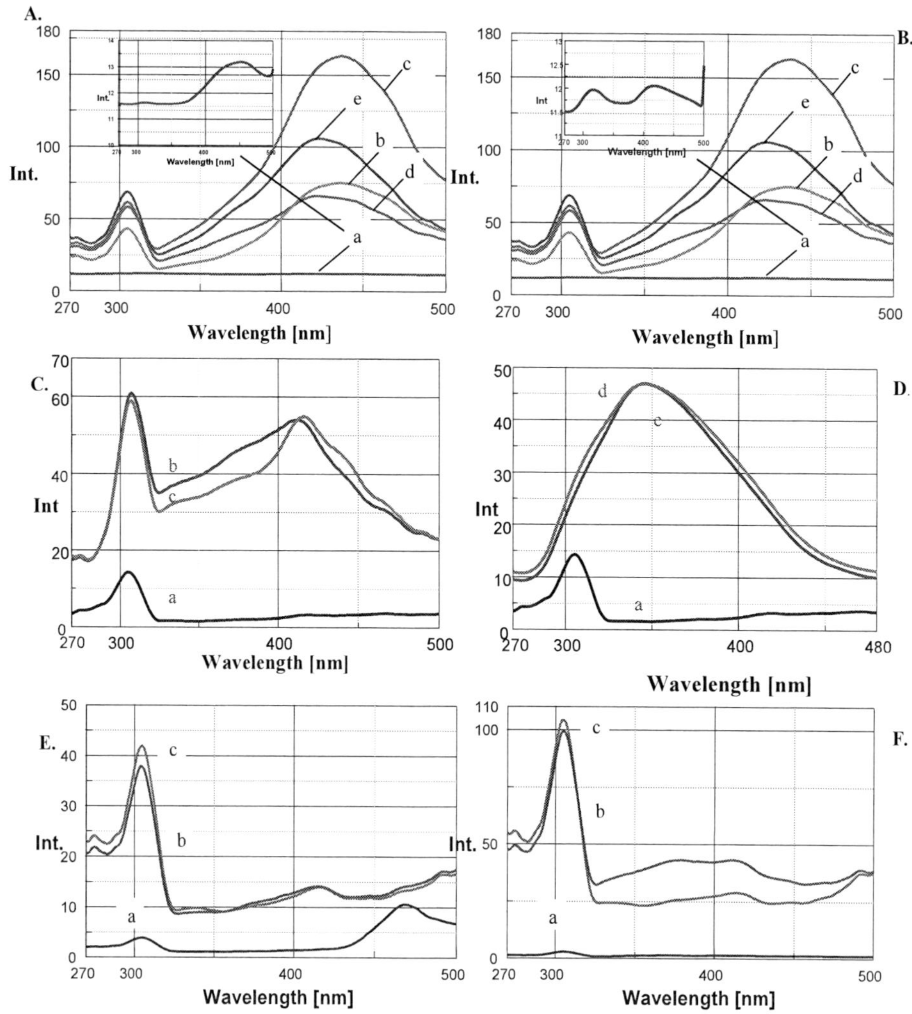

Figure 13. The fluorescence properties of diferent vegetal extracts immobilized in different matrices [180-183]: A. ornamental bush extract encapsulated samples (a = native OBE; b = OBE–PEG–SiO_2; c = OBE–PEG–SiO_2-Sq; d = OBE–Gl–SiO_2; e = OBE–Gl–SiO_2-Sq;); B. orange peel extract encapsulated samples (a = native OPE; b = OPE–PEG–SiO_2; c = OPE–PEG–SiO_2-Sq; d = OPE–Gl–SiO_2; e = OPE–Gl–SiO_2-Sq; C and D. green tea extract encapsulated samples (a = native GTE; b = GTE–ASA–SiO_2; c = GTE–ASA–SiO_2-Sq; d = GTE–PEG–SiO_2; e = GTE–PEG–SiO_2-Sq E. rutin encapsulated samples (a = native rutin; b = R–C12-2–SiO_2; c = R–C18-3–SiO_2); F. Begonia extract encapsulated samples (a = native Begonia extract; b = Beg–C12-2– SiO_2; c = Beg–C18-3–SiO_2).

All these results that underline an important fluorescence enhancing may be explained having in view, mainly the role of silica-based matrices as appropriate host for encapsulation process and the adequate encapsulation process conditions [182]. These aspects are based, firstly, on the physical adsorption of polyphenolic molecules from extract which does not affect the basic chemical structures. Secondly, due to the well known optical properties of silica glasses, this optical transparence of silica leading to the enhancement of the transmission of optical signal within the matrix. On the other side, silica matrix assures an appropriate homogeneity and pores size inside the entire network. Also, the three-dimensional network of silica may assure an optimum conformational arrangement of flavonoidic molecules, without allowing a mobility of organic compounds and neither side reactions occurring from chemical point of view. The last process may result in apparition of quenching fluorescent effect.

The synthesized hybrid materials exhibited better fluorescent properties and their application as optical functional materials would be potentially promising owing to the good optical functions itself and the simple encapsulation process for the preparation.

Another aspect that needs to be analysed was the effect of these weak interactions on the morphology and the porous structure of the silica-derived matrices. In order to evaluate the size distribution and morphology of the synthesized hybrid fluorescence materials, dynamic light scattering and electronic microscopy analyses were performed. In figures 14 ÷16 are exemplified DLS measurements and TEM images for two kinds of hybrid polymeric materials with entrapped orange peel extract and ornamental bush extract [180, 181]. The average diameter size of about hundreds nanometers for silica-based materials containing immobilized extracts have been evidenced by DLS measurements, excepting some samples templated with D-glucose. From the figures 14 A and B is observed that for all materials synthesized either by using inorganic silica or hybrid matrices, only a single peak area has been observed, with both kinds of template (PEG and D-glucose). The particle size distribution and estimation of average diameters of hybrid OPE-silica-derived materials were: 587 nm and 979 nm – in immobilized samples templated with D-glucose, respectively 788 nm and 392 nm – for PEG templated immobilized samples), with a polydispersity ranging between 0.110 and 0.440 (figure 14 A).

In case of OBE encapsulated samples, the average diameter is slightly increased (figure 14 B): 1060 nm and 211 nm – for samples with D-glucose template and 1330 nm and 531 nm – for samples with PEG template, respectively. These results evidenced that for all the four OBE entrapped samples, a narrow size distribution was observed, with a polydispersity ranging between 0.022 and 0.426 [180]. As expected, the samples based on silsesquioxane exhibit a lower size, comparative to samples prepared by using an inorganic silica matrix, as can be seen in figure 14 B. The lower size of hybrid materials prepared by using silica-silsesquioxane matrix is in good accordance with optical signals of these materials that exhibit a higher intensity as comparing to those prepared in inorganic silica matrix. A possible explanation for this decreasing in size may be assigned to a change in porosity of final materials, due to the content of silsesquioxane oligomer and its structure that presents channels and pores of a few nanometers, with the main role of decreasing the pores size owing to their cage structure.

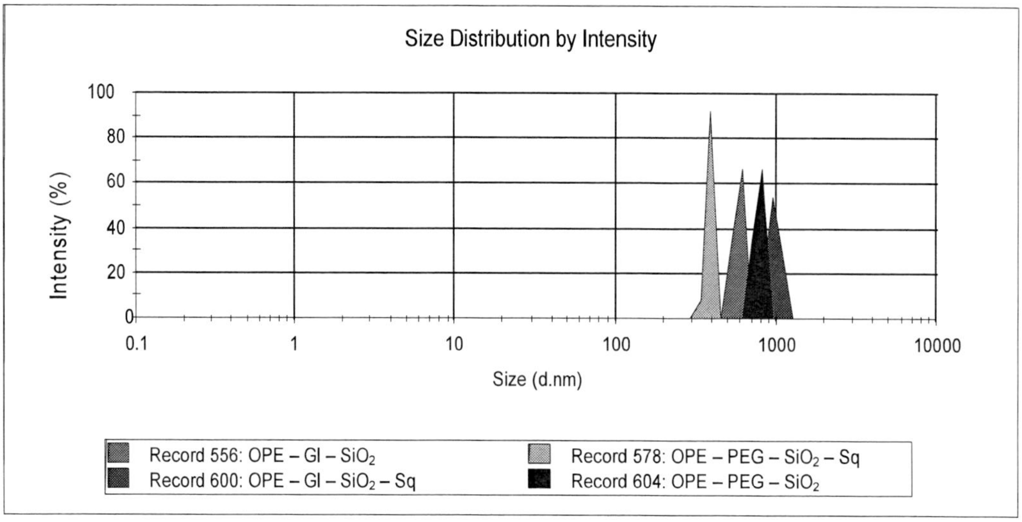

A

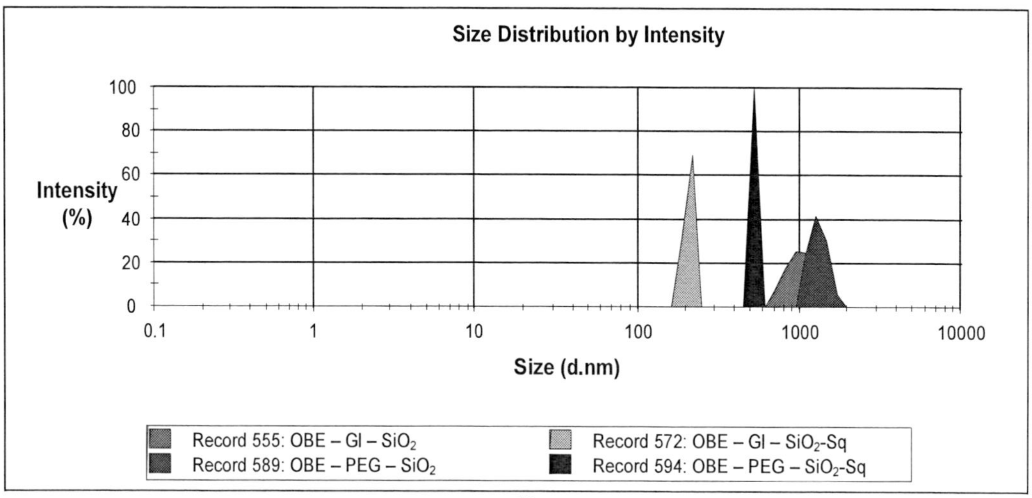

B

Figure 14. DLS measurements of entrapped orange peel extract samples (A) and entrapped ornamental bush extract samples (B) into polymeric silica and silica – silsesquioxane matrices [181, 180].

The immobilization of orange peel and ornamental bush extracts in both selected polymeric matrices has been also demonstrated by transmission electronic microscopy [181, 180]. The average diameters of synthesized polymeric materials obtained by DLS technique were in good agreement to those obtained by TEM analysis. In figure 15 A and B are comparatively presented the microscopy images of orange peel extract immobilized with D-glucose as template in inorganic silica and hybrid silica-silsesquioxane networks [181], and for OBE only in hybrid silica – silsesquioxane network derived by using D-glucose and PEG as template agents (figure 16 A and B) [180].

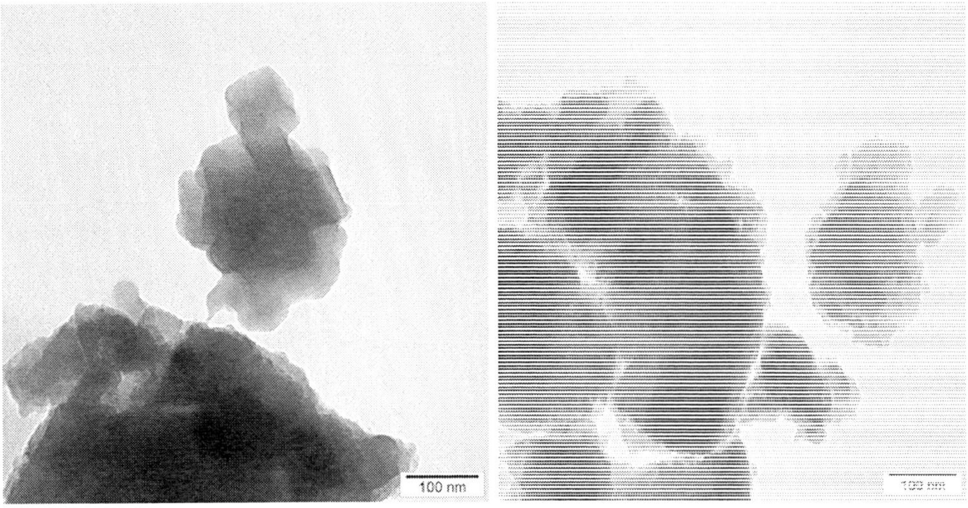

A B

Figure 15. TEM imagines of orange peel extract templated with D-Glucose and immobilized in: (A) polymeric silica matrix, (B) hybrid silica-silsesquioxane network [181].

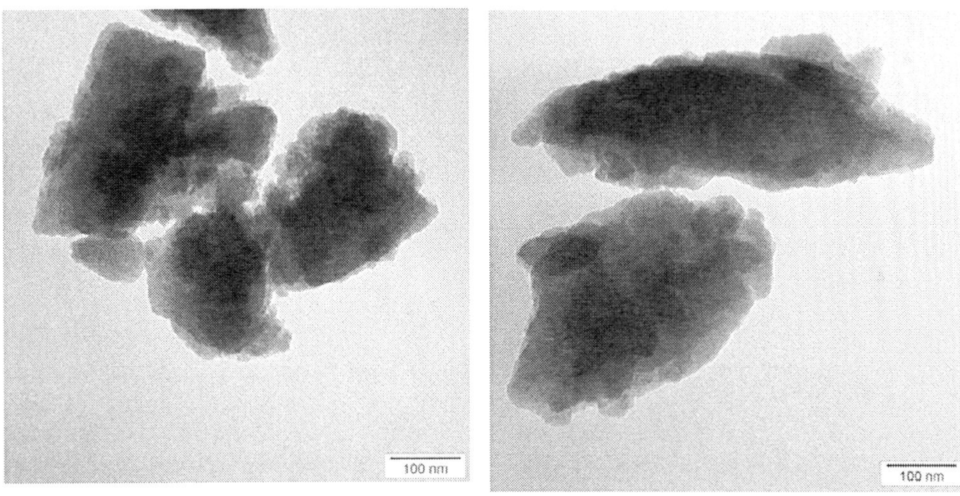

A B

Figure 16. TEM images of entrapped ornamental bush extract in hybrid SiO_2-Sq network: A. samples templated with D-glucose; B. samples templated with PEG [180].

According to TEM images, the average sizes of hybrid nanomaterials are of about hundred nanometers, but some larger aggregates are observed, more evident in the case of D-glucose template, since the synthesis process appears to cluster the organosilsesquioxane subunits into hydrogen-bonded aggregates that further are grouped into higher clusters (figure 15). Nevertheless, there are individual smaller subunits in case of ornamental bush extract samples, less than 400 nm, when the cages of silsesquioxane may be observed. As expected, the polymeric silica matrix does not confer a polyhedral uniformity (figure 15 A) as may be observed in hybrid silica-silsesquioxane network, where the polyhedral structure of silsesquioxane units is clearly outlined (figure 15 B and 16 A, B) [181, 180]. The dispersion and immobilization of vegetal extract at molecular level are confirmed by advanced transparency degree of silsesquioxane units. Therefore, the result of the synthesis is the vegetal extract entrapped into silica-based structure containing repeated octahedral forms originated from silsesquioxanic unit.

3. Methods for Physico-Chemical Characterisation of Nanostructured Materials Obtained by Sol Gel Procedure

Beside the preparation of hybrid materials, much attention has been paid to investigation of the physical-chemical properties of organic-inorganic hybrid materials obtained by the method of sol-gel synthesis and their nano- and microstructure formation. Interest in this problem is motivated by the relative ease of obtaining new nanomaterials that controllably combine the thermal stability and high optical properties of inorganic glasses with the plasticity of organic polymers.

The range of characterization methods used in the analysis of the composition, the molecular and nanometer structure as well as the physical properties of hybrid materials is quite large. Due to the great structural diversity of nanomaterials, the physical properties of nanomaterials strongly depend on their size distribution, shape and chemical composition. The size of nanostructures constrains the applications of well-established measurement techniques, therefore appropriate methods and approaches must be used in order to obtain the best characterization information and to detect the properties of the desired materials. The heterogeneous nature of hybrid materials supposes that generally a variety of analytical methods has to be used to get a satisfactory answer to structure-property relationships. Many of these methods are specific for particular compositions of materials, therefore a complete list of these techniques is not the aim of this chapter, but only several of them often used for the investigation of hybrid materials have been introduced in this chapter.

3.1. IR Spectroscopy

One of the most challenging tasks facing chemists from a variety of disciplines (e.g. biochemistry, chemistry, materials science) is to determine the identity of unknown compounds and materials, the mechanisms of reactions and the nature and stability of intermediates [184]. For example, understanding the mechanisms of interaction can allow the design of properties. Powerful analytical tools, such as FT-IR spectroscopy are often used to study such features (e.g. adsorbed species, intermediates and products).

Infrared spectroscopy is certainly one of the most important analytical techniques available to the scientists. Fourier transform infrared spectroscopy, or simply FT-IR analysis, is a technique that provides information about the chemical bonding or molecular structure of materials, whether organic or inorganic. It is used in failure analysis to identify unknown materials present in a specimen, and is usually conducted to complement EDX analysis.

Infrared spectroscopy is a technique based on the atoms vibrations from a molecule. The technique works on the fact that bonds and groups of bonds vibrate at characteristic frequencies. During FT-IR analysis, an infrared spectrum is usually obtained by passing infrared radiation through a sample and determining what fraction of the incident radiation is absorbed at a particular energy [185]. The energy at which any peak in an absorption spectrum appears corresponds to the frequency of a vibration of a part of a sample molecule. In other words, a molecule that is exposed to infrared rays absorbs infrared energy at frequencies which are characteristic to that molecule. The transmittance of the specimen and reflectance of the infrared rays at different frequencies is translated into an IR absorption plot consisting of reverse peaks, with obtaining of a FT-IR spectral pattern.

FT-IR spectroscopic methods are well known techniques for the structural characterisation of materials. In case of a hybrid material, this technique evidenced the presence of organic compounds in silica matrix in three forms: free in the hybrid matrix pores, hydrogen bonded with the silanol groups and chemically bonded (by covalent bonds or other condensation reactions) in the silica network. The interactions between sol-gel matrices and different organic compounds include electrostatic, hydrogen bonding and hydrophobic interactions. They play key roles in the ability of a host matrix to entrap organic molecules and are important in determining the arrangement of entrapped molecules. Moreover, the study of these interactions is very important science they significantly affect the specific properties of entrapped molecules, particularly, if their properties are preserved or enhanced. In cases where some weak interactions exist between the matrix and the molecule to be entrapped, than the physical entrapment of organic molecule in a sol-gel host matrix preserves original structure and functionality of the encapsulated molecule. Organic molecule – matrix interactions by hydrogen bonds allow the adsorption of the organic molecule on the microporous surface thus, preventing its dimerization (e.g. the dye molecules usually manifest this effect), protecting them from physico-chemical perturbations and enhancing the thermal stability of organic molecule entrapped. This fact is mainly due to the sol-gel matrix "cages" which provide an appropriate environment and a good compatibility with other materials. For example, a physical interaction by H bonds between the functional substituting groups (e.g. hydroxyl and carbonyl groups) of an organic component (retinyl palmitate) and residual OH groups of silica precursor species (hybrid silica-silisesquioxane network) was proved by FT-IR analysis [186]. The IR spectra of both retinyl palmitate encapsulated samples (variant a = 25% and b = 50% from the total content of silica-derived matrix) are presented in figure 17. The weak interactions between the components of hybrid materials have emerged from the slight shifting of characteristic vibrations. The band specific for Si–O–Si asymmetric vibration is significantly shifted (1079 cm^{-1} for variant a, and 1081 cm^{-1} for variant b) as referring to its position in the sol prepared with glycoside template agent (1173 cm^{-1}). The same behaviour is observed in the case of O–Si–O bending vibration at 458 cm^{-1} in encapsulated samples and 468 cm^{-1} in silica sol. These aspects represent a clear indication for the modification of silica network after retinyl palmitate encapsulation. Also, the evidence of retinyl palmitate presence inside silica-derived network is provided by the appearance in encapsulated spectra of some characteristic bands of retinyl palmitate: ν_{C-O} band (at 1161 cm^{-1}), $\nu_{C=C}$ (at 2983 cm^{-1}), the bands from 1450 cm^{-1} and 1399 cm^{-1} assigned to δ_{CH2}, $\delta_{CH=CH}$, γ_{CH} , γ_{OH} at 575 cm^{-1}. Moreover, in the region of C–H stretching vibrations, both encapsulated samples present the characteristic bands at 2873 cm^{-1} ($\nu_{C\text{-}H\ sym}$), 2950 cm^{-1} ($\nu_{C\text{-}H\ asym}$). Not the last, the hydrogen bond interactions between Si-OH and O=C< groups during the gel formation are clearly shown by the broad band from 3400 cm^{-1} and by shifting of Si-OH bond at 958 cm^{-1} in encapsulated samples [186].

One of the great advantages of infrared spectroscopy is that virtually any sample in any state may be studied. Liquids, solutions, powders, films, fibres and surfaces can be examined with a judicious choice of sampling technique. As a consequence of the improved instrumentation, a variety of new sensitive techniques have been developed in order to examine formerly intractable samples.

3.2. Fluorescence Analysis

Fluorescence spectroscopy or spectrofluorimetry is a useful tool to analyze the fluorescence properties from a variety of samples. Molecular fluorescence involves the using of a beam of light, usually ultraviolet light, that excites the electrons in molecules of certain compounds and causes them to emit light of a lower energy which represent the fluorescence emission. The fluorescence emission constitutes a highly sensitive readout signal, with a large signal-to-noise ratio and excellent spatial resolution [160].

The wavelengths of the emitted radiation are different from those absorbed and are useful in the identification of the molecule. The intensity of emitted radiation can be used for quantitative analysis. A considerable number of compounds present fluorescence and it may be detected by this very sensitive method. Fluorescence compounds contain multiple conjugated bonds with the associated delocalized π electrons, provided by the electron-donating groups, such amine and hydroxyl groups. Such multiple donating groups increase the apparition of fluorescence effect.

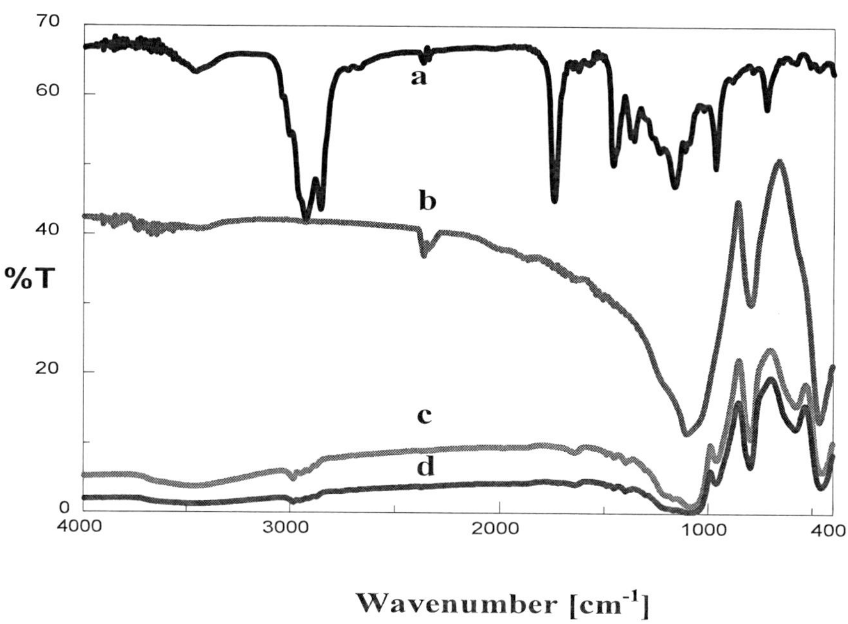

Figure 17. FT-IR spectra of: a – native retinyl palmitate, b – templated calcinated sol, c – encapsulated retinyl palmitate (variant *a*), d – encapsulated retinyl palmitate (variant *b*) [186].

In case of hybrid nanomaterials, fluorescence is typically generated through incorporation of either inorganic or organic fluorescencent dyes into the particle of an organic or inorganic material. While the inorganic dyes are typically more stable, their number and compatibility is rather restricted. Instead, the large variety of organic dyes makes them attractive for creating fluorescent particles which have broad applications in tagging, tracing and labelling [161]. However, there is a problem in the stability of these organic dyes. Incorporation of dyes into silica-derived matrices seems to be one of the most promising approaches due to the excellent sealing ability of silica and wide compatibility of silica with other materials, as have been previously demonstrated. The dopant entrapped inside the sol-gel matrix can interact with other molecules or surrounding physical environment. These interactions are easily monitored by appropriate optical methods (absorption and fluorescence).

Fluorescence was first by used as an analytical tool to determine the concentrations of various species, either neutral or ionic. When the analyte is fluorescent, direct determination is possible; otherwise, a variety of indirect methods using derivatization, formation of a fluorescent complex or fluorescence quenching have been developed. Fluorescence sensing is the method of choice for the detection of analyte with a very high sensitivity, and often has an outstanding selectivity thanks to specially designed fluorescent molecular sensors. For example, clinical diagnosis based on fluorescence has been the object of extensive development, especially with regard to the design of optodes, i.e. chemical sensors and biosensors based on optical fibers coupled with fluorescent probes (e.g. for measurement of pH, pO_2, pCO_2, potassium, etc. in blood) [187]. Fluorescence is also used for investigating the structure and dynamics of matter or living systems at a molecular or supramolecular level. Polymers, solutions of surfactants, solid surfaces, biological membranes, proteins, nucleic acids and living cells are well-known examples of systems in which estimation of local parameters such as polarity, fluidity, order, molecular mobility and electrical potential is possible by means of fluorescent molecules playing the role of probes [188].

The fluorescence technique has some advantages such as its speed of analysis, it is reagentless and small amounts of sample are required, non-invasive and non-destructive. Fluorescence spectroscopy can be used to identify and analyze fluorescent compounds at very low concentration (in parts per billion range) while providing information about structure, formulation, and stability. All these advantages have allowed the utilization of this technique in different domains: physics and chemistry [189, 190], biology [191], medicine [192] and even for the authentication of food products [193].

This technique has been intensively used by the authors [180 – 183] in order to demonstrate the enhancement of fluorescence intensity after encapsulation of various plant extracts within silica or silica-silsesquioxane matrices by sol-gel procedures using different types of templated agents. As it is clearly illustrated in figure 13, this is a versatile technique to select the appropriate synthesis conditions, matrix and the best template agent in order to maximize the fluorescence emission of the nanostructured hybrid material with an encapsulated flavonoid extract.

3.3. Chemiluminescence Method

Aerobic organisms produce a wide range of oxygen radicals and other reactive oxygen species (ROS), for useful purposes (e.g. defence, redox signalling) and also, by "accidents of chemistry". Free radicals are highly charged and active particles which are made of unstable molecules or atoms due to their odd electrons. Reactive oxygen species such as superoxide radical, hydroxyl radical, and peroxyl radical, can damage vital biomolecules (DNA, lipids, proteins) in cells and body fluids. ROS distroy the cells in the body, making it vulnerable to disorders and diseases such as, atherosclerosis, coronary heart disease, hypertension, diabetes, rheumatoid arthritis, Alzheimer disease, cancer, AIDS etc [194]. These ROS are metabolised by a series of antioxidant defence agents, some of them synthesized in vivo and other diet-derived mechanisms. The purpose of an "antioxidant defence network" is not to remove all ROS, but to control their levels so as to allow useful functions whilst minimising oxidative damage [195, 196]. The antioxidant systems are among the variety of defence mechanisms against oxidative stress induced by free radicals.

An appropriate technique largely used for evaluation of antioxidant activity of different compounds/systems is the chemiluminiscence method. Chemiluminesence may be defined as the emission of light as a result of the generation of electronically excited states formed within of chemical reaction [197]. The measurements of chemiluminescence are often used to determine initial radical products by employing luminol or isoluminol based assays [198]. The chemiluminescence of luminol (5-amino-2,3-dihydrophthalazine-1,4-dione) was firstly described by Albrecht [199], and involves the oxidation of luminol in a basic solution, generating an energy rich intermediate with subsequent light emission of the aminophtalic acid, in present one oxidizing agent (H_2O_2). The maximum of luminol chemiluminesence in aqueous medium is at 425 nm in aprotic media (DMF, DMSO) at 480 nm, and in DMSO/water mixtures are observed both maximum. The determination of antioxidants is based on the decrease of chemiluminescence intensity derived from luminol and superoxide anion radical ($O^{-}{}_2$). The constant chemiluminescence intensity, that is recorded as background (baseline) is required for an accurate and precise detection before injection of sample solutions containing tested antioxidants.

This method was recently used to evaluate the antioxidant activity of a hybrid nanomaterial obtained by encapsulation of one of the highly hydrophobic and photosensitive and labile vitamin – retynil palmitate [186]. Although retinyl palmitate and its derivatives are appropriate vitamins for pharmaceutical and cosmetic area, with many useful and important roles on human skin and mucous membranes (they induce thickening of the epidermis and are effective for treatment of skin diseases [200] it manifests serious deficiency, being very unstable against light, heat, pH and oxygen [201]. Nanoencapsulation by sol-gel technique is considered to be an ideal approach since it seems to solve the problem of losing of specific properties such as antioxidant activity. Thus, the stability of retinyl palmitate is one of the most important properties to be followed and maintained after encapsulation process and storage at room temperature. The antioxidant behaviour of the encapsulated samples with moderate and high content of silsesquioxane precursor, in which the antioxidant concentration doped was as high as $1 \cdot 10^{-5}$ M, is presented in figure 18, along with that of retinyl palmitate alcoholic solution with the same concentration for comparison purpose [186].

In order to evaluate the antioxidant activity and drug photo-stability after encapsulation process, retinyl palmitate and both encapsulated samples were subjected to an UVA and UVB irradiation process in two stages (figure 18) [186]:

- a short irradiation (for UVA: 2x26 J/cm^2, 160 min; for UVB: 32 J/cm^2 · 200 min);
- a prolonged irradiation (for UVA: at 3x26 J/cm^2, 230 min; for UVB: 93 J/cm^2, 600 min).

In case of a short irradiation (irradiation I), retinyl palmitate and encapsulated samples present a superior antioxidant activity due to the formation of some activated compounds (for a short period of time) acting as antioxidants and scavengers for free peroxy radicals. After a prolonged irradiation period (irradiation II) retinyl palmitate was almost completely decomposed since a high number of free radicals is generated in the system. As a consequence, its antioxidant activity became insignificant. A different behaviour is observed in case of the two encapsulated samples after a prolonged UVA and UVB irradiation, when the antioxidant activity is diminished by only cca. 30%. This behaviour demonstrated the effective protective action of silica – silsesquioxane network surrounding the retinyl palmitate which assures a remarkable antioxidant activity even in such aggressive photo-degradation conditions. Therefore, the vitamin entrapped into sol-gel matrices is isolated in an individual cage, and is not in direct contact with external environment. As a consequence, encapsulation of retinyl palmitate in this silica-silsesquioxane network ensures much better protection of the molecules, allowing safer disposal and higher stability than the free molecule [186].

The chemiluminescence method offers numerous advantages such as high sensitivity, wide linear range, safety and controllable emission rate, and the use of simple and inexpensive instrumentation for monitoring emission. All these advantages have allowed its utilization in determination of many organic and inorganic compounds in food samples [202], medicine [203], biology and environmental protection [204]. In present, beside chemiluminescence method, photochemiluminesence, radiochemiluminescence [205], radiostorage- and photostorage-chemiluminescence [206] may be used for determination of antioxidant activity.

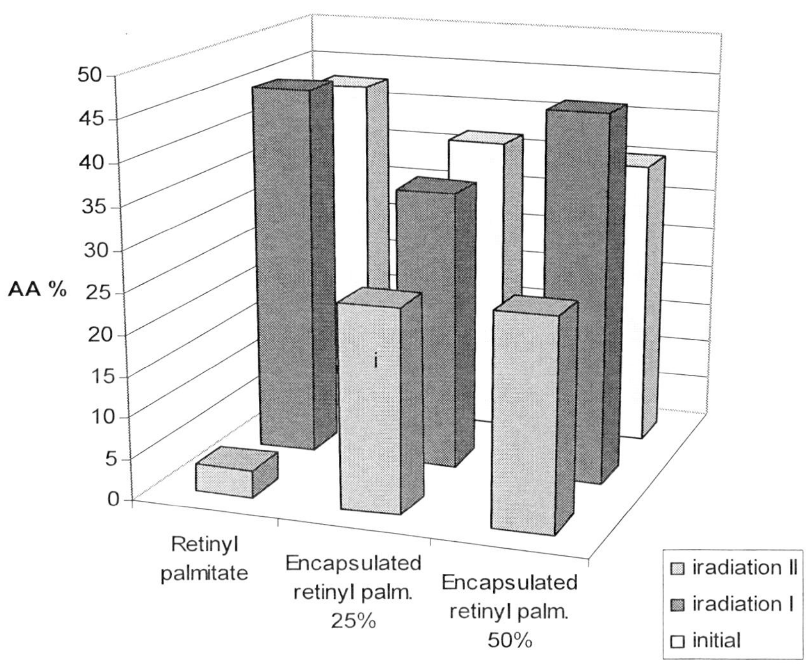

Figure 18. The behaviour of retinyl palmitate encapsulated samples at UVA and UVB irradiation [186].

3.4. Dynamic Light Scattering

Dynamic light scattering (DLS), also known as photon correlation spectroscopy (PCS) and quasi-elastic light scattering (QUELS), is one of the most popular methods designed for particle analysis, which provides a rapid detection of the size distribution of solid and agglomerated materials with nano- and micron range particles. This technique is used to measure hydrodynamic sizes, polydispersities and aggregation effects of various samples [207].

DLS technique measures the laser light scattered from dissolved molecules or suspended particles. Particles, emulsions and molecules in suspension undergo Brownian motion. This is the motion induced by the bombardment by solvent molecules that themselves are moving due to their thermal energy. Due to the Brownian motion of the molecules and particles in solution, some fluctuations of the scattering intensities may be observed. The shining of a monochromatic light beam onto a solution with spherical particles in Brownian motion causes a Doppler shift when the light hits the moving particle, changing the wavelength of the incoming light. This change is related to the size of the particle. These fluctuations increase as particle size decreases. The intensity of the scattered light fluctuates at a rate that is dependent upon the size of the particles, as smaller particles are hinted further by the solvent molecules and move more rapidly. Analysis of these intensity fluctuations yields the velocity of the Brownian motion and hence determines the particle size using the Stokes-Einstein relationship [207, 208].

Due to the fact that large molecules or particles move slower than small molecules, a defined correlation function results. From the correlation function the diffusion coefficient (D) of the molecules can be calculated by fitting the data. The translational diffusion coefficient will depend not only on the size of the particle "core", but also on any surface structure, as well as the concentration and type of ions in the medium. This means that the size can be larger than measured by electron microscopy, for example, where the particle is removed from its native environment. By DLS can be determined the hydrodynamic radius (R_h) of the sample (particles and molecules). This diameter refers to how a particle diffuses within a fluid. The hydrodynamic diameter obtained by this technique is that of a sphere that has the same translational diffusion coefficient as the particle being measured [209].

Typical applications of DLS are the measurement of the size (*z*-average) and size distribution of particles emulsions and molecules dispersed or dissolved in a liquid (e.g. proteins, polymers, micelles, carbohydrates, nanoparticles, colloidal dispersions, emulsions, microemulsions). DLS is also used in the nanotechnology research for the accurate and fast size measurement of nanoparticles made of different materials [210]. An example in this context is the application of DLS measurements on the encapsulated samples of retinyl palmitate inside polymeric silica – silsesquioxane network, in order to determine the particle size distribution of the hybrid material [186]. As shown in figure 19 the mean diameter of encapsulated retinyl palmitate particles were 577 nm, 100% (for variant *a*) and 507 nm, 100% respectively (for variant *b*), with a very narrow size distribution (polydispersity for variant *a* is 0.057 and 0.259 for variant *b*).

The particle size is an important parameter for in-process control and particularly in quality assurance, because the physical stability of studied systems depends on particle size distribution. In case of polydisperse systems, calculation of the particle size distribution using special transformation algorithms is also possible. For this purpose certain requirements need to be fulfilled, e.g. a spherical particle shapes sufficient dilution and a large difference between the refractive indices of the inner and the outer phase. Since usually not all requirements can be fulfilled, z-average as a directly accessible parameter is preferred to the distribution function.

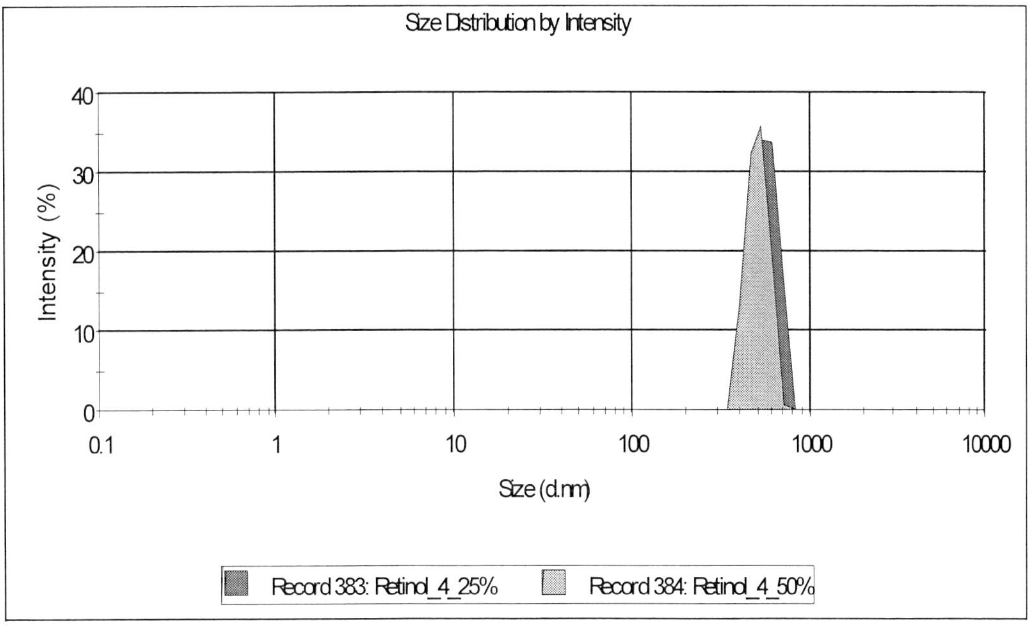

Figure 19. DLS measurements of encapsulated retinyl palmitate [186].

This method has several advantages [207-211]: the possibility to analyze samples containing broad distributions of species of widely differing molecular masses; measurement in the native environment of the material; detection of very small amounts of the higher mass species (<0.01% in many cases); measurements the scattered light of a sample in very short time periods (one or two minutes); the experiment duration is short and it is almost all automatized so that routine measurements an extensive experience is not required; has modest development costs; typical time scale for a measurement is less than one minute; a low volume of samples is required; a filtration step before measurement it is not needed. The DLS technique is especially impressive because there is no sample information needed, only the solvent viscosity and the refractive index of the solvent must be known. For nanoparticles the density of the particles might be important. Moreover, with this technique it is also possible to obtain absolute measurements of several parameters of interest, like molecular weight, radius of gyration, translational diffusion constant and so on.

The commercial particle sizing systems mostly operate at only one angle (90°) and use red light (675 nm). Usually in these systems the dependence on concentration is neglected. Using more sophisticated experimental equipment (projector, short wavelength light source), the methods can be not only considerably extended, but also more complicated and expensive [211].

3.5. Transmission Electronic Microscopy Analysis

The synthesis process with template agents leads to the formation of highly controlled inorganic-organic materials, in which the organic template exerts a substantial control on the entire process, which results in formation of materials of uniform size, novel crystal morphology, specific crystallographic orientation and not the last, remarkable properties [42]. In this respect, a technique that may evidence the nanostructure, as well as density characteristics of organic-inorganic hybrids is transmission electron microscopy (TEM).

The electron microscopy became one of the most important techniques to characterize the materials morphology on the nanometer and nowadays even on the atomic scale. The combination of direct imaging as well as local structural and chemical analyses makes transmission electron microscopy a powerful tool to study materials down to the atomic level. This method provides a direct image of the sample and by coupling the microscopy with analytical techniques some supplementary information may be obtained (e.g. elemental distribution) [212].

Transmission electron microscopy comprises a complete repertoire of imaging, diffraction, and analysis techniques. The combination of imaging contrast, high resolution, 1 nm spatial chemical resolution, make this instrument unique in the arsenal of characterizing methods [45]. This technique is similar to optical microscopy, except that the photons are replaced by electrons. Since electrons have a much smaller wavelength than photons, the diffraction barrier is less of an issue and much higher resolution can be achieved. TEM is able to examine fine detail, even as small as a single column of atoms, which is tens of thousands times smaller than the smallest resolvable object in a light microscope. Based on similar principles to the conventional light microscope, the transmission electron microscope is capable of magnifications of 100,000 times and extreme resolution, far beyond the possibilities afforded by visible light.

Transmission electron microscopy is a technique in which a very thin sample of the material to be tested (e.g. an ultra thin specimen) is exposed to a beam of electrons that passes through specimen and interacts with him. An image that carries information about the inner structure of specimen is formed from the interaction of the electrons transmitted through the specimen [213]. At smaller magnifications TEM image contrast is due to absorption of electrons in the material, due to the thickness and composition of the material. At higher magnifications complex wave interactions modulate the intensity of the image, requiring expert analysis of observed images. Alternate modes of use allow for the TEM to observe modulations in chemical identity, crystal orientation, electronic structure and sample induced electron phase shift as well as the regular absorption based imaging [213]. The image is recorded, magnified and focused onto an imaging device, such as a fluorescent screen, on a layer of photographic film, or to be detected by a light sensitive sensor such as a CCD camera. The latter has the advantages that the image may be displayed in real time on a monitor or computer. As a result, the atomic arrangement of the sample may be observed [214].

The transmission electron microscope has become an indispensable tool to characterize the microstructure of materials (e.g. the constitution of phases and nanoparticles, the local arrangement of atoms), to observe the microstructure of colloidal systems that can be visualized with the high-magnification power of the electron microscopy. It is also used in crystallography, to study crystal defects such as grain boundaries, dislocations, precipitates, and their impact on the physical properties of the solid, to identify imperfections in the atomic-level structure of material by analysis of microscopic surfaces, in solving structural problems and optimization of the functional properties of hybrid nanomaterials, etc. For example, in order to observe the effect of encapsulation process of retinyl palmitate on the morphology of a hybrid silica-silsesquioxane network, a transmission electron microscopy image was recorded [186]. By TEM analysis it has been shown that there is a significant influence of silsesquioxane molecules on the porous structure of silica matrix. One can clearly see in figure 20, that the size of polymeric hybrids which contain encapsulated retinyl palmitate decreased with increasing in silsesquioxane content: in the sample with 25% silsesquioxane the diameter of all polymeric aggregates is less than 600 nm, while in the sample with 50% silsesquioxane the diameter is less than 500 nm. Retinyl palmitate has been encapsulated in silica – silsesquioxane network and its size seems to be less than 10 nm, this suggests that the retinyl palmitate was entrapped in the cage structure of polyhedral silsesquioxane, recognized by their typical shape texture [186].

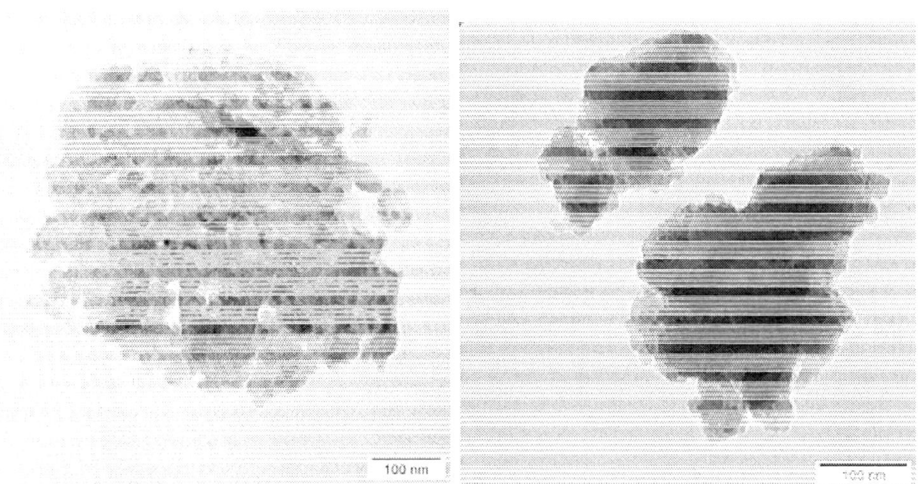

Figure 20. TEM image of encapsulated retinyl palmitate with various silsesquioxane content in the silica matrix (A. 25% and B. 50%) [198].

The increasing need to study materials on the atomic scale, which is primarily given by the scaling down of electronic devices, the development of nanostructures, the optimization of catalysts, and also by the goal to obtain a basic understanding of physical and mechanical properties of solids, can in many cases be met by applying the experimental TEM techniques.

As the resolution continues and the instrument is improved, new techniques such as high-resolution transmission electron microscopy (HR-TEM) become powerful tools indispensable for nanotechnology, largely used for characterizing of nanomaterials. This microscopy forms a major analysis method in a range of scientific fields, chemical, physical and biological sciences. TEMs find application in materials science, cancer research, virusology, as well as pollution and semiconductor research. The use of HR-TEM may covers a wide interest areas: the study of dynamic shape transformation of nanocrystals, in-situ nanoscale property measurements on the mechanical, electrical and field emission properties of nanotubes or nanowires, environmental microscopy for direct observation of surface reactions, aberration-free angstrom-resolution imaging of light elements, imaging of atom clusters with atomic resolution chemical information; electron holography of magnetic materials, etc. [215].

3.6. Atomic Force Microscopy

In recent years, the applications and demand for solid surfaces with micrometer to nanometer patterns have increased because of extensive developments with many potential applications in various fields of nanoscience and nanotechnology. An adequate series of tools to image, measure and manipulate atoms and structures on a variety of surfaces at the nanoscale, are well developed and include atomic force microscopy.

The atomic force microscope (AFM) or scanning force microscope (SFM) is a very high-resolution type of scanning probe microscopy, with demonstrated resolution of fractions of a nanometer, more than 1000 times better than the optical diffraction limit. By AFM, a small probe is scanned across the sample to obtain information about the surface of sample [216-219]. The AFM probe has a very sharp tip, often less than 100Å diameter, at the end of a small cantilever beam. The cantilever is typically silicon or silicon nitride with a probe radius of curvature on the order of nanometers [217, 218]. Interatomic forces between the probe tip and the sample surface lead to a deflection of the cantilever (according to Hooke's law) as the sample's surface topography (or other properties) changes. A laser light reflected from the back of the cantilever measures the deflection of the cantilever. This information gathered from the interaction of probe with the surface is feedback to a computer, which generates a map of topography and/or other properties of interest. Additional properties of the sample, such as thermal and electrical conductivity, magnetic and electric field strength, and sample compliance can simultaneously be obtained using a specialty probe. Since the images are collected in digital format, a wide variety of image is available for AFM data. Quantitative topographical information, such as lateral spacing, step height, and surface roughness are readily obtained. Images can be presented as two-dimensional or three-dimensional representations in hard copy or as digital image files for electronic transfer.

The basics modes of operation for an AFM are contact mode, tapping mode or non contact mode, and dynamic contact mode. In the contact mode operation, by maintaining a constant deflection, the force between the probe and the surface is keeping constant during scanning. By using this contact mode, the AFM probe is scanned at a constant force between the probe and the sample surface to obtain a 3D topographical map. In the non-contact mode, the probe cantilever is externally oscillated at or near its resonant frequency of the cantilever. These changes in oscillation provide information about the properties of the samples. By applying this mode of operation, the system monitors the probe position and vibrational amplitude to obtain topographical and other property information. Accurate topographical information can be obtained even for very fragile surfaces. In the dynamic contact mode, the cantilever is oscillated such that it comes in contact with the sample with each cycle, and then enough force is applied to detach the probe from the sample.

Atomic force microscopy is a surface imaging technique that can be applied at sub-nanometre resolution in liquids and gases [217-219]. The usual applications of AFM include: analysis of surfaces and nanoobjects, three dimensional topography of samples, roughness measurements for chemical mechanical polishing, materials evaluation (e.g. surface roughness on implanted silicon wafers; thermal properties such as thermal conductivity, glass transition temperature, and melting temperature of various phases of a blended polymer measured down to the nanometer scale). Other applications are on analysis of microscopic phase distribution in polymers, surface profiles for thin film deposition and coatings, the mechanical and physical property measurements for thin films, respectively, imaging magnetic domains on digital storage media, magnetic field mapping of recording media, imaging of submicron phases in metals, microscopic imaging of fragile biological samples, the measurement of force-distance curves, etc. This technique has been used in chemistry, biophysics and extended recently to cell biology. For instance, the atomic force microscopy (AFM) was recently applied in biophysics and clinics to the field of molecular recognition as well as to the evaluation of dynamic parameters (e.g. unbinding force, dissociation rate and energy barrier) for various interactions between proteins and ligands in their native environment [220].

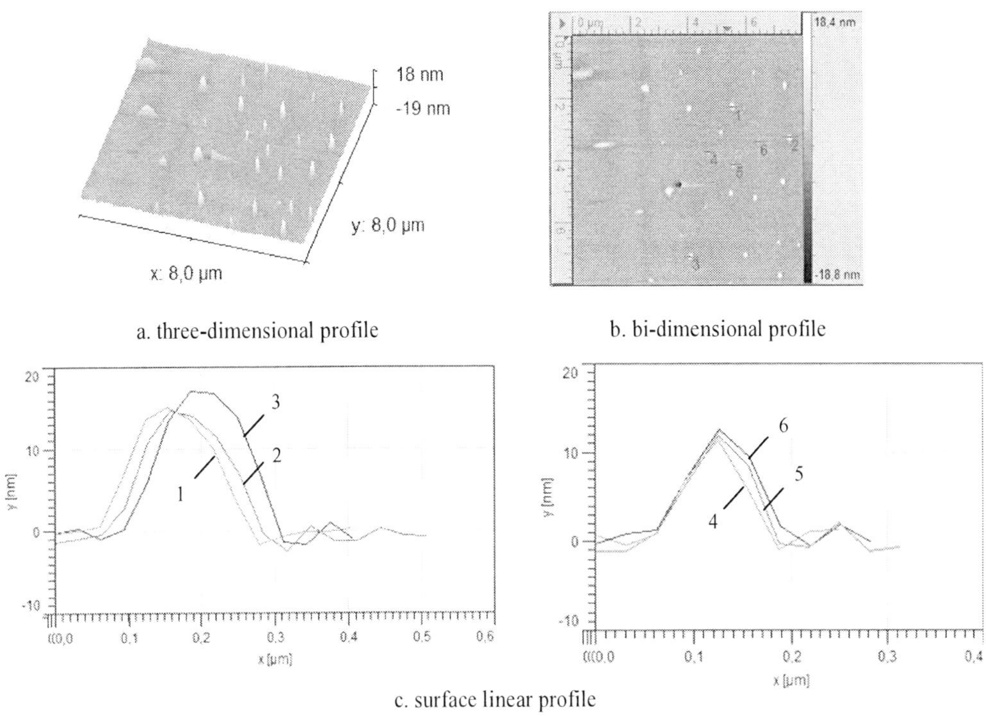

Figure 21. AFM images of the surface of a dried glass materials; the scanning size is 8µm x 8µm [183].

Thus, there are many promising areas in nanotechnology which can be examined by AFM. The atomic force microscopy is known as a versatile technique that may be applied to a large number of research topics. It has the advantage of imaging almost any type of surface, including polymers, ceramics, composites, glass, and biological samples. An example is the evaluation of morphology and size of a hybrid glass material obtained by encapsulation of green tea extract into a polymeric silica network [183]. Figure 21 shows the AFM patterns of the surface morphologies of silica glass derived from an acid one-step catalyzed sol which contains a green tea extract. AFM and particle size distribution analysis indicated a uniform size distribution pattern of green tea extract within the glass materials, the incorporated organic molecules being well dispersed and homogeneously distributed in the silica host matrices [183]. According to AFM data, the glass has a porous like structure and the size of aggregates from pores is highly uniform and about 200 nm on average diameter, for 1-3 aggregates and about 130 nm for 4-6 aggregates (figure 21 b, c).

The AFM has several advantages over other techniques used for the analysis of surface: it is a quite simple technique, provides a true three-dimensional surface profile, samples viewed by this technique do not require any special treatments (such as metal/carbon coatings) that would damage the sample and most AFM modes can work in ambient air or even in a liquid environment. However, this technique presents a disadvantage compared to the other microscopy techniques since, it may only image a maximum height on the order of micrometers and a maximum scanning area of around 150 by 150 micrometers, as comparing to SEM that can image an area on the order of millimetres by millimetres with a depth of field on the order of millimetres.

Conclusion and Perspectives

Research in nanomaterials faces out many challenges in materials synthesis, property characterization, and device fabrication which represent the three key steps in the development of nanoscience and nanotechnology. Synthesis of hybrid materials represents one of the most fascinating developments in materials chemistry in recent years. The tremendous possibilities of combination of different properties in one material initiated an explosion of ideas about potential hybrid nanomaterials and applications. The combination of organic and inorganic components into a complex material may promote and optimise the functionalities and properties of final materials, such as optical, electrical, magnetical, mecanical, catalytic properties, etc. These combinations offer exciting new opportunities for scientists to develop new strategies and techniques for the synthesis and applications of these materials. Hybrid materials not only integrate the elementary properties and the composition of component materials, but also they bring new functionalities that one component does not possess, owing to the interactions of organic and inorganic materials. Investigations in this field, in particular to anticipate the importance of interface between the inorganic and organic materials which has to be tailored to overcome serious problems in the preparation of hybrid materials, are crucial. This chapter has shown how appropriate self-assembling and building block approaches were used for the preparation of novel fluorescence hybrid nanomaterials loaded with bioactive principles, and adapted in order to bridge the differences between the inorganic and organic materials. By using this encapsulation process, a homogeneous and stable distribution of silsesquioxane building blocks in hybrid nanomaterials with active principles was achieved, without separation of some undesired microphases.

This research realised in the field of bioencapsulation process, may bring new contributions to the development of the sol-gel procedure by physical entrapment of a complex vegetable mixture in polymeric matrices as integral component of silica and hybrid silica-silsesquioxane networks which leads to a significant enhancement of the functional properties of natural products (e.g. flavones, vegetal extracts and vitamins) such as fluorescence intensity and antioxidant capacity, thus diversifying the potential applications of organic encapsulated sol-gel nanomaterials. Moreover, synthesis of some novel fluorescence nanomaterials loaded with bioactive polyphenols which are present in most plants (concentrated in seeds, fruit skin or peel) with a high spectrum of biological activity, by replacing synthetic chemicals, may open new opportunities for optical and bio-medical applications. However, a concerted multi-team effort has to be focused on studying the possibility of replacing the chemical compounds with natural compound that present multifunctional properties. In this respect, a much advanced generation of scientists which may facilitate the problems of human civilization, can be the result of a future research in this area.

The chapter was intended to give an overview of issues on fundamental aspects of the sol-gel process, templating effect of sol-gel microenvironment for synthesis of novel fluorescent nanomaterials and some selected methods for the physico-chemical characterization of nanostructured materials, and to guide the reader towards more specialized topics. By far away the most attention has been paid to the self-assembling process of hybrid organic-inorganic nanomaterials, by distinct synthesis strategies. The synthesis of hybrid materials with controlled hydrophobicity and functionality may open a new way for the guest-host organic –inorganic chemistry. The main driving forces within the framework of intensive activities in this area are the new and different properties of these materials that do not exist in traditional materials at macroscale level. However, on the future some issues are still expected to be developed in order to give a more attention of this active research area. Firstly, it is a challenge to improve these techniques by a rigorous control of size, shape, structure and defect distribution of hybrid materials. Also, the self-assembling mechanism of hybrid materials have to be still careful explored, especially for understanding the role of components in the final structure. Ultimately, it is important to enhance the main properties of component materials in order to obtain attractive applications with better performances. The complexity of researches on self-assembling approach makes necessary the multidisciplinary efforts, involving in this case experts in chemistry, physics, biology, medicine and material science. Therefore, the interdisciplinary knowledge is becoming more and more important to circumvent the newly emerging problems in biomedical, environmental and industrial sectors.

REFERENCES

[1] Nakanishi, K., Sato, Y., Ruya, Y. & Hirao, K. (2003). Supramolecular Templating of Mesopores in Phase-Separating Silica Sol-Gels Incorporated with Cationic Surfactant. *J. Sol-Gel Sci. Technol.*, 26, 567 – 570.

[2] Avnir, D. & Braun, S. (1996). *Biochemical Aspects of Sol-gel Science and Technology*, Kluwer Academic Publishers, Boston.

[3] Livage, J. (1996). Bioactivity in sol-gel glasses, C.R. *Acad. Sci.*, 322, 417 – 427.

[4] Loy, D.A., Baugher, B.M., Baugher, C.R., Schneider, D.A. & Rahimian, K. (2000). Substituent Effects on the Sol-Gel Chemistry of Organotrialkoxysilanes, *Chem. Mater.*, 12, 3624 – 3632.

[5] Koch, C.C. (2002). *Nanostructured Materials, Processing, Properties and Potential Applications,* William Amdrew Publishing, New York, 20 – 22.

[6] Gupta, R. & Kumar, A. (2008). Molecular imprinting in sol-gel matrix. *Biotechnology Advances,* 26, 533 – 547.

[7] Zarzycki, J. (1997). Past and present of sol-gel science and technology. *Journal of Sol-Gel Science and Technology,* 8, 17 – 22.

[8] Uhlmann, D.R. & Teowee, G. (1998). Sol-gel science and technology: current state and future prospects. *Journal of Sol-Gel Science and Technology,* 13, 153 – 162.

[9] Stachs, O., Gerber, Th. & Petkov, V. (1999). The structure formation of Zirconium oxide gels in alcoholic solutions. *J. Sol-Gel Sci. Technol.,* 15, 23 – 30.

[10] Brinker, C.J. & Scherer, G.W. (1990). Sol-Gel Science. *The Physics and Chemistry of Sol-Gel Processing,* Academic Press, New York.

[11] Fireman-Shoresh, S., Avnir, D., & Marx, S. (2003). General method for chiral imprinting of sol-gel thin films exhibiting enantioselectivity. *Chem. Mater.,* 15, 3607 – 3613.

[12] Marx, S., & Liron, Z. (2001). Molecular imprinting in thin films of organic – inorganic hybrid sol-gel and acrylic polymers. *Chem. Mater.,* 13, 3624 – 3630.

[13] Jones, R.G., Corriu, R.J.P., Douglas, W.E. et al. (2000). Silicon-Containing Polymers. *The Science and Technology of their Synthesis and Aplications,* Kluwer Academic Publishers., 667 – 695.

[14] Podbielska, H. & Ulatowska-Jarza, (2005). A. Sol-gel technology for biomedical engineering. *Bulletin of the Polish Academy of Sciences Technical Sciences,* 53(3), 261 – 271.

[15] Brinker, C.J., Keefer, K.D., Schaefer, D.W. & Ashley, C.S.J. (1982). Sol-gel transition in simple silicates. *J. Non-Cryst. Solids,* 48, 47 – 64.

[16] Bartlett, J.R., & Woolfrey, J.L. (1998). Ceramic Interfaces: *Properties and Applications,* London, 461 – 482.

[17] Lakshmi, B.B., Patrissi, C.J., & Martin, C.R. (1997). Sol-gel template synthesis of semiconductor oxide micro- and nanostructures. *Chemistry of Materials,* 9, 2544 – 2550.

[18] Crişan, M., Jitianu, A., Crişan, D., Bălăşoiu, M., Drăgan, N., & Zaharescu, M. (2000). Sol-gel monocomponent nano-sized oxide powders. *Journal of Optoelectronics and Advanced Materials,* 2(4), 339 – 344.

[19] Diaz-Garcia, M.E., & Badıa, L.R., (2005). Molecular imprinting in sol–gel materials: recent developments and applications. *Microchem. Acta,* 149, 19 – 36.

[20] Gupta, R., & Kumar, A. (2008). Bioactive materials for biomedical applications using sol–gel technology. *Biomed. Mater.,* 3(3), 034005, doi: 10.1088/1748-6041/3/3/034005.

[21] Chaudhury, N.K., Gupta, R., & Gulia, S. (2007). Sol-gel technology for sensor applications. *Def. Sci. J.,* 57, 241 – 253.

[22] Shilova, O.A. (2003). Spin-on Glass Films for Semiconductor Technology. *Surf. Coat., Part B: Coat. Trans.,* 86(3), 195 – 202.

[23] Shilova, O.A., Hashkovsky, S.V., & Kuznetsova, L.A. (2003). Sol-Gel Preparation of Coatings for Electrical, Laser, Space Engineering, and Power. *J. Sol-Gel Sci. Technol.,* 26 (1–3), 687 – 691.

[24] Garcia-Sancez, M.A., De la Luz, V., Estrada-Rico, M.L., Murillo-Martinez, Coahuila-Hernandez, M.I., Sosa-Fonseca, R., Tello-Solis, S.R., Rojas, F. & Campero, A. (2009). Fluorescent porphyrins covalently bound to silica xerogel matrices. *J. of Non-Crystalline Solids,* 355, 120 – 125.

[25] Khamova, T.V., Shilova, O.A., Movchan, T.G., Sazhnikov, V.A., & Rusanov, A.I. (2008). Sol-Gel Synthesis and Fluorescence Properties of Hybrid Nanocomposite Materials Doped with the Nile Red Dye. *Glass Physics and Chemistry,* 34(1), 63 – 67.

[26] Radin, S., Chen, T. & Ducheyne, P. (2009), The controlled realease of drugs from emulsified, sol gel processed silica microspheres. *Biomaterials,* 30, 850 – 858.

[27] Stephanson, C.J. & Flanagan, G.P. (2003). Synthesis of a novel anionic hydride organosiloxane presenting biochemical properties. *International Journal of Hydrogen Energy,* 28, 1243 – 1250.

[28] Gallardo, J., Galliano, P., Moreno, R., Duran, A. (2000). Bioactive sol-gel coatings for orthopaedic prosthesis. *J Sol-Gel Sci Tech.* 19, 107 – 111.

[29] Livage, J. (1997). Sol-gel processes. Curr. Opin. Solid State Mater. Sci., 2, 132 – 138.

[30] Li, J., Tan, S.N., Ge H. (1996). Silica sol-gel immobilized amperometric biosensor for hydrogen peroxide. *Anal. Chim. Acta,* 335, 137 – 145.

[31] Prokopowicz, M., Przyjazny, A. (2007). Synthesis of sol-gel mesoporous silica materials providing a slow release of doxorubicin. *J. Microencapsul.,* 24(7), 694 – 713.

[32] Kortesuo, P., Ahola, M., Karlsson, S., Kangasniemi, I., Kiesvaara, J., & Yli-Urpo, A., (1999). Sol-gel processed sintered silica xerogel as cartier in controlled drug delivery. *J Biomed Mater Res.,* 44, 162 – 167.

[33] Wu, Z., Joo, H., Lee, T.G., & Lee, K. (2005). Controlled release of lidocaine hydrochloride from the surfactant-doped hybrid xerogels. J. Control Release, 104, 497 – 505.

[34] Hench, L.L. (1994). Bioactive ceramics theory and clinical applications. *Bioceramics,* 7, 3 – 14.

[35] Peltola, T., Jokinen, M., Veittola, S., Rahiala, H., & Yli-Urpo, A. (2001). Influence of sol and stage of spinnability on in vitro bioactivity and dissolution of sol-gel derived SiO_2 fibers. *Biomaterials,* 22, 589 – 598.

[36] Gallardo, J., Galliano, P.G., & Porto Lopez, J.M. (2002). Preparation and in vitro evaluation of porous silica gels. *Biomaterials,* 23, 4277 – 4284.

[37] Coradin, T., & Livage, J. (2007). Effects of relative humidity on enzyme activity immobilized in sol-gel-derived silica nanocomposites. Acc. Chem. Res., 40(9), 819 – 826.

[38] Kortesuo, P., Ahola, M., Kangas, M., Leino, T., Laakso, S., Vuorilehto, L. & et al. (2001). Alkyl substituted silica gel as carrier in the controlled release of dexmedetomidine. *J. Controlled Release,* 76 (3), 227 – 238.

[39] Radin, S., El Bassyouni, G., Vresilovic, E.J. Schepers, E., & Ducheyne, P., (2005). In vivo tissue response to resorbable silica xerogels as controlled release materials. *Biomaterials, 26*(9), 1043 – 1052;

[40] Radin, S., Ducheyne, P., Kamplain, T., & Tan B.H. (2001). Silica sol gel for the controlled release of antibiotics. *J. Biomed. Mater. Res.,* 57, 313 – 320.

[41] Aughenbaugh, W.B., Radin, & S, Ducheyne, P. (2001). Silica sol gel for the controlled release of antibiotics. *J. Biomed. Mater. Res.,* 57, 321 – 326.

[42] Liu, Q., Zhang, J., Liu, Q., Zhu, Z., & Chen, J. (2009). J. Sol-gel synthesis and characterization of silica film with two oposite structures: nano-porous and protuberant. *Materials Chemistry and Physics,* 114, 309 – 312.

[43] Kortesuo, P., Ahola, M., Kangas, M., Yli-Urpo, A., Kiesvaara, J., & Marvola, M. (2001). In vitro release of dexmedetomidine from silica xerogel monoliths: effect of sol-gel synthesis parameters. *Int. J. Pharm* . 221, 107 – 114.

[44] Kawakami, Y., (2007). Structural control and functionalization of oligomeric silsesquioxanes, *Reactive & Functional Polymers* 67, 1137 – 1147.

[45] Kickelbick, G. (2007). *Hybrid Materials. Synthesis, Characterization, and Applications,* Wiley-VCH Verlag GmbH & Co. KGaA, Weinheim.

[46] Tereshchenko, T.A., (2008). Synthesis and Application of Polyhedral Oligosilsesquioxanes and Spherosilicates. *Polymer Science*, Ser. B, 50(9–10), 249 – 262.

[47] Li, G., Wang, L., Ni, H., & Pittman, U.C. (2002). Polyhedral oligomeric silsesquiosane polymers and copolymers: A Review. *Journal of Inorganic and Organometallic Polymers,* 11(3), 123 – 154.

[48] do Carmo, R.D., Filho, N.L.D., & Stradiotto N.R., (2007). Encapsulation of titanium (IV) silsesquioxane into the NH4USY zeolite: Preparation, characterization and application. *Materials Research Bulletin, 42*, 1811 – 1822.

[49] Wada, K., & Mitsudo, T. (2005). Preparation of Novel Materials for Catalysts Utilizing Metal-Containing Silsesquioxanes, *Catalysis Surveys from Asia,* 9(4), 229 – 241.

[50] Vrastil J., Matejka L., Spacek V., Vecera M., & Prokupek L. (2005). Organization in sol-gel polymerization of methacrylate co-oligomers containing trimethoxysilylpropyl methacrylate. *Polymer,* 46, 11232 – 11240;

[51] Choi, J., Harcup, J., Yee A.F., Zhu, Q., & Laine, R.M. (2001). Organic/inorganic hybrid composites from cubic silsesquioxanes. *J. Am. Chem. Soc.* 123, 11420.

[52] Guizhi, L., Lichang, W., Hanli, N., & U. Pittman Jr. (2001). C., Polyhedral Oligomeric Silsesquioxane (POSS) Polymers and Copolymers: A Review. *Journal of Inorganic and Organometallic Polymers,* 11(3), 123 – 154.

[53] Mori, H., Lanzendörfer M. G., & Müller A. H. E. (2004). Silsesquioxane-based nanoparticles formed via hydrolytic condensation of organotriethoxysilane containing hydroxy groups. *Macromolecules,* 37, 5228 – 5238.

[54] Abe, Y. & Gunji, T. (2004). Oligo- and Polysiloxanes. *Prog. Polym. Sci.* 29, 149 – 182.

[55] Cho, H.J., Hwang, D.H., Lee, J.I., Jung, Y.K., Park. J.H., Lee, J., & et al. (2006). Electroluminescent polyhedral oligomeric silsesquioxane-based nanoparticle. *Chem. Mater.,* 18, 3780 – 3787.

[56] Lin, W.J., Chen, W.C., Wu, W.C., Niu, Y.H., & Jen A.K.Y. (2004). Synthesis and Optoelectronic Properties of Starlike Polyfluorenes with a Silsesquioxane Core. *Macromolecules,* 37, 2335 – 2341.

[57] Adachi, K., Tamaki, R. & Chujo, Y. (2004). Synthesis of Organic-Inorganic Polymer Hybrids from Ammoniumpropyl-Functionalized Polyhedral Oligomeric Silsesquioxane. Bull. *Chem. Sos. Jpn.,* 77, 2115 – 2119.

[58] McCusker, C., Carroll, J.B. & Rotello, V.M. (2005). Cationic Polyhedral Oligomeric Silsesquioxane (POSS) Units as Carriers for Drug Delivery Processes. *Chem. Comm.,* 996 – 998.

[59] Watanabe, A. & Miyashita, T. (2006). Micropatterning of organic-inorganic hybrid film using photosensitive sol-gel system consisting of double-decker-shaped multifunctional silsesquioxane. *Chemistry Letters,* 35, (10), 1130 – 1131.

[60] Kim, M.S., Seok, S.I, Ahn, BY., Koo, S.M., Paik, S.U., (2003). Encapsulation of Water-Soluble Dye in Spherical Sol-Gel Silica Matrices. *J. of Sol-Gel Sci. and Techn.,* 27, 355 – 361.

[61] Tamaki, R., Tanaka, J., Asuncion, M. Z., & et al., (2001). Octa(aminophenyl) silsesquioxane As A Nanoconstruction Site. *J. Am. Chem. Soc.* 123, 12416 – 12417.

[62] Zhang, C. & Laine, R.M. (2000). Hydrosilylation of allyl alcohol with $[HSiMe_2OSiO_{1.5}]_8$. Octa (3-hydroxypropyldimethylsiloxy)octasilsesquioxane and its octamethacrylate derivative as potential precursors to hybrid nanocomposites. *J. Am. Chem. Soc.* 122, 6979 – 6980.

[63] Pyun, J., Matyjaszewski, K., Wu, J., Kim, G.M., Chun, S.B. & Mather P.T., (2003). ABA triblock copolymers containing polyhedral oligomeric silsesquioxane pendant groups: synthesis and unique propertie. *Polymer,* 44(9), 2739 – 2750.

[64] Lam, C.W., James, J.T., McCluskey, R., & Hunter, R.L. (2004). Pulmonary Toxicity of Single-Wall Carbon Nanotubes in Mice 7 and 90 Days After Intratracheal Instillation. *Toxicol. Sci.,* 77, 126 – 134.

[65] Hanssen, R.W.J.M., Meetsma, A., van Santen, R.A., & Abbenhuis, H.C.L. (2001). Synthesis, structural characterization and transmetalation reactions of a tetranuclear magnesium silsesquioxane complex. *Inorg. Chem.,* 40, 4049-4052

[66] Asuncion, M.Z., Laine, R.M. (2007). Silsesquioxane barrier materials. *Macromolecules*, 40(3), 555 – 562.

[67] Feher, F.J., und Weller, K.J., (1991). Synthesis and characterization of vanadium-containing silsesquioxanes. *Inorg. Chem.,* 30(8), 1689 – 1694.

[68] Zeng, K., Wang, L., Zheng, S., Qian, X. (2009). Self-assembly behavior of hepta(3,3,3-trifluoropropyl) polyhedral oligomeric silsesquioxane-capped poly(ε-caprolactone) in epoxy resin: nanostructures and surface properties. *Polymer,* 50, 685 – 695.

[69] Enqvist, J. and Enqvist, M., European Patent 1 791 000 A1, Date of publication 30.05.2007, Bulletin 2007/22 (21) *Application number* 05397022.4.

[70] Xu, H., Yang, B., Wang, J., Guang, S., & Li, C. (2007). Preparation, T_g improvement, and thermal stability enhancement mechanism of soluble poly(methyl methacrylate) nanocomposites by incorporating octavinyl polyhedral oligomeric silsesquioxanes. *J. Polym. Sci., Part A: Polym. Chem*., 45, 5308 – 5317.

[71] Yang, B., Xu, H., Wang, J., Gang, S., & Li, C. (2007). Preparation and thermal property of hybrid nanocomposites by free radical copolymerization of styrene with octavinyl polyhedral oligomeric silsesquioxane. *J. Appl. Polym. Sci.,* 106, 320 – 326.

[72] MuCusker, C., Carroll, J.B., & Rotillo, V.M. (2005). Cationic polyhedral oligomeric silsesquiosane (POSS) units as carriers for drug delivery processes. *Chem. Com.,* 996 – 998.

[73] Siddiqui, W.H., & York, R.G. (1993). Quaternary Silsesquioxane: A Developmental Toxicity Study in Rats. *Toxicological Science,* 21, 66 – 70.

[74] Cassidy, S.L., Ryan, B.M., Mallett, E., Shamoun, D., Siddiqui, W.H. (2000). Phenylsilsesquioxane fluid: Developmental toxicity studies in rats and rabbits following oral administration. *Journal of Toxicology and Environmental Health Part A.,* 60(5), 305 – 316.

[75] Mousavi, L., Keshtgar, M., Darbyshire, A., Cannon, R., Seifalian, A. (2008). Development of new generation of breast implants using silsesquioxane nanocomposite shell. *European Journal of Surgical Oncology,* 34, 10, 1196 – 1197.

[76] Braun, S., Rappoport, Zusman, S., Avnir D., & Ottolenghi M. (1990). Biochemically active sol-gel glasses: the trapping of enzymes. *Mater. Lett.,* 10, 1 – 8.

[77] Jin, W., Brennan, J. D. (2002). Properties and applications of proteins encapsulated within sol-gel derived materials. *Analytica Chimica Acta,* 461, 1 – 36.

[78] Dunn, B., Miller J.M., Dave B.C., Valentine J.S., & Zink J.I. (1998). Strategies for encapsulating biomolecules in sol-gel matrixes. *Acta Materialia,* 46, 737 – 741.

[79] Miller J.M., Dunn B., Dave B.C., Valentine J.S., Zink J.I., (1996). Synthesis conditions for encapsulating cytochrome C and catalase in SiO_2 sol-gel materials. *Journal of Non-Crystalline Solids, 202*, 279 – 289.

[80] Zink, J.I., Dunn, B., Valentine J.S., Yamanaka, S., & Akbarian F. (1995). Optical sol-gel materials based on binding and catalysis by biomolecules. *Polymeric Materials Science and Engineering, 73*, 264 – 265;

[81] Valentine, J.S., Dunn, B., Zink, J.I., Fukuto, J.M., El Sayed, M.A., Ellerby, L.M., Chung, K.E., Dave, B.C., & Nishida, C.R. (1994). Sol-gel encapsulation of metalloproteins for biosensor", *Polymeric Materials Science and Engineering,* 71, 666 – 667.

[82] Braun, S., & Avnir D. (1996). Special issue on biochemical aspect of sol-gel processing, J. *Sol-Gel Sci. Technol.,* 7, 5 – 143;

[83] Gill, I., & Ballesteros, A. (2000). Bioencapsulation within synthetic polymers (Part 1): sol–gel encapsulated biologicals. *Tibtech*, 18, 282 – 296.

[84] Dave, B.C., Dunn, B., Valentine, J.S., & Zink, J. I. (1994). Sol–gel encapsulation method for biosensors. *Analytical Chemistry* 66, 1120A – 1127A.

[85] Lin, J. & Brown, C.W. (1997). Sol-gel glass as a matrix for chemical and biochemical sensing. *Trends Anal. Chem.* 16, 200 – 211.

[86] Gill, I. & Ballesteros, A. (1998). Encapsulation of biologicals within silicate, siloxane and hybrid sol-gel polymers: an efficient and generic approach. *J. Am. Chem. Soc.* 120, 8587 – 8598.

[87] Kandimalla V. B., Tripathi V. S., & Ju H. (2006). Immobilization of Biomolecules in Sol-Gels: Biological and Analytical Applications, *Critical Reviews in Analytical Chemistry,* 36, 73 – 106.

[88] Innocenzi, P., Abdirashid, M.O., & Guglielmi, M. (1994). Structure and Properties of Sol-Gel Coatings from Methyltriethoxysilane and Tetraethoxysilane. *J. Sol-Gel Sc. Tech., 3,* 47 – 55.

[89] Gill, I., & Ballesteros, A. (2000). Bioencapsulation within synthetic polymers (Part 2): non-sol–gel protein–polymer biocomposites. *Trends Biotechnol.* 18(11), 469 – 479.

[90] Grod, I., Mista, W., Strek, W., Bukowska, E., Hermanowicz, K., & Maruszewski, K., (2004). Synthesis and properties of an inorganic-organic hybrid preparedby the sol-gel method. *Optical Mat.,* 26, 207 – 211.

[91] Pereira, A.P.V., Vasconcelos, W.L., & Orefice R.L., (2000). Novel multicomponent silicate/poly (vinyl alcohol) hybrids with controlled reactivity. *J. Non-Crystalline Sol.,* 273, 180 – 185.

[92] Chen H., Brook M.A. & Sheardown H. (2004). Silicone elastomers for reduced protein adsorption. *Biomaterials,* 25, 2273 – 2282.

[93] Mackenzie, J.D., Bescher, E.P. (1998). Structures, Properties and Potential Applications of Ormosils. *J. Sol-Gel Sci. Technol.,* 13(1–3), 371 – 377.

[94] Chujo, Y. (1996). Organic-inorganic hybrid materials. *Curr. Opin. Solid-State Mater.* Sci.,1, 806 – 811.

[95] Arkles, B. (1999). Hybrid Polymers in the Marketplace. *CHEMTECH,* 29(12), 7.

[96] Grandi, S. Tomasi, P., Mustarelli, C., Clemente & F., Carbonaro, C.M. (2007). Characterisation of a new sol-gel precursor for a SiO_2-rhodamine 6G hybrid class II material. *J. Sol-Gel Sci. Technol.* 41, 57 – 63.

[97] Avnir, D., Levy, D., Reisfeld, R. (1984). The nature of silica glass cage as reflected by spectral changes and enhanced photostability trapped of rhodamine 6G. *J. Phys. Chem.* 88, 5956 – 5960.

[98] Makishima, A. & Tani, T. (1986). Preparation of Amorphous Silicas Doped with Organic Molecules by the Sol-Gel Process. *J. Am. Ceram. Soc.* 69(4), C72 – C74.

[99] Takahashi, Y., Kitamura, T., Nogami, M., Uchida, K., Yamanaka, T. (1994). Luminescence of rhodamine dyes doped in amorphous silica glasses. *J Luminescence* 60&61, 451 – 453.

[100] Costa, T.M.H., Stefani, V., Balzaretti, N., Francisco, L.T.S.T., Gallas, M.R., de Jornada J.A.H. (1997). High pressure loading of organic dyes into a silica matrix. *J Non-Cryst Solids* 221, 157 – 162.

[101] Hungerford, G., Suhling, K., Ferriera, J.A. (1999). Comparison of the fluorescence behaviour of rhodamine 6G in bulk and thin film tetraethylorthosilicate derived sol-gel matrices. *J Photochem Photobiol A: Chem* 129, 71 – 80.

[102] Zhu, X-L., Lam, S-K., Lo, D. (2000). Distributed-feedback dye-doped solgel silica lasers. *Appl Opt* 39(18), 3104 – 3107.

[103] Rao, A.P. & Rao, A.V. (2003). Studies on the effect of organic additives on the monolithicity and optical properties of the rhodamine 6G doped silica xerogels. *Material Letters,* 57, 3741 – 3747.

[104] Sieminska, L. & Zerda, T.W. (1996). Diffusion of steroids from sol-gel glass. *J. Phys. Chem.,* 100, 4591-4597.

[105] Harrison, C.C., McGiveron, J.K., Li, X. (1994). Spectroscopic studies of post-doped silica gel monoliths. *J Sol-Gel Sci Technol.* 2, 855 – 859.

[106] Gall, G.J., Li, X., King, A.T. (1994). Optical characterisation of **sol-gel** optical composites *J. Sol-Gel Sci Technol.* 2, 775 – 778.

[107] Chernev, G.E., Samuneva B.I., Djambaski, P.R., Salvado I.M.M., & Fernandes H.V. (2006). Silica hybrid nanocomposites. *Central European Journal of Chemistry,* 4(1), 81 – 91.

[108] Dash, S., S. Mishra, Patel S., &Mishra B. K. (2008). Organically modified silica: Synthesis and applications due to its surface interaction with organic molecules. *Advances in Colloid and Interface Science,* 140, 77 – 94.

[109] Price, P.M., Clark, J.H. & Macquarrie, D.J. (2000). Modified silicas for clean technology. *J. Chem. Soc. Dalton Trans.,* 101 – 110.

[110] Cong, H.P. & Yu, S.H. (2009). Self-assembly of functionalized inorganic–organic hybrids. *Current Opinion in Colloid & Interface Science,* 14, 71 – 80.

[111] Monnier, V., Dubuisson, E., Sanz, N. & Ibanez A. (2008). Organic nanocrystals grown in sol-gel matrices: new hybrid organic-inorganic materials for optics. *Res. Chem. Intermed.*, 34 (2–3), 155 – 167.

[112] Que, W. & Hu, X. (2007). Optical switch and luminescence properties of sol-gel hybrid organic-inorganic materials containing azobenzene groups and doped with neodymium ions. *Appl. Phys. B* 88, 557 – 561.

[113] Deng, T-S., Zhang, Q-F., Zhang, J-Y., Shen, X., Zhu, K-T., & Wu, J-L. (2009). One-step synthesis of highly monodisperse hybrid silica spheres in aqueous solution. Journal of Colloid and Interface Science. 329, 292 – 299.

[114] Feng, P., Xu, Y., & Wu, D. (2007), Non-surfactant synthesis of mesoporous silica with dye as template. *Chinese Chemical Letters,* 18, 1182 – 1186.

[115] Samuneva, B., Kabaivanova, L., Chernev, G., Djambaski, P., Kashchieva, E., Emanuilova, E.,. Salvado, I. M. M, Fernandes, M. H. V., & Wu, A. (2008). Sol-gel synthesis and structure of silica, hybrid materials. *J. Sol-Gel Sci. Technol.* 48, 73 – 79.

[116] Malins, C., & Maccraith, B.D., (1998). Dye-doped organically modified silica glass for fluorescence based carbon dioxide gas detection. *Analyst,* 123, 2373 – 2376.

[117] Levy, D. (1996). Organic doped sol - gel glasses for electro-optics and display applications. *Pure Appl. Opt*, 5, 621 – 629.

[118] Levy, D., Einhorn, S., & Avnir, D. (1989). Applications of the sol-gel process for the preparation of photochromic information-recording materials: synthesis, properties, mechanisms. *J. Non-Cryst. Solids* 113, 137 – 145.

[119] Vazquez-Duran, A., Araujo-Andrade, G., Martinez, C., Ortega-Zarzosa, G., Ruiz, F., Martinez, J.R. (2006). Spectral characterization of chlorophyll fluorescence in extract of barley leves embedded in silica xerogel matrix. *J Sol-Gel Sci Technol.* 39, 223 – 227.

[120] Dire, S., Tagliazucca, V., Brusatin, G., Bottazzo, J., Fortunati, I., Signorini, R., Dainese, T., Andraud, C., Trombetta, M., Vona Di, M.L., & Licoccia, S. (2008). Hybrid organic/inorganic materials for photonic applications via assembling of nanostructured molecular units. *J Sol-Gel Sci Technol.* 48, 217 – 223.

[121] Buestrich, R., Kahlenberg, F., Popall, M., Dannberg, P., Muller-Fiedler, R., & Rösch, D. (2001). ORMOCERRS for Optical Interconnection Technology. *J Sol-Gel Sci Technol* 20, 181.

[122] Haas, K.H., & Rose K. (2003). Hybrid inorganic/organic polymers with nanoscale building blocks: precursor, processing, properties and application. *Rev. Adv. Mat. Sci.,* 5, 47 – 52.

[123] Castelvetro, V. & Vita, C. (2004). Nanostructured hybrid materials from aqueous polymer dispersions", *Adv. Coll. Interface Sci.,* 108 –109, 167 – 185.

[124] Ceccato, R., Dire, S., & Lutterotti L., (2003). Pyrolysis pathway of sol-gel derived organic/ inorganic hybrid nanocomposites. *J. Non-Crystalline Solids,* 322, 22 – 28.

[125] Sanchez, C., De Soler-Illia, A.A.G.J., Ribot, F.& Grosso, D. (2003). Design of functional nano-structured materials through the use of controlled hybrid organic-inorganic interfaces, *C.R. Chimie,* 6, 1131 – 1151.

[126] Houbertz, R., Frohlich, L., Poall, M., Streppel, U., Dannberg, P., Brauer, A., Serbin, J., & Chichkov, B.N. (2003). Inorganic-organic hybrid polymers for information technology: from planar technology to 3D nanostructures. *Adv. Eng. Mater.* 5, 551.

[127] Holthoff, E.L., & Bright, F.V. (2007). Molecularly templated materials in chemical sensing. *Analytica Chimica Acta,* 594, 147 – 161.

[128] Eggins, B.R. (2002). (Ed.) *Chemical Sensors and Biosensors,* John Wiley & Sons, Chichester, UK.

[129] Yan, M., & Ramstrom O. (2005). (Eds.) Molecularly Imprinted Materials: *Science and Technology,* Marcel Dekker, New York.

[130] Ahola, M., Kortesuo, P., Kangasniemi, I. Kiesvaara, J., & Yli-Urpo, A. (2000). Silica xerogel carrier material for controlled release of toremifene citrate. *Int. J. Pharm.,* 195(12) 219 – 227.

[131] Radin, S., & Ducheyene, P. (2004). Nanostructural controlof implantable xerogels for the controlled release of biomolecules. In Reis RL, Weiner S. editors. *Learning from nature how to design new implantable materials: from biomineralization fundamentals to biomimetic materials and processing route.* Netherland: Kluwer Academic Publishers, 59 – 74.

[132] Santos, E. M., Radin, S. & Ducheyene, P. (1999). Sol gel derived carrier for the controlled release of proteins. *Biomaterials,* 20, 1695 – 1670.

[133] Chen, T. & Ducheyene, P. (2009). The controlled release of drugs from emulsified sol-gel procesed microspheres. *Biomaterials*, 30, 850 – 858.

[134] Ahola, M., Sa¨ilynoja, E., Raitavuo, M., Vaahtio, M., Salonen, J., & Yli-Urpo, A. (2001). In vitro release of heparin from silica xerogels. *Biomaterials,* 22, 2163 – 2170.

[135] Teoli, D., Parisi, L., Realdon, N., Guglielmi, M. Rosato, & A., Morpurgo, M. (2006). Wet sol-gel derived silica for controlled release of proteins. *Journal of Controlled Release,* 116, 295 – 303.

[136] Lopez, T., Basaldella, E.I., Ojeda, M.L., Manjarrez, J., & Alexander-Katz, R. (2006). Encapsulation of valproic acid and sodic phenytoin in ordered mesoporous SiO_2 solids for the treatment of temporal lobe epilepsy, *Optical Materials,* 29, 75 – 81.

[137] Wang, S. (2009). *Ordered mesoporous materials for drug delivery, Microporous and Mesoporous Materials,* 117, 1 – 9.

[138] Prokopowicz, M. (2009). Correlation between physicochemical properties of doxorubicin-loaded silica/polydimethylsiloxane xerogel and in vitro release of drug. *Acta Biomaterialia,* 5, 193 – 207.

[139] Prokopowicz, M. (2007). In vitro controlled release of doxorubicin from silica xerogels. *J. Pharm Pharmacol.,* 59, 1365 – 1373;

[140] Ying, J.Y., Mehnert, C., & Wong, M.S. (1999). Synthesis and applications of supramolecular templated mesoporous materials. *Angew Chem. Int. Ed. Eng.*l, 38, 56 – 77.

[141] Stein, A., Melde, B.J., & Schroden, R.C. (2000). Hybrid inorganic–organic mesoporous silicatesnanoscopic reactors coming of age. *Adv Mater.,* 12, 1403 –1419.

[142] Soler-Illia, G.J.A.A., Rozes, L., Boggiano, M.K., Sanchez, C., Turrin, C.O., Caminade, A.M., & et al. (2000). New mesotextured hybrid materials made from assemblies of dendrimers and titanium(iv)-oxo-organo clusters. *Angew Chem Int Ed Eng*, 39, 4249 – 4254.

[143] Holmberg, K. (2004). Surfactant-templated nanomaterials synthesis. *J. of Colloid and Interface Science,* 274, 355-364.

[144] Smith, G., Kumar, P., Nguyen, D. (2004). Formulating Cleaning Products with Microemulsions, *CD Proceedings 6th World Surfactant Congress CESIO,* Berlin, Germany, June 21 – 23.

[145] Marques, E. F., Regev, O., Khan, A., & Lindman, B. (2003). Self-organization of double-chained and pseudodouble-chained surfactants: counterion and geometry effects. *Advances in Colloid and Interface Science,* 100-102, 83 – 104.

[146] Langevin, D. (1998). Structure and dynamic properties of surfactant systems. *Studies in Surface Science and Catalysis,* 117, 129 – 134.

[147] Yang, J.Y., Mehnert, C.P. & Wong, M.S. (1999). Synthesis and Applications of Supramolecular-Templated Mesoporous Materials. *Review. Angew. Chem., Int. Ed. Engl.,* 38, 56 – 77.

[148] Beck, J.S., Vartuli, J.C., Roth, W.J., Leonowicz, M.E., Kresge, C.T., Schmitt, K.D., Chu, C.T.W., Olson, D.H., Sheppard, E.W. & et al. (1992). A new family of mesoporous molecular sieves prepared with liquid crystal templates. *Journal of the American Chemical Society,* 114, 10834 – 10843.

[149] Dong, H. (2002). *Organic-Inorganic Hybrid Mesoporous Silica Materials and Their Application as Host Matrix for Protein Molecules,* Ph.D. Thesis.

[150] Štangar, U.L., Hüsing, N., *Formation of Mesostructured SiO_2 Films in the Presence of Sugar-Based Surfactants.* http://www.mariecurie.org/annals/volume4/chem3.pdf.

[151] Wei, Y. and Qiu, K.Y. (2004). Chapter 28. A novel Nonsurfactant route to nanoporous materials and its biological applications, 873 – 887, *in Nonporous Materials – Science and Engineering*, Ed. Lu, G.Q., & Zhao, X.S.

[152] Dutta, J. & Hofmann, H. (2003). *Nanomaterials,* 17 – 18, 38; http://www.nano.ait.ac.th.

[153] Gou, Z.R., Weng, W.J., Zou, C., Du, P.Y., & Han G.R. (2007). Effect of Inorganic Ions on Gelation, Particle Morphology, and Texture of Solo-Gel-Derived Silica Nanospheres. *Key Engineering Materials,* 330-332, 203 – 206.

[154] Lacatusu, I., Badea, N., & Meghea A. (2009). Effect of surfactants on stability of colloidal sols as precursors in sol-gel encapsulation. Key Materials Engineering, 415, 57 – 60.

[155] Arunkumar, E., Forbes, C.C., & Smith, B.D. (2005). Improving rhe properties of organic dyes by molecular encapsulation, *Eur. J. Org. Chem.,* 4051 – 4059.

[156] Yao, Y., Zhang, M., Shi, J., Gong, M., Zhang, H., & Yang Y. (2001). Encapsulation of fluorescein into MCM-41 mesoporous molecular sieve by a sol-gel method. *Materials Letters,* 4844 – 4848.

[157] Veith, S. R, Perren, M. & Pratsinis, S. E. (2005). Encapsulation and retention of decanoic acid in sol-gel-made silicas. *J of Coll and Interface Sci,* 283(2), 495 – 502.

[158] Cintora-Gonzales, O., Estournes, C., Guille J.L., Grob, J.J., Honerlage, B., Lemoigne, J., Levy, R., Lutz, T., Merle, J.C., Muller D., Richard, M., Rehspringer, J.L., Schell, J., Viart, N. (2000). Aggregates in silica based matrices. *ANALUSIS,* 28(2), 109 – 113.

[159] Ahn, B.Y., Seok, S.I. & Baek, I.C. (2008). Sol-gel microencapsulation of hydrophilic active compounds from the modified silicon alkoxides: The control of pore and particle size. *Mat Sci and Eng,* 28(7), 1183 – 1188.

[160] Fölling, J., Polyakova, S., Belov, V., Alfons van Blaaderen, Bossi, M.L. & Hell S.W. (2008). Synthesis and Characterization of Photoswitchable Fluorescent Silica Nanoparticles, *Small,* 4(1), 134 – 142.

[161] Sokolov, I., Kievsky, Y.Y., & Kaszpurenko, M. (2007). Self-Assembly of Ultrabright Fluorescent Silica Particles, *Small,* 3(3), 419 – 423.

[162] Han, X.M., Lin, J., Xing, R.B., Fu, J., & Wang, S.B. (2003). Patterning and Optical Properties of Rhodamine Bdoped Organic-Inorganic Silica Films Fabricated by Sol-Gel Soft Lithography, *Mater. Lett.,* 57(7), 1355 – 1360.

[163] Stalikas, C.D. (2007). Journal of Separation Science, 30, 3268-3295.

[164] Walle, T., Ta, N., Kawamori, T., Wen, X., Tsuji, P.A., & Walle, U.K. (2007). *Biochem Pharm,* 73(9), 1288 – 1296.

[165] Stalikas, C.D. (2007). Extraction, separation, and detection methods for phenolic acids and flavonoids. *Journal of Separation Science,* 30, 3268 – 3295.

[166] Malešev, D., & Kuntič, V. (2007). Investigation metal - flavonoid chelates and the determination of flavonoid via metal flavonoid complexing reactions. *J. Serb. Chem. Soc.* 72(10), 921 – 939.

[167] Hatziantoniou, S., Demetzos, C. & Wagner, M. (2005). Phase transitions of lipids and liposomes. *User. Com.,* 1, 16 – 19.

[168] Halliwell, B. (2008), Are polyphenols antioxidants or pro-oxidants? What do we learn from cell culture and in vivo studies? *Archives of Biochemistry and Biophysics,* 476, 107 – 112.

[169] Georgetti S.R., Casagrande, R., Mambro Di V.M., Azzolini, A.E.C.S., & Fonseca, M.J.V. (2003). Evaluation of the antioxidant activity of different flavonoids by the chemiluminescence method. *AAPS PharmaSci,* 5(2), 1-5.

[170] Fusco, D., Colloca, G., Lo Monaco M.R. & Cesari M. (2007). Effects of antioxidant supplementation on the aging process. *Clin. Interv. Aging.* 2(3), 377 – 387.

[171] Es-Safi, N-E., Ghidouche S., & Ducrot, P.H. (2007). Flavonoids: Hemisynthesis, reactivity, characterization and free radical scavenging activity. *Molecules,* 12, 2228 – 2258.

[172] Fan, K., Kurihara, N., Abe, S., Ho C.T., Ghai, G., & Yang, K. (2007). Chemopreventive effects of orange peel extract (OPE). I: OPE inhibits intestinal tumor growth in ApcMin/+ mice. *J Med Food,* 10 (1), 11 – 17.

[173] Abe, S., Fan, K., Ho C.T., Ghai, G., & Yang, K. (2007). Chemopreventive effects of orange peel extract (OPE). II: OPE inhibits atypical hyperplastic lesions in rodent mammary gland. *J Med Food,* 10 (1), 18 – 24.

[174] Dornoff, J.M., & Davies, M.B. (1999). *United States Patent 5916576, Application Number* 08/865913, Publication Date 06/29/1999.

[175] Sergeev, I.N., Ho, Chi-Tang, Li, S., Colby J., & Dushenkov S. (2007). Apoptosis-inducing activity of hydroxylated polymethoxyflavones and polymethoxyflavones from orange peel in human breast cancer cells, *Mol. Nutr. Food Res.,* 51, 1478 – 1484.

[176] Huang, M. T., Chi-tang, Ho, Rosen, R. T. Ghai, G., Lipkin, M., Chen, K. Y., Telang, N. Boyd, C., & Csiszar, K. (2007). *U.S. Patent 7,*201,928, Application Number 10/088664, Publication Date 04/10/2007.

[177] Moon, H. S., Byung Ryeul, L., & Key Hyun, L. (2001*). United States Patent 6241975,* Application Number 09/478356, Publication Date 06/05/2001.

[178] Ma, Q. H., Xia Q., Lu, Y.Y., Hao X.Z., Gu N., Lin X.F., & Luo D. (2007). *Solid State Phenom.,* 121-123, 705 – 708.

[179] Ermakova, S., Choi, B.Y., Choi H.S., Kang, B.S., Bode A.M., & Zigang D. (2005). The intermediate filament protein vimentin is a new target for epigallocatechin gallate *J. Biol. Chem.,* 280(17), 16882 – 16890.

[180] Lacatusu, I., Badea, N., Bojin, D., Iosub, S., & Meghea, A. (2009). Novel fluorescence nanostructured materials obtained by entrapment of an ornamental bush extract in hybrid silica glass. *Journal of Sol-Gel Science and Technology,* 51(1), 84 – 91.

[181] Lacatusu, I., Badea, N., Nita, R., Giurginca, M., Bojin, D., Iosub, I., & Meghea A. (2009). Synthesis of high fluorescent silica hybrid materials by immobilization of orange peel extract in silica-silsesquioxane matrix. *Journal of Organic-Physics Chemistry,* 21, 1 – 8.

[182] Lacatusu, I., Badea, N., Nita R., Murariu, A., Miculescu, F., Iosub, I. & Meghea, A. (2009). *Encapsulation of fluorescence vegetable extracts within a templated sol – gel matrix, Optical Materials,* in press, doi:10.1016/j.optmat.2009.09.001.

[183] Lacatusu, I., Badea, N., Giurginca, M., Garcia, C.V., Sahraoui, B. & Meghea, A. (2009). Synthesis and properties of hybrid glass nanomaterials prepared by entrapping of green tea extract in sol-gel templated matrices, *Journal of Optoelectronics and Advanced Materials,* accepted for publication.

[184] Denev, I. & Markova (2008). FT-IR spectroscopy investigations of Co nanostructured materials, *Journal of the University of Chemical Technology and Metallurgy,* 43(4), 427-432.

[185] Stuart, B., (2004). *Infrared Spectroscopy: Fundamentals and Applications,* John Wiley Publication.

[186] Lacatusu, I., Nita, R., Badea, N., Bojin, D. & Meghea, A. (2009). The role of silsesquioxane compounds used as „building blocks" in sol – gel nanoencapsulation of retinyl palmitate, *Materials Research Innovations,* 13(3), 330 – 333.

[187] Drezek, R., Sokolov, K., Utzinger U., Boiko I., Malpica, A., Follen, M. & Richards-Kortum, R. (2001). Understanding the contributions of NADH and collagen to cervical tissue fluorescencen spectra: Modeling, measurements, and implications, *Journal of Biomedical Optics 6*(4), 385 – 396.

[188] Maurice, Eftink, R. (2002). *Topics in Fluorescence Spectroscopy, Intrinsic Fluorescence of Proteins,* Kluwer Academic Publishers.

[189] Krishna Murthy, N.V., Reddy, A.R. & Bhudevi, B. (2008). Wavelength dependant quenching of 2,5-diphenyloxazole fluorescence by nucleotides, *J Fluoresc.,* 18, 29 – 34.

[190] Gutzeit, H.O., Henker Y., Kind, B. &, Franz, A. (2004). Specific interactions of quercetin and other flavonoids with target proteins are revealed by elicited fluorescence, *Biochemical and Biophysical Research Communications,* 318, 490 – 495.

[191] Lakowicz, J.R., Shen, B., Gryczynski, Z., D'Auria, S. & Gryczynski, I. (2001). Intrinsic fluorescence from DNA can be enhanced by metallic particles, *Biochemical and Biophysical Research Communications,* 286, 875 – 879.

[192] Sokolov, K., Galvan, J., Myakov, A., Lacy, A., Lotan, R., Richards-Kortum, R. (2002). Realistic three-dimensional epithelial tissue phantoms for biomedical optics, *Journal of Biomedical Optics* 7(1), 148 –156.

[193] Dufour, E., Letort, A., Laguet, A., Lebecque, A. & Serra, J.N. (2006). Investigation of variety, typicality and vintage of French and German wines using front-face fluorescence spectroscopy, *Analytica Chimica Acta* 56, 292 – 299.

[194] El-Sayed Saleh Abdel-Hameed. (2009). Total phenolic contents and free radical scavenging activity of certain Egyptian Ficus species leaf samples, *Food Chemistry,* 114, 1271 – 1277.

[195] Halliwell, B. & Gutteridge, J.M.C., (2007). *Free Radicals in Biology and Medicine,* fourth ed., Clarendon Press, Oxford, UK.

[196] Rhee, S.G., (2006). Cell signaling:H_2O_2, a necessary evil for cell signaling. *Science,* 312(5782),1882 – 1883.

[197] Gundermann, K.D. & McCapra, F. (1987). *Chemiluminescence in Organic Chemistry,* Berlin Heidelberg, Springer-Verlag.

[198] Visioli, F. & Galli, C. (1997). Evaluation of Antioxidant Capacity by Chemiluminescence, *Analitycal Biochemistry,* 249(2), 244 -246.

[199] Albrecht, H.O. (1928). Chemiluminescence of aminophtalic hydrazide, *Z. Phys. Chem.*, 136, 321.

[200] Schottner. G. (2001). Hybrid Sol-Gel Derived Polymers: Applications of Multifunctional Materials. *Chem. of Mater.*, 13, 3422 – 3435.

[201] Hwang, Y.J., Oh C. & Seong-Geun, O. (2005). Controlled release of retinol from silica particles prepared in O/W/O emulsion: The effects of surfactants and polymers *J. Control. Release*, 106, 339 – 349.

[202] Kranl K., Schlesier K., Bitsch R., Hermann H., Rohe M. & Böhm V. (2005). Comparing antioxidative food additives and secondary plant products – use of different assays. *Food Chemistry,* 93 (1), 171 – 175.

[203] Popov, L. & Lewin, G. (1999). Antioxidative Homeostasis: Characterisation by means of chemiluminescent Tehniques. Methods of Enzimology, vol. 300, *Oxidants and Antioxidants, Part.* B, Ed. By L. Parker, Academic Press.

[204] Lesser, M.P. (2006). Oxidative stress in marine environments: biochemistry and physiological ecology, *Annual review of physiology,* 68, 253-278.

[205] Papadopoulos, K., Triantis, T., Dimotikali, D. & Nikokavouras, J. (1999). Photo- and radiochemiluminescence: reductive chemiluminescence of lucigenin by photo- or radiooxygenated amines and amides, *Journal of Photochemistry and Photobiology A: Chemistry,* 124(1-2), 85 – 90.

[206] Papadopoulos, K., Triantis, T., Dimotikali, D. & Nikokavouras, J. (2000). Radiostorage- and photostoragechemiluminescence: analytical prospects, *Analytica Chimica Acta,* 423 (2), 239 – 245.

[207] Berne, B.J. & Pecora, R. (1975). *Dynamic Light scattering,* John Wiley Publication.

[208] Malvern Instruments. (2005). *Zetasizer Nano Series User Manual,* Zeta Potential Theory.

[209] Berne, B.J. & Pecora, R. (2000). *Dynamic light scattering with applications to chemistry, biology and physics,* Dover Publication.

[210] *Dynamic Light Scattering – an Important Tool for Protein Crystallography and Nanotechnology,* web: www.chemeurope.com/articles/e/61632/.

[211] Marta Sartor, *DLS to determine the radius of small beads in Brownian motion in a solution, web*: www.fys.ku.dk/~nielshj/samlet_final_AFLEVERET_udgave.pdf.

[212] Wang, Z.L., Poncharal, P. & de Heer, W.A. (2000). Nanomeasurements in Transmission Electron Microscopy, *Microsc. Microanal.* 6, 224 – 230.

[213] Williams, D.B. & Carter, C.B. (2009). *Transmission Electron Microscopy*, 2nd Edition.

[214] Erni, R. & Browning, N.D. (2005). *Transmission Electron Microscopy,* published in „Encyclopedia of Chemical Processing", Taylor and Francis, Editor: Sunggyu Lee.

[215] Wang, Z.L. (2000). Transmission Electron Microscopy of Shape-Controlled Nanocrystals and Their Assemblies, J. *Phys. Chem.* B, 104 (6), 1153 – 1175.

[216] Sugimoto, Y., Pou, P., Abe, M., Jelinek, P., Pérez, R., Morita, S. & Custance, O. (2007). Chemical identification of individual surface atoms by atomic force microscopy. *Nature* 446 (7131), 64 – 67.

[217] Bowen, W.R. (2009). Atomic force microscopy in process engineering, *Elsevier,* ISBN-13: 978-1-85617-517-3.

[218] Giessibl, F.J. (2003). Advances in atomic force microscopy. Reviews of Modern Physics 75, 949. doi:10.1103/*RevModPhys*.75.949.

[219] Braga, P.C. & Ricci, D. (2004). *Atomic force microscopy: Biomedical methods and applications.*

[220] Lin, S. (2008). *Atomic force microscopy: Determination of dynamic parameters for protein-ligand interactions,* web: www.scitopics.com/Atomic_force_microscopy_Determination_of_dynamic_parameters_for_protein_ligand_interactions.html.

In: Natural Products: Structure, Bioactivity and Applications ISBN 978-1-62081-728-5
Editors: Ramiro E. Goncalves and Marcos Cunha Pinto ©2012 Nova Science Publishers, Inc.

Chapter 6

NATURAL PRODUCTS AS A SOURCE OF POTENTIAL DRUGS FOR THE TREATMENT OF FUNGAL INFECTIONS

Fernanda F. Campos*[*], *Susana Johann and Vera L. dos Santos
Departament of Microbiology, Instituto de Ciências Biológicas,
Universidade Federal de Minas Gerais, Belo Horizonte, MG, Brazil

ABSTRACT

The increasing numbers of immunocompromised people and the advances in medical technology, such as organ and bone marrow transplants, have been accompanied by an increasing occurrence of invasive fungal infections. In addition, the indiscriminate use of antifungal agents has resulted in the development of resistance to the drugs currently in use, which emphasizes the need for the discovery of new classes of antifungal compounds to treat fungal infections. Thus, there is an urge need for the discovery of new antifungal drug therapies that are safe and effective. Moreover, the increase of azole-resistant strains emphasizes the importance of identifying lead compounds that act on novel targets. Microorganisms and plants are sources of various natural products with antimicrobial activity. Biomolecules, such as biosurfactants and secondary metabolites belonging to different chemical classes including terpenoids, saponins, alkaloids and flavonoids, have been evaluated for their anti-adhesive, fungistatic and fungicidal properties in vitro and in vivo. These molecules can be used directly or as precursors to the development of new antifungal drugs. Formerly, combinatorial chemistry appeared to be the future for drug discovery, but in the late 1990s, synthetic chemists realized that combinatorial libraries lacked the "complexity" usually associated with natural compounds. Thus, with the advent of the concept of diversity-oriented synthesis, research into biologically active natural products has had a reprise. This review attempts to summarize the current status of the importance of natural products of microorganisms and plants as sources of drugs to treat fungal infections that are more effective and less toxic.

[*] E-mail address: fcampos@cpqrr.fiocruz.br; Tel: +55-31-3409 2741; Fax: +55-31-3409 2730; Address: Departamento de Microbiologia, Instituto de Ciências Biológicas, Universidade Federal de Minas Gerais, Belo Horizonte, MG, P. O. Box 486, CEP 31270-901, Brazil. (Corresponding author)

1. Introduction

Natural products or secondary metabolites comprise low-molecular-weight compounds that are regarded as dispensable for sustaining life, but indispensable for the survival of the producing organism (Hartmann, 1996; Hadacek, 2002). For centuries, plants have provided mankind with medicines and food-conserving additives. Still, plant- or microbial-derived compounds are regarded as a substantial source for the development of medicines and biocides. Recently, the focus has been on microorganisms that live as endophytes in plants or occur in deep-sea habitats (Clardy and Walsh, 2004).

Furthermore, the application of combinatorial chemistry in generating structural diversity has affected the significance of natural products in biological activity screening. However, concerning the identification of novel mode of actions, natural products are still regarded as a valuable pool for medicines and biocides. Furthermore, during the last decade, costs have increased dramatically due to additionally imposed conditions to elucidate the modes of action and the side effects of novel antimycotic drugs and fungicides; an increasing chemistry-critical public has also fuelled this process. As a result, today the cost estimates for the development of a new drug or biocide and the formulation of their application amount to approximately 150 million US Dollars. Taken together, these issues might impose a constraint on drug development; and even more, some representatives from the industry regard the quality of existing biocides as sufficient to combat the recognized pest organisms (Stetter and Lieb, 2000).

Biosurfactants are compounds produced by a wide variety of bacteria and fungi that exhibit pronounced surface and emulsifying activities. These molecules comprise a wide range of chemical structures, e.g., glycolipids, lipopeptides and lipoproteins, fatty acids, neutral lipids, polysaccharide-protein complexes, phospholipids and polymeric and particulate structures (Desai and Banat, 1997). Although most biosurfactants are considered secondary metabolites, some play essential roles in the survival of biosurfactant-producing microorganisms by facilitating nutrient transport or microbe-host interactions, or by acting as biocide agents (Rodrigues et al. 2006, Fernandes et al., 2007). Biosurfactants participate in several activities whenever a microbe encounters an interface; motility, including gliding and swarming motility and de-adhesion from surfaces; cell–cell interactions, occurring during formation, maintenance and maturation of biofilm, quorum sensing, amensalism and pathogenicity; cellular differentiation; substrate accession via direct interfacial contact and pseudosolubilization of substrates; and avoidance of toxic elements and compounds (Hamme et al., 2006). They might also be used as carbon and energy storage molecules, as protection from high ionic strength, or simply be by-products that are released in response to environmental change as extracellular coverings (Fisher, 1992, Healy et al. 1996, Kearns and Losick, 2003, Hoffman et al., 2010, Kavamura and Esposito, 2010, Simões et al., 2010). Besides their application in various industrial processes, biosurfactants have also been found to possess several properties of therapeutic and biomedical importance (Rodrigues et al. 2006, Eshrat Gharaei-Fathabad, 2011). They have antibacterial, antifungal and antiviral properties, (Banat et al. 2000, Singh and Cameotra, 2004) inhibit fibrin clot formation and possess anti-adhesive action against several pathogenic microorganisms (Rodrigues et al. 2006, Monteiro et al. 2011). In recent years, the biosurfactants produced by bacteria and yeasts are getting more attention for their antimicrobial activity, lower toxicity for plants and animals, high biodegradability, and low irritancy and compatibility with human skin (Cameotra and Makkar, 2004). The only factor that works against the widespread use of biosurfactants is the economics of their production; however, much effort has been given to ease the optimization process at both the engineering and biological levels.

Fungal diseases and antifungal agents have received less attention. However, because they are eukaryotic microorganisms, fungi are insidious pathogens; they are more closely related to human cells, and thus are more difficult to target with antifungal chemotherapy. Until 1970, fungal infections had generally been considered curable, and thus the demand for new antifungal agents was relatively low; however, they have emerged worldwide as an increasingly frequent cause of healthcare-associated infections (Vicente et al., 2003, Chakrabarti et al., 2005, Di Santo, 2010). Fungal diseases can be divided into two types: superficial and systemic. The first group includes cosmetically unpleasant, but less serious, infections of the skin, nails, and hair shafts; the majority of these can be treated successfully with existing antifungal agents. However, systemic infections, such as cryptococcosis and aspergillosis, are potentially fatal and are a growing danger for human health.

A recent epidemiological survey in the US reported that the incidence of fungal sepsis increased three-fold between 1979 and 2000. An alarming rise in life-threatening systemic fungal infections has been reported in AIDS patients who suffer suppression of the immune system. Other factors that have caused an increase in invasive fungal infections include bone marrow and organ transplants, the use of antineoplastic agents, long-term use of corticoids, and even the indiscriminate use of antibiotics. Invasive medical procedures, such as the use of vascular catheters, peritoneal dialysis, hemodialysis and parenteral nutrition, have also contributed to the rise of fungal infections. The factors influencing the outcome of fungal infections include the impact of the course of the underlying disease, and the ineffectiveness of diagnostic tools, which are poor and usually permit the detection of the infection only at a late stage (Vicente et al., 2003, Chakrabarti et al., 2005, Di Santo, 2010).

There are increasing concerns regarding the rapid emergence of the resistance phenomena in pathogens toward specific applied fungicides, especially in antimycotic therapy (Knight et al., 1997; Henningsen, 2003; Baddly and Moser, 2004). This approach generally utilizes fungicides that affect non-target mechanisms less than their synthetic analogues, and here natural products certainly constitute potential candidates. Consequently, there is interest in finding low-molecular-weight compounds with antifungal activities within various research approaches that reflect themselves in the choice of assays. So far, key areas for the application of antifungal bioassays include the control of crop pathogens in phytopathology and human pathogenic fungi in antimycotic chemotherapy.

Further, the current opinion is that resource utilization and susceptibility to secondary metabolites will attract more attention in the future to address basic ecological questions, such as the functioning and structuring of plant communities. In this context, the impact of microbial symbionts and pathogens will attract more attention.

A small number of antifungal classes are available for the treatment of serious mycoses, based on their chemistry and antifungal action (Sanglard and Odds, 2002). Since 1950, polyenes, such as amphotericin B, have been used to bind to the ergosterol units of the fungal cell membrane. Changes in the sterol content decrease the binding of the drug and contribute to resistance. Furthermore, an alteration in the fungal cell wall restricts the ability of amphotericin B to reach its target site. Although amphotericin B is highly effective against a wide range of fungi, it must be administered intravenously and is somewhat toxic, producing a range of side effects (Clemons et al., 2002, González et al., 2004, Laniado-Laborı´n and Cabrales-Vargas, 2009).

The azoles (e.g., fluconazole, itraconazole, and ketoconazole) interfere with the biosynthesis of sterols that comprise the fungal cell membrane by inhibiting cytochrome P450-dependent lanosterol 14-demethylase, which is responsible for converting lanosterol to ergosterol (Chakrabarti, 2005). Consequently, the lack of ergosterol in the cell membrane results in cell permeability and death. Each of the azoles has a different spectrum of effectiveness and defined limitations. For example, fluconazole is highly effective against *Cryptococcus*, which is responsible for serious infections and is common in AIDS patients, but is ineffective against *Aspergillus* and has limited effectiveness against certain *Candida* species. Itraconazole has the broadest range of activity and the fewest side effects among the azoles, but has certain drawbacks like unpredictable bioavailability, variation between patients, and frequent drug interactions.

Allylamines (e.g., terbinafine) inhibit squalene epoxidase, another enzyme that is converted to ergosterol. Fluoropyrimidines (e.g., 5-fluorocytosine) are pyrimidine analogs that are selectively converted by a fungal enzyme to an active nucleoside, which interferes with DNA synthesis in fungi. Their spectrum of activity is fairly limited, and drug resistance develops if they are used alone. For that reason, 5-fluorocytosine is utilized in combination with amphotericin B or fluconazole. Toxicity is frequent, however, which includes mucositis and myelosuppression. Usage of these antifungal compounds is becoming limited due to the development of resistance to the azoles and toxicity of the polyenes (Vicente et al., 2003).

The cyclic lipopeptides, known as candins (or echinocandins or pneumocandins), inhibit (1,3)-β-glucan synthase and thus the biosynthesis of the (1,3)-β-glucan layer of the *Candida albicans* cell wall; they are relatively nontoxic. Although the previous derivative, papulocandin, failed due to a spectrum restricted to *Candida* and lack of *in vivo* activity, the parenteral semisynthetic caspofungin of Merck (also known as pneumocandin, L-743,872, MIC 991, or Cancidas), produced by *Zalerion arboricola,* inhibits the same enzyme and was approved in 2000 (Franzot and Casadevall, 1997). It is active against many species of *Candida*, including *C. albicans*, *Aspergillus*, and *Histoplasma* and can be administered as an aerosol for prophylaxis against *P. carinii*, a major cause of death in HIV patients. Pneumocandin is more active and less toxic than is amphotericin B. Another semisynthetic candin is micafungin (FK-463), which was recently approved in Japan. Lilly's anidulafungin (V-echinocandin, LY-303366A), produced by *Aspergillus nidulans* var. *echinulatus*, was licensed to Versicor (now Vicuron Pharmaceuticals) and is awaiting FDA approval. In contrast to the currently used azoles, which are fungistatic, candins are fungicidal (Franzot and Casadevall, 1997, Jekosch et al., 2000).

However, many populations do not benefit from antimycotic drugs, especially in developing countries, where people have to rely on traditional medicines from plant sources. This fact, together with the emerging resistance phenomena, the increased occurrence of fungal strains with multiple antibiotic resistance, and new emerging fungal diseases, fuels interest in developing screening studies for antimycotic natural products (Ficker et al., 2004).

2. Natural Products from Microorganism as Antifungal Agents

2.1. Natural Products Require More-advanced Studies for the Treatment of Fungal Infections

Polyene antibiotics were the first class of antifungal agents that were vital for the treatment of systemic fungal diseases. Within this class, the first clinically relevant compounds were amphotericin B and nystatin, which were isolated from the *Streptomyces* species. Amphotericin B has been used since 1956 and is still considered the gold standard of therapy for systemic fungal infections due to its broad spectrum of activity and fungicidal properties (Clemons et al., 2002, González et al., 2004, Laniado-Laborı´n and Cabrales-Vargas, 2009).

Amphotericin B acts as a fungicide that binds to ergosterol, a steroid present in the membrane of fungi, changing permeability and causing the loss of cytoplasmic constituents. Additionally, it causes oxidative injury that result in metabolic changes that prevent cell survival. To a lesser degree, amphotericin B also binds to the membrane cholesterol of human cells, altering them and causing adverse effects (Hossain and Ghannoum, 2000). The spectrum of etiological agents sensitive to the antifungal effects of amphotericin B includes those of the major endemic mycoses: Paracoccidioides brasiliensis, Histoplasma capsulatum, Coccidioides immitis, Blastomyces dermatitidis, Cryptococcus neoformans, Cr. Gattii, and Sporothrix schenckii. Amphotericin B is effective against the Candida species, although a few samples of the non-albican species can be resistant. It is also effective against Aspergillus fumigatus, but the sensitivity of other species of Aspergillus is variable. Varying susceptibility is also observed in zygomycetes (Mucor and Rhizopus) Fusarium spp. Some microorganisms that cause opportunistic fungal infections are usually resistant to amphotericin B, such as the Trichosporon spp. Pseudallescheria boydii, Cladosporium spp., and Phialophora spp (Andes, 2003).

Oxford, Raistrick and Simonart (1939) isolated a natural product called griseofulvin from the culture filtrates of *Penicillium griseofulvin*. Studies on the griseofulvin-sensitivity of different organisms showed that dermatophytes were very sensitive, whereas bacteria and yeasts were insensitive (Rothf et al., 1959). The spectrum of etiological agents is restricted to *Trichophyton rubrum, T. mentagrophytes, T. interdigitale, T. verrucosum, T. schoenleini, T. sulfureum, Epidermophyton floccosum, Microsporum audouini, M. canis, and M. gypseum.* The mechanism of griseofulvin action is believed to involve the selective inhibition of fungal cell mitosis in association with its accumulation in the keratin layers of the epidermis. Early studies demonstrated that griseofulvin inhibits mitosis in sensitive fungi in a manner resembling the actions of colchicine and other antimitotic drugs that disrupt spindle microtubule (MT) function in mammalian cells; however, its precise mechanism of action in sensitive fungi remains unclear (Panda et al, 2005).

Another natural product shown to be an antifungal agent was echinocandin B, which was isolated from *Aspergillus nidulans*, whereas an additional natural product was pneumocandin B isolated from *Zalerion arboricola*, which was later reclassified as *Glarea lozoyensis* (Schwartz et al., 1995). However, only some semisynthetic derivatives, such as caspofungin, micafungin, and anidulafungin, showed sufficiently favorable solubility, pharmacokinetic profiles and tolerability to be introduced into clinical use (Denning et al, 2002). Echinocandins are fungal secondary metabolites comprising a cyclic hexapeptide core with a lipid side-chain responsible for antifungal activity against (1,3)-β-glucan synthesis (Nyfeler and Keller, 1974). The echinocandins are concentration-dependent, noncompetitive inhibitors of (1,3)-β and (1,6)-β-glucan synthase, an enzyme complex composed of two subunits, encoded by the genes *FKS1* and *FKS2.* Glucan synthase is involved in the synthesis of (1,3)-β-glucan, a polysaccharide composed of 3 helically entwined polymers of glucose. Glucan is an essential carbohydrate component of all fungal cell walls, comprising 30%–60% of the fungal cell wall in *Candida* and *Saccharomyces* (Denning et al., 2002). Changes in the characteristics of the cell wall can lead to osmotic instability and eventual cell lysis. Importantly, human cells do not contain (1,3)-β-glucan, thus avoiding direct human cell toxicity (Denning et al, 2002).

The proportion of the fungal cell wall composed of glucan varies widely between different species of fungi. (1,3)-β-glucan is more predominant in the cell walls of *Candida* and *Aspergillus* species (especially *C. albicans* and *A. fumigatus*) than in yeast forms of dimorphic fungi. Likewise, the cell walls of mycelial forms of *H. capsulatum, B. dermatitidis,* and *P. brasiliensis* contain significant amounts of (1,3)-β-glucan, while zygomycetes lack this target component. However, these characteristics do not always predict echinocandin activity. For example, the cell wall of *Cr. neoformans* contains (1,3)-β-glucan, but the echinocandins demonstrate little activity against this pathogen (Denning, 2003). This observation suggests that there are likely additional (or alternate) components of the mechanism of action of the echinocandins (Feldmesser et al., 2000, Marybeth et al., 2005).

Aureobasidin A (AbA), an antifungal depsipeptide antibiotic produced by *Aureobasidium pullulans*, has strong fungicidal activity against many pathogenic fungi, including *C. albicans*, *Cr. neoformans*, and some *Aspergillus* spp (Takesako et al., 1991, Takesako et al., 1993). *In vitro* aureobasidin A is effective when administered orally to mice with candidiasis with little toxicity to the mice (Takesako et al., 1993). This drug kills only the growing cells of *C. albicans*; the resting cells are unaffected. AbA does not inhibit the synthesis of DNA, RNA, or protein nor does it decrease glucose incorporation into insoluble cell wall glucans of *C. albicans* (Heidler and Radding, 1995). These results suggest that the action of AbA is different from the actions of amphotericin B and other antifungal agents and that AbA inhibits some intracellular processes essential for fungal growth (Endo et al., 1997). Its mechanism of action remained unknown until it was shown that aureobasidin A inhibits the inositol phosphorylceramide synthase from *S. cerevisiae* with an IC50 of approximately 0.2 nM (Zhong et al., 2000).

The crocacins are a group of compounds that are regularly found in the extracts of the bacteria *Chondromyces crocatus* and *Chondromyces pediculatus* strains and which represent a second novel group of modified peptides from *C. crocatus* (Dias et al., 2005, Di Santo, 2010). Crocacin D shows higher biological activity against *S. cerevisiae* and higher toxicity in L929 mouse fibroblast cell cultures when compared to crocacins A-C. Crocacin A showed a minimal inhibitory concentration (MIC) of 0.625 μg/mL against both *Ustilago maydis* and *S. cerevisiae,* whereas crocacin D was more potent against *S. cerevisiae*, showing an MIC of 1.4 ng/mL. The crocacins A and D are novel natural products that inhibited the electron transport chain at complex III in a mitochondrial respiration assay (Dias et al., 2005, Di Santo, 2010).

Another metabolite, cispentacin, was isolated from the culture broth of a *Bacillus cereus* strain (L450-B2) (Konishi et al., 1989). This antibiotic demonstrated weak activity against certain fungi *in vitro*, but it effectively protected mice from lethal infection by *C. albicans*. The structure–activity relationship (SAR) studies of this lead compound led to the development of (1R, 2S)-2-amino-4-methylene-cyclopentane carboxylic acid (icofungipen). Icofungipen showed high antifungal activity against *C. albicans* (IC50 0.13 mg/mL) *in vitro* and was shown to target protein biosynthesis through the inhibition of isoleucyl tRNA synthetase. Phase IIb studies of efficacy indicate its potential for salvage therapy in the oral treatment of azole-resistant oropharyngeal and esophageal candidiasis (Konichi et al., 1989, Di Santo, 2010).

Sordarin was isolated in 1969 from the fungus *Sordaria araneosa* by scientists at the Sandoz Co., in Switzerland, and it was patented as SL 2266. The degradation of sordarin with concentrated aqueous hydrochloric acid in acetone released a diterpenoid aglycone called sordaricin (Sigg and Stoll, 1969, Hauser et al., 1971). Sordarin production by fermentation was optimized to simplify purification and to increase the yield. Interestingly, early antifungal screens in the 1970's excluded sordarin, but two decades later, a renewed appreciation of this natural product arose as a consequence of its potent inhibition of protein synthesis in *C. albicans in vitro* (Hauser et al., 1971, Harris et al., 2001, Tully et al., 2007).

Unlike traditional antifungal agents, which target only the integrity of the cell membrane through binding of ergosterol or inhibition of its biosynthesis, sordarin acts on elongation factor 2 (EF2) (Justice et al., 1998, Shastry et al., 2001, Anderson et al., 2003). This enzyme catalyzes the translocation of the ribosome along mRNA during elongation of the emerging polypeptide chain. Sordarin inhibits this translocation by stabilizing the EF2/ribosome complex. Strong activity against *S. cerevisiae*, *C. albicans*, and a number of pathogenic fungi make sordarin a promising antimycotic agent (Liu et al., 2006). These compounds are endowed with relatively high antifungal activities and are considered an interesting class of antifungal agents due to their new mechanism of action (Di Santo, 2010).

Pyrrolnitrin, a chloropyrrole, was isolated from *Pseudomonas pyrrocinia* (Arima et al., 1964). Pyrrolnitrin exhibited high antibiotic activity against dermatophytic fungi, particularly members of the genus *Trychophyton,* many soil-borne fungal plant pathogens, such as *Rhizoctonia solani* and *Fusarium sambucinum,* and foliar fungal pathogens, such as *F. graminearum, F. culmorum* and *Pyrenophora tritici repentis*. This compound is marketed in Japan under the name PYRO-ACE for the treatment of superficial dermatophytic infections. Its sensitivity to light prevented the use of pyrrolnitrin as a fungicide in plant protection. Recently, antimycobacterial activity was reported for related compounds (Burkhead et al., 1994).

At low concentrations, the biological activity of pyrrolnitrin was demonstrated to be due to the uncoupling of oxidative phosphorylation in *Neurospora crassa*; at higher concentrations, biological activity was due to the inhibition of electron transport in the flavin region and through cytochrome coxidase. However, it was recently reported that the use of pyrrolnitrin leads to glycerol accumulation and the stimulation of triacylglycerol synthesis, resulting in leaky cell membranes and a concomitant breakdown of biosynthetic activity, followed by the cessation of cell growth (Burkhead et al., 1994).

Rustmicin (galbonolide A) is a 14-member macrolide produced by the bacteria *Micromonospora chalcea* and *Streptomyces galbus* (Di Santo, 2010). Rustmicin was previously identified as an inhibitor of plant pathogenic fungi by an undefined mechanism. Mandala et al. (1998) discovered that rustmicin inhibits inositol phosphoceramide synthase, resulting in the accumulation of ceramide and the loss of complex sphingolipids. Rustmicin has potent fungicidal activity against clinically important human pathogens that are correlated with its sphingolipid inhibition (Mandala et al., 1998). It is especially potent against *C. neoformans*, where it inhibits growth and sphingolipid synthesis at concentrations < 1 ng/mL and inhibits the enzyme with an IC50 of 70 nM. This inhibition of the membrane-bound enzyme is reversible moreover, rustmicin is nearly equipotent against the soluble enzyme. Rustmicin was efficacious in a mouse model for cryptococcosis, but it was less active than its potency against this pathogen *in vitro*. Stability and drug efflux were identified as two factors limiting rustmicin's activity. In the presence of serum, rustmicin rapidly epimerizes at the C-2 position and is converted to g-lactone, a product that is devoid of activity (Mandala et al., 1998).

Khafrefungin, which is composed of aldonic acid linked via an ester to a C22 modified alkyl chain, has fungicidal activity against *C. albicans*, *Cr. neoformans*, and *S. cerevisiae* (Mandala et al., 1997). Sphingolipid synthesis is inhibited in these organisms at the step in which phosphoinositol is transferred to ceramide, resulting in the accumulation of ceramide and loss of complex sphingolipids. *In vitro*, khafrefungin inhibits the inositol phosphoceramide synthase of *C. albicans* with an IC50 of 0.6 nM. Khafrefungin does not inhibit the synthesis of mammalian sphingolipids (Mandala et al., 1997).

Another natural product that was isolated from *Fusarium larvarum* is the Parnafungins (Figure 1), which contain an isoxazolidinone ring and have been shown to be potent inhibitors of fungal polyadenosine polymerase (Xu et al., 2007, Bills et al., 2008).

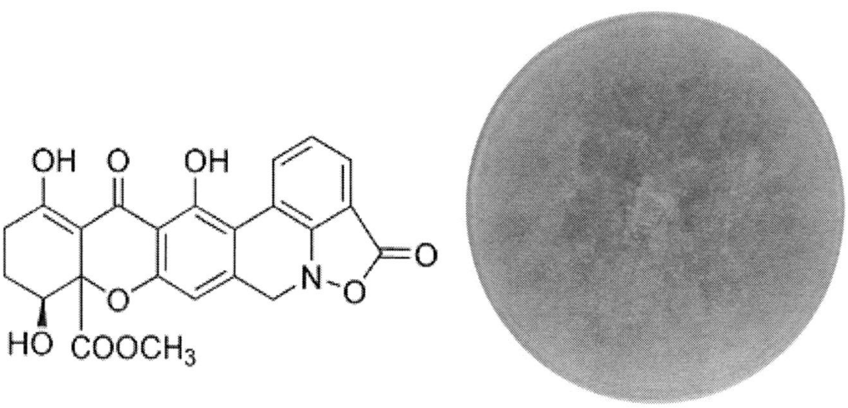

Figure 1. Chemical structure of the parnafungins and *Fusarium larvarum*.

The extraction and analysis of fermentation broths of taxonomically-related organisms identified as closely related *Fusarium* spp. produce not only parnafungin A and B but also significant quantities of two related components (Xu et al., 2007, Bills et al., 2009). These members of the paranfungin family of natural products have been isolated, and the structure of each has been elucidated. While structurally analogous to parnafungin A, parnafungin C is further elaborated by the methylation of a phenolic hydroxyl group, and parnafungin D has both the methyl phenol ether and an epoxide in the xanthone ring. Parnafungins C and D have potent broad-spectrum antifungal activity and have been shown to target fungal mRNA cleavage and polyadenylation (Overy et al., 2009).

Table 1. Source, compound and mechanism of action of the antifungal agents

Source	Compound	Mechanism of action	Reference
Streptomyces nodosus	Amphotericin B	Acts as fungicide binding to ergosterol, a steroid present membrane of fungi	Gold et al., 1956
Penicillium griseofulvin	Griseofulvin	Inhibition of fungal cell mitosis, disrupting spindle microtubule (MT) function	Oxford, Raistrick and Simonart (1939)
Aspergillus nidulans	Echinocandin B	Activity against b-(1,3)-D-glucan synthesis, lead to osmotic instability and eventual cell lysis	Nyfeler and Keller (1974)
Glarea lozoyensis	Pneumocandin B	Activity against b-(1,3)-D-glucan synthesis, lead to osmotic instability and eventual cell lysis	Nyfeler and Keller (1974)
Aureobasidium pullulans	Aureobasidin A	Inhibits the IPC synthase	Takesako et al., (1991)
Chondromyces crocatus *Chondromyces pediculatus*	Crocacins	Inhibits the electron transport chain at complex III in a beef heart mitochondrial respiration assay	Jansen et al., (1999)
Bacillus cereus	Cispentacin	Target protein biosynthesis through inhibition of isoleucyl tRNA synthetase	Konichi et al. (1989)
Sordaria araneosa	Sordarin	Inhibits the translocation of the ribosome along mRNA during elongation of the emerging polypeptide chain	Sigg and Stoll. (1969)
Micromonospora chalcea *Streptomyces galbus*	Rustmicin	Inhibits inositol phosphoceramide synthase, resulting in the loss of all of the complex sphingolipids.	Mandala et al. (2000)
Unidentified sterile endophyte fungus	Khafrefungin	Inhibits sphingolipid synthesis, resulting in accumulation of ceramide and loss of all of the complex sphingolipids	Mandala et al. (1997)
Fusarium larvarum	Parnafungins	Inhibitors of the fungal polyadenosine polymerase	Xu et al. (2007)

2.2. Novel Natural Products with Antifungal Activities

The work described below is in the initial phase; thus, we did not find any reference to the mechanism of action of the isolated compound. The mechanism for some compounds was already evaluated in *vivo*, but the majority of the information collected cites only the isolation and antifungal activity.

Coronamycin, a complex of novel peptide antibiotics with activity against pythiaceous fungi and the human fungal pathogen *Cr. neoformans*, is produced by a verticillate *Streptomyces* spp. It was also active against the malarial parasite, *Plasmodium falciparum*, with an IC50 of 9.0 ng/mL (Ezra et al., 2004). Recently, two antibiotics were reported by chemical studies of an organic extract from maize kernel fermentations of *Acremonium zeae* (NRRL 13540), which displayed significant antifungal activity against *Aspergillus flavus* and *Fusariwn verticillioides*. Wicklow and Poling (2009) revealed that the metabolites accounting for this activity were the two antibiotics pyrrocidines A and B.

Some publications reported the presence of chlorflavonin, a new antifungal antibiotic, in the cultures of *Aspergillus candidus*. Chlorflavonin is a flavonoid isolated from fungi with demonstrated biological activities against *Candida* and the *Aspergilus* sp (Burns et al., 1979, Burns et al., 1981, Munden et al., 1970).

The crude extract of the endophytic fungus *Fusarium* sp. was able to inhibit several clinical strains of *P. brasiliensis* and was active in a bioautographic assay against *Cladosporium sphaerospermum* (Campos et al., 2010). The endophytic fungus was isolated from the plant *Piptadenia adiantoides* J.F. Macbr (Fabaceae). The fungus was identified as *Fusarium* sp. based on its macro- and micro-morphology (Figure 2) and on the sequence of the internally transcribed spacer regions (ITS) of its rRNA gene (Campos et al., 2010).

The chromatographic fractionation of the fungal extract, guided by the bioautographic assay, produced three known trichothecene mycotoxins (Figure 3). The MIC of the these compounds against eleven clinical strains of *P. brasiliensis* was evaluated and found to be in the range between 75 and 640 nmol for *1* and 160–640 nmol for the mixture of *2* and *3* (Campos et al., 2010).

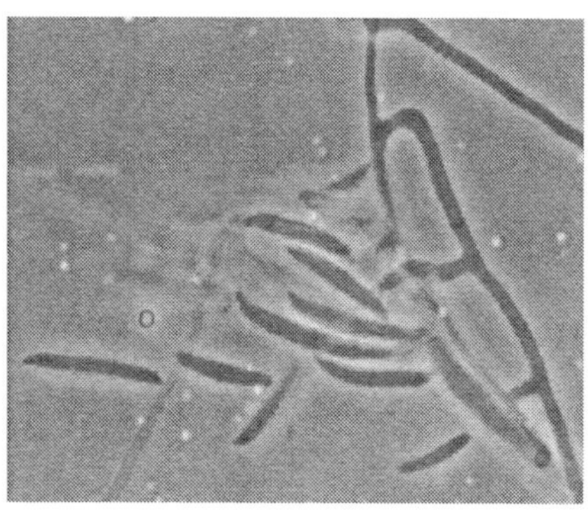

Figure 2. Micrograph of macroconids from the *Fusarium* sp. in Czapek medium after 14 days culture at 28 ± 2°C.

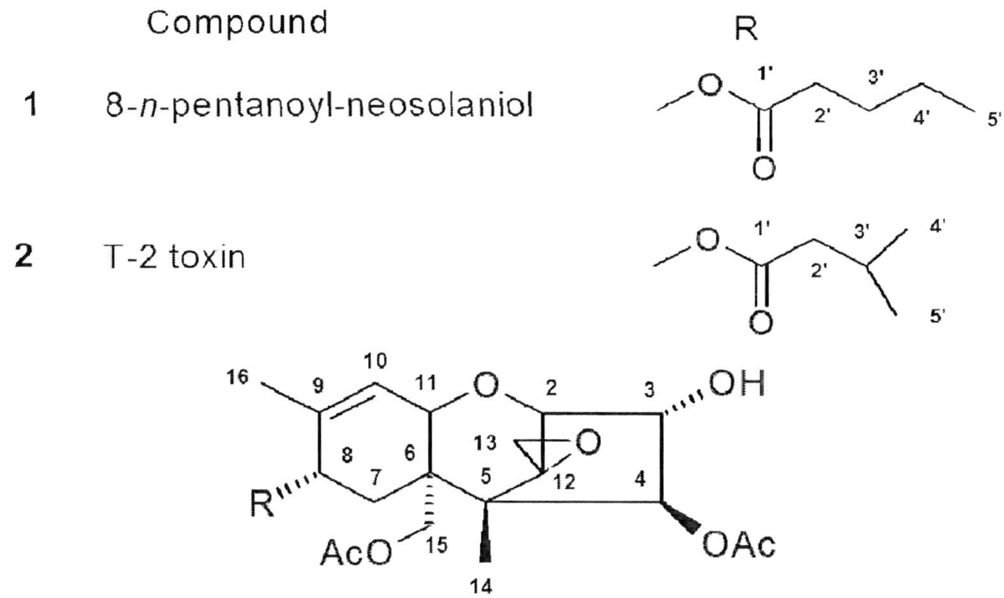

Figure 3. Compounds obtained from *Fusarium* sp.

Cryptosporiopsis cf. quercina is the imperfect stage of *Pezicula cinnamomea*, a fungus commonly associated with hardwood species in Europe. This fungus and related species occur as endophytes in many parts of the world. It was isolated as an endophyte from *Tripterigeum wilfordii*, a medicinal plant native to Eurasia (Strobel et al., 1999). *In vitro*, *C. quercina* demonstrated excellent antifungal activity against some important human fungal pathogens *C. albicans* and the *Trichophyton* spp. A unique peptide antimycotic, termed cryptocandin, was isolated and characterized from *C. quercina* (Strobel et al., 1999).

This compound contains a number of peculiar hydroxylated amino acids and a novel amino acid-3-hydroxy-4-hydroxy methyl proline. The bioactive compound is related to the known antimycotics, echinocandins and pneumocandins. Cryptocandin is also active against a number of plant pathogenic fungi, including *Sclerotinia sclerotiorum* and *Botrytis cinerea* (Strobel et al., 1999). Cryptocandin is currently being tested and developed by several companies for use against a number of fungi that cause diseases of the skin and nails (Strobel et al., 1999, Strobel, 2003).

The ecomycins are produced by *Pseudomonas viridiflava* (Miller et al., 1998). *P. viridiflava* is a member of a group of plant-associated fluorescent bacteria that are either weakly pathogenic or saprophytic. *P. viridiflava* is generally associated with the leaves of many grass species and is located on and within the tissues. The ecomycins represent a family of novel lipopeptides and have molecular weights of 1153 and 1181. Besides common amino acids, such as alanine, serine, threonine, and glycine, some unusual amino acids are also involved in the structure of the ecomycins, including homoserine and b-hydroxyaspartic acid. The ecomycins are active against such pathogenic fungi as *Cr. neoformans* and *C. albicans*. Another group of antifungal compounds is the pseudomycins, produced by a plant-associated pseudomonas (Harrison et al., 1991, Strobel, 2003).

3. Biological Activity and Therapeutic Applications of Biosurfactants

The biosurfactants present different natural roles in microbial physiology and behavior. The study of these functions is important because they are the basis for the development of biotechnologies employing these biomolecules in the medical and pharmaceutical area. Although biosurfactants are considered less toxic and more environmentally favorable, compared with chemical surfactants in industrial applications, these molecules can have antimicrobial activity. It is likely that microorganisms produce molecules such as antagonistic agents to gain a competitive advantage during colonization and cell–cell competition in the natural environment (Ron and Rosenberg, 2001, Hamme et al. 2006). Due to these properties and an increasing concern over the emergence of various multi-drugs resistant pathogens, these molecules have emerged as potential drug molecules. It was also found that, in addition to their detergent activity, many surfactants could cause membrane disruption, leading to cellular lysis, by increasing membrane permeability and causing metabolite leakage, by altering the physical membrane structure or disrupting protein conformations, thus interfering with important membrane functions, such as energy generation and transport (Thimon et al., 1995, Hutchison and Gross, 1997). In the case of enzymes, such interactions can modify enzyme specificity/activity and/or stability characteristics.

3.1. Glycolipids

The glycolipids are the most common biosurfactants that have been isolated and studied. They are carbohydrates in combination with long-chain aliphatic acids or hydroxy aliphatic acids. The best-studied glycolipids are rhamnolipids synthesized by *Pseudomonas* sp. (Jarvis and Johnson, 1949); mannosylerythriol lipids synthesized by *Pseudozyma* (previously *Candida) antarctica* (Kitamoto et al., 1990); trehalose lipids synthesized by *Rhodococcus* sp., *Nocardia* sp. and *Mycobacterium* sp. (Lang and Wagner, 1987); and sophorolipids synthesized by yeasts (Cooper and Paddock, 1984).

Abdel-Megeed et al. (2011) described that glycolipids produced by *Rhodococcus erythropolis* exhibited high inhibitory activity against the bacteria *Aspergillus niger* and *A. flavus*; the cells presented a considerable reduction of wrinkling and deformation. The fungal cells were observed in damaged pellets and somewhat resembled pseudohyphae that were highly irregular and aborted.

Mannosylerythritol lipid (MEL) contains a 4-O-β-D-mannopyranosylmeso-erythritol hydrophilic group and a hydrophobic fatty acid and/or an acetyl moiety (Kitamoto, 2008). The fatty acid chains can be short (2 to 8 carbon atoms) or long (10 to 18 carbon atoms). MEL is reported to be secreted by *Ustilago* sp. as a minor component along with cellobiose lipid (Haskins et al., 1955), *Schizonella melanogramma* (shizonellin) (Deml, 1980), *Geotrichum candidum* (Kurz et al., 2003) and *Pseudozyma* (previously Candida) *antarctica* as a major component (Kitamoto et al., 1990). The type of MEL produced varies with the strains and the species of *Pseudozyma*. Hence, its production can be used as an important taxonomic index to identify *Pseudozyma* yeasts. Based on the degree of acetylation at the C4 and C6 positions and their order of appearance by thin layer chromatography (TLC), the MELs are classified as MEL-A, -B, -C and -D. MEL-A represents the diacetylated compound, whereas MEL-B and MEL-C are monoacetylated at C6 and C4, respectively. The completely deacetylated structure is known as MEL-D (Rau et al., 2005).

These compounds are believed to facilitate the dissolution of organic hydrophobic compounds for consumption; to present antimicrobial, immunological and neurological properties; and to induce the differentiation of certain cancer cells (Kitamoto et al., 1993, Arutchelv et al. 2008). Mixtures of these four MELs (particularly MEL-A and MEL-B) isolated from *Pseudozyma* (previously Candida) *antarctica* T34 have shown excellent surface and antimicrobial activity, including activity against *C. albicans* (Kitamoto et al., 1993). Other MELs, accumulated in the cell of *S. melanogramma,* also exhibit antibacterial and antifungal activities (Deml al., 1980).

In addition to MEL production, *Ustilago maydis* secretes a second class of glycolipids, the cellobiose lipid ustilagic acid. Haskins, who observed the formation of an insoluble compound with antibiotic activity in the submerged culture of *U. maydis,* first described ustilagic acid in 1950. The sugar moiety of ustilagic acid is the disaccharide cellobiose, which is O-glycosidically linked to the ω-hydroxyl group of the unusual long-chain fatty acid 15,16-dihydroxyhexadecanoic acid or 2,15,16-trihydroxyhexadecanoic acid. In addition, the cellobiose is esterified at several positions with either acetyl or medium-chain hydroxy fatty acids. Cellobiose-derived glycolipids have since been detected in other fungal species. The fungal biocontrol agent, *P. flocculosa* produces flocculosin, a rare cellobiose lipid with antifungal activity (Cheng et al., 2003); *Metschnikowia pulcherrima* produces pulcherrimic acid (Nguyen, 1998); and the cellobiolipids of *Cryptococcus humicola* (Puchkov et al., 2002), *P. fusiformata* (Kulakovskaya et al., 2009) and *Trichosporon porosum* (Kulakovskaya et al., 2010), are responsible for fungicidal action. Taxonomically close *T. porosum* and *Cr. Humicola* of the order Trichosporonales produce cellobiose lipids of a similar structure, which differs from the structure of cellobiose lipids produced by representatives of the genus *Pseudozyma* and *Ustilago* of the order Ustilaginales. *T. porosum* secreted mainly 16-(tetra-*O*-acetyl-β-cellobiosyloxy)-2-hydroxyhexadecanoic acid and 16-(tri-*O*-acetyl-β-cellobiosyloxy)-2-hydroxyhexadecanoic acid. *Cr. Humicola* secreted 16-(tetra-*O*-acetyl-β-cellobiosyloxy)-2,15-dihydrohexadecanoic acid and 16-(tetra-*O*-acetyl-β-cellobiosyloxy)-2-hydroxyhexadecanoic as major products, whereas *P. fusiformata* secreted 16-[6-*O*-acetyl-2′-*O*-(3-hydroxyhexanoyl)-β-cellobiosyloxy)-2,15-dihydroxyhexadecanoic acid (Puchko et al. 2002, Kulakovskaya et al. 2005, 2007 and 2009). These compounds exhibit similar fungicidal activities against many species of yeasts and mycelial fungi, including known candidosis and cryptococcosis agents, such as *Candida albicans* and *Filobasidiella (previously Cryptococcus) neoformans* (Golubev et al., 1994, Puchkov et al., 2002, Cheng et al., 2003, Kulakovskaya et al. 2005). The cellobiose lipids can induce the cell death of yeast and mycelial fungi. This effect is conditioned by their membrane-damaging properties. The cellobiose lipids were shown to intercalate into the liposomal lipid matrix (Puchkov et al. 2002). This intercalation results in the permeabilization of the lipid bilayer of the cytoplasmic membrane of yeast cells, which results in ATP leakage and the susceptibility of the treated cells to staining with bromocresol purple (Puchkov et al., 2002, Kulakovskaya et al., 2003). These compounds present maximal activity at acidic pH values, thermal stability and protease-resistance (Golubev et al., 1994 and 2001, Puchkov et al., 2002, Kulakovskaya et al. 2005, and low toxicity for animal cells (Cheng et al., 2003, Mimee et al., 2005). The acidic pH optimum may be explained by the necessity of carboxyl group ionization for this activity activity (Kulakovskaya et al. 2009).

Golubev et al. (2001) reported that the glycolipid produced by *Pseudozyma fusiformata* VKM Y-2821 was active against more than 80% of the 280 yeast and yeast-like species tested under acidic conditions at 20–30°C, including medically important species of the genera *Filobasidiella, Malassezia and Trichosporon.*

In the study conducted by Kulakovskaya et al. (2006), Basidiomycetous yeasts were more sensitive to the glycolipids produced by *P. fusiformata* than were ascomycetous ones. The minimal effective glycolipid concentration was 0.13 and 0.26 mg ml^{-1} for *Cr. terreus* and *Filobasidiella (previously Cryptococcus) neoformans*, whereas for *C. albicans* and *Saccharomyces cerevisiae,* it was 1.0 and 1.6 mg ml−1. In addition, the cells of *S. cerevisiae* lost the ability to acidify the medium. It has been suggested that a difference in the cytoplasmic membrane lipid composition among ascomycetous and basidiomycetous yeasts is probably an important factor that influences the level of sensitivity (Kaneko et al., 1976, Kim et al., 1975). In addition, a difference in the cell envelope composition and surface charge might also cause variations in the sensitivity of a yeast cell to membrane-damaging compounds (Ivanov et al., 1996).

The yeast *T. porosum* suppresses the growth of ascomycetes and basidiomycetes belonging to 52 genera (Kulakovskaya et al., 2010). Additionally, the suppression was maximal at pH 3.5–4.0. Fungicidal preparation obtained from the culture broth was shown to be a mixture of cellobiosides of dihydrodecane acid with different degree of acetylation of cellobiose residue. The preparation caused the death of *C. albicans* and *Filobasidiella (previously Cryptococcus) neoformans,* cells at concentrations of 0.2 and 0.03 mM, respectively. These studies conducted with cellobiose lipids of different structures, which present different spectra of activities, suggested that the structures of both the carbohydrate and fatty acid aglycon moieties are important for the fungicidal activity of these bio-molecules.

Rhamnolipids (RLs) are a mixture of a number of analogous molecules. The four most important molecules comprise either one or two rhamnose units linked glycosidically to a 3-hydroxy acid, thence by an ester bond to a further 3-hydroxy acid (Syldatk et al., 1985). 3- or β-Hydroxydecanoic acid is the most common fatty acid constituent, but other fatty acids may be found depending on the *Pseudomonas* species or growth conditions, including 12:0, 12:1, 12:2 and 8:2 (each with a 3-hydroxyl group), resulting in a number of distinct molecular species. All of these lipids have antifungal and antiviral properties, and they exhibit bactericidal properties to Gram-positive bacteria. However, they are considered as one of the virulence factors in *Pseudomonas sp.* (Abdel-Mawgoud et al. 2010). They also found inhibitory activity against a range of fungal species, including the fungi *Chaetomium globosum, Aureobasidium pullulans*, and *Gliocladium virens*, and the phytopathogens *Botrytis cinerea, Pythium sp., Phytophora sp, Plasmopara sp* and *Rhizoctonia solanii* (Kim et al. 2000, Abalos et al. 2001, Haba et al. 2003, Benincasa et al. 2004). In these studies, the biosurfactant was observed to cause zoospore lysis, inhibition of zoospore and spore germination, and hyphal growth inhibition. Although the studies evaluating antifungal activity have been conducted with plant pathogenic fungi, the mechanisms of action provided here suggest that they might be effective against fungal animal pathogens.

Sophorolipids, which consist of the sugar sophorose linked to a long-chain hydroxy fatty acid, are synthesized by a phylogenetically diverse group of yeasts., including *C. apicola* (Gorin et al., 1961), *C. bombicola* (Spencer et al., 1970), *C. batistae* (Konoshi et al., 2008) and *C. riodocensis, C. stellata* and *Candida sp.* NRRL Y-27208 (Kurtzman et al. 2010) of the *clade Starmerella, Wickerhamiella domercqiae* (Chen et al., 2006) and the basidiomycetous yeast *Rhodotorula bogoriensis* (Tulloch et al., 1968). These glicolipids have antimicrobial activity against certain yeasts (Ito et al., 1980), plant pathogenic fungi (Lang and Wagner, 1993, Yoo et al., 2005) and bacteria (Lang et al., 1989) and have antiviral and anticancer activity (Shah et al., 2005, Chen et al., 2006). A lactonic sophorolipid isolated from the culture broth of *Torulopsis bombicola* by extraction with EtOAc was identified as 17-L-[(2'-O-b-D-glucopyranosyl-b-D-glucopyranosyl)oxy] octadecenoic acid 1,4''-lactone 6',6''-diacetate. This biomolecule inhibited the growth of 19 species of 5 genera of yeasts *Candida, Pichia, Debaryomyces, Saccharomycopsis* and *Lodderomyces* on the hexadecane, even at a very low concentration (0.4 or 0.8 mM) (Ito et al., 1980).

3.2. Lipopeptides

Lipopeptides are low molecular mass biosurfactants, including surfactin, iturin, lichenysins, mycosubtilin, arthrofactin and fengycin (Hsieh et al., 2008), produced by a wide spectrum of microorganisms including species of *Bacillus*, *Streptomyces*, *Pseudomonas* and *Arthrobacter* (Lee et al., 2006, Kim et al., 2010). The characteristic structural element of lipopeptides is a specific fatty acid, which is combined with an amino-acid moiety (Banat et al., 2000). The length of the fatty acid chains can vary from C_{13} to C_{16} for surfactins, from C_{14} to C_{17} for iturins, and from C_{14} to C_{18} for fengycins, giving different homologous compounds and isomers (n, iso, anteiso) for each lipopeptide. Beside tenso-actives properties, lipopeptides play an important role in biological activities, such as antibacterial, antifungal or antiviral activity, cytolytic activity, inhibition of fibrin clot formation, macrophage stimulating activity and enzyme inhibitors. The mode of action of most of these compounds has not been clarified in detail, although it is likely that their surface- and membrane-active properties play an important role. Surfactin, the best-known lipopeptide exhibits a large antibacterial spectrum, and other lipopeptides exhibit antifungal properties, including iturins (Delcambe and Devignat, 1957), bacillomycins (Peypoux et al., 1981), and mycosubtilins (Besson and Michel, 1990). By scanning electron and optical microscopies, Thimon et al. (1995) discovered iturin A produced by *B. subtilis,* is an antifungal pore-forming cyclic heptapeptide with a 14–16 carbon atom side chain that also forms small vesicles in the plasma membrane of yeast resulting in metabolite leakage. Iturin A is able to penetrate the cell wall without causing damage, whereas the membranes of cytoplasmic organelles are affected. Moreover, it also significantly increases the electrical conductance of biomolecular lipid membranes. Iturin A has been proposed as an effective antifungal agent for profound mycosis (Tanaka et al., 1997). Other members of the iturin group, including bacillomycin D and bacillomycin Lc, were also found to have antimicrobial activity against *Aspergillus flavus*, but the different lipid chain length apparently affected the activity of the lipopeptide against other fungi (Moyne et al., 2001). Bechard et al. (1998) isolated an antimicrobial lipopeptide from a strain of *B. subtilis* found in apples that demonstrated a broad spectrum of activity against Gram-negative bacteria, little activity against Gram-positive organisms and was active against the fungi *Penicillium expansum*.

Hino et al. (2001) discovered a novel series of lipopeptide compounds structurally related to but highly superior to echinocandin B. Echinocandin-type compounds are lipopeptides consisting of a cyclic hexapeptide nuclear structure acylated with a long fatty acid chain. The producing strains were classified into two groups, the *Coleomycetes* group and the *Hyphomycetes* group. Compound FK463, a derivative of WF11899A, is currently in Phase 3 clinical development as a novel antifungal antibiotic.

3.3. Anti-adhesive Activities of Biosurfactants

Biosurfactants are amphipathic molecules with both hydrophilic and hydrophobic moieties that partition preferentially at the interface between fluid phases that have different degrees of polarity and hydrogen bonding, such as oil and water or air and water interfaces. These molecules can spontaneously adsorb to surfaces forming a film at the interfaces with the environment through the orientation of the polar and nonpolar groups, according with the hydrophilic/hydrophobic character of the surface. This mechanism is called conditioning film (Neu, 1996). During its formation, microbial biosurfactant can be excreted into the environment outside the cell and form a microbial film on a solid surface, interacting with the surface by hydrophobic or hydrophilic portion, resulting in a hydrophobic and hydrophilic interface, respectively. This interaction of surfactants with a substratum surface interferes with the microbial adhesion and desorption processes (Rodrigues et al. 2006a). An example of this mechanism was observed for *Serratia marcescens*, which produces lipopeptides capable of producing a hydrophobic interface on hydrophilic surfaces, causing their detachment from the micro areas (Matsuyama et al., 1989). In addition, biosurfactants can form a film on the cell connected to the hydrophobic groups in the free cell surface, altering its hydrophobic character (Neu, 1996). Due to their affinity to surfaces, biosurfactants are being used as a new strategy for inhibiting the adhesion of pathogenic microorganisms, including fungal to non-biological surfaces or to infection sites (Rodrigues et al., 2006c, d).

For example, *C. albicans* possesses various virulence attributes that may represent potential targets for new antifungal treatments. Many authors have suggested that the virulence of *C. albicans* could be attributed, at least in part, to the filamentous form following a yeast-hyphal transition (Jones et al., 1994, Soll, 2002). A correlation was established between the degree of antifungal drug resistance of *C. albicans* strains and their capacity to form hyphae in the presence of azole antifungal drugs (Ha and White, 1999). Many studies have reported that a higher frequency of hyphae phenotype of *C. albicans* exists in fungal infection sites, which facilitates soft tissue invasion and allows the microorganisms to hide from the host defense system sites (Soll et al., 1987, Jones et al. 1994), Biofilm formation is also an important property that helps *C. albicans* to cause many types of infections (Seneviratne et al., 2008). *Candida albicans* is the fungal microorganism most associated with biofilm formation, and these biofilms have been reported to be 30–2000 times more resistant to various antifungal agents, compared with their planktonic counterparts (Hawser and Douglas, 1995). Thus, control of infections caused by yeasts of the genus *Candida* requires strategies that combine the use of antifungal agents with substances that block the adhesion to host cells.

It has been suggested that the early events of *C. albicans* attachment may involve both nonspecific factors, like hydrophobicity of the surface and electrostatic forces between cells and surface, and specific adhesin molecules at the surface of *Candida*, recognizing their ligands in the conditioning film, and/or chemical signaling between cells. The initial attachment to a surface is followed by cell division, proliferation and the development of biofilm (Ramage et al., 2006). Several authors have shown that the formation of germ tubes is related to increased cell hydrophobicity of the strains of *C. albicans* and represent a mechanism of adhesion to biological surfaces, but also to inert surfaces, such as a variety of plastic materials, including acrylic used to make dental prostheses. Monteiro et al. (2011) demonstrated that the incubation of the *C. albicans* CC cells in the presence of glycolipid polymeric biosurfactant inhibited the germ tubes or hyphae formation during the biofilm formation on polystyrene plate surface, and the cells showed lower values of hexadecane adhesion (MATH), compared with the cells incubated without biosurfactant. Biofilm formation was reduced up to 87.4 % with use of the biosurfactant at a concentration of 16 mg/ml. The lipid portion of glycolipid contains16.03% (9Z)-octadec-9-enoic acid, 14.92% 35 hexadecanoic acid, and 9.63% (E) octadec-9-enoic acid and the carbohydrate 36 portion containing mannose (35.29%), xylose (41.99%), arabinose (17.47%), and 37 glucose (5.25%) (Monteiro et al., 2009). Monteiro et al. (2011) observed that *C. albicans* CC cells with negative charges form biofilms in the untreated polystyrene plates (hydrophobic surfaces and uncharged), which can bind microbial cells mainly through passive interactions. The addition of glycolipid produced by *Trichosporon montevideense* CLOA72 to plates decreased the hydrophobicity of their surfaces, and the authors suggested that biosurfactant connects to the polystyrene surface forming a hydrophilic interface, thus inhibiting the adhesion of *C. albicans* CC. ITC and zeta potential study of the interaction of glycolipid with *C. albicans* CC cells showed that biosurfactants bind to *C. albicans* CC cells by an entropically driven mechanism. The inhibition of adhesion can be result of both free surfactant molecules and micelles that might be adsorbed on the cell surface of microorganisms and solid surfaces blocking or facilitating microbial adherence to hydrophobic and hydrophilic interfaces (Neu, 1996). This phenomenon can be been attributed also to the ability of biosurfactant to disrupt negatively charged cell wall surface; our zeta potential analysis shows a negative effect of biosurfactant on the negative charge of the *Candida* cell surface. Our group also demonstrated also that the adhesion of the clinical isolate of *C. albicans* to epithelial bucal cells was reduced up to 87.4% with use of biosurfactant at a concentration of 16 mg/ml. Measurements of the zeta potential of the cell surface showed a partial neutralization of net surface charge; the value of zeta potential ranged from -20,1 mV to -8,3 mV during the titration with biosurfactant.

Some studies have been conducted with dairy *Streptococcus thermophilus* strains producing biosurfactants that cause reduction of yeasts adhesion to silicone rubber in a parallel plate flow chamber system and interference with the formation of a mixed fungal/bacterial biofilm on silicone rubber voice prostheses in the modified Robbins device (Busscher et al. 1997 and 1999, Rodrigues et al., 2004a,b).

The effect of biosurfactants obtained from probiotic bacteria, *Lactobacillus lactis* 53 and *S. thermophilus* A, and also a rhamnolipid obtained from *P. aeruginosa* DS10–129, on the adhesion of bacteria and yeast strains isolated from explanted voice prostheses to silicone rubber with and without an adsorbed biosurfactant layer in a parallel plate flow chamber has been studied (Rodrigues et al., 2004a,b and 2006e). Both biosurfactants reduced the numbers of bacteria adhering to silicone rubber after 4 h by approximately 90%, except for *C. albicans* GBJ 13/4A, *C. tropicalis* GB 9/9 and *R. dentocariosa* GBJ 52/2B, which showed less reductions ranging between 56 and 70%. Additionally, Rodrigues et al. (2006a, e) demonstrated the activity of inhibiting biofilm formation by biosurfactants produced *S. thermophilus,* suggesting that this biosurfactant could be used as a biodetergent solution for prostheses cleaning, prolonging their lifetime and directly benefiting laryngectomized patients.

Another group of biomolecules with the potential to inhibit adhesion is the hidrophobins, formed by small, relatively hydrophobic fungal proteins with the ability to convert hydrophobic surfaces to hydrophilic ones and to convert hydrophilic surfaces to hydrophobic ones by self-assembly into an amphipathic protein membrane (Wessels, 1997). They also belong to the most surface-active molecules known (Wösten and de Vocht, 2000). The surface activity of hydrophobins is comparable to that of the commercial synthetic surfactants and other biosurfactants. In nature, these exceptional characteristics serve different functions. The hydrophobin membrane covers emergent fungal structures, making the cell wall hydrophobic, which facilitates the attachment of hyphae to hydrophobic surfaces, aerial growth of the hyphae, spore dispersal, and proper gas exchange in fungal air channels (Wösten and Wessels 1997). The high surface activity of hydrophobins enables the fungal hyphae to escape from liquid media to the air (Wösten et al. 2000). For medical applications, hydrophobins could be used in combination with antimicrobial compounds to prevent microorganisms from adhering to surfaces such as catheters and to improve the growth of human fibroblasts (Hektor and Scholtmeijer, 2005) (for example, SC3 fused with the cell-binding domain of fibronectin, RGD, supports the growth of human fibroblasts on a coated Teflon surface). Neither soluble-state nor assembled SC3 showed toxicity to the cultured cells, nor did they seem to be immunogenic. For nanotechnological applications, hydrophobins, carrying different functional groups after certain modifications, might be used to pattern different molecules on a surface with nanometre accuracy, given that the hydrophobin rodlets are approximately 10 nm wide and are chemically highly stable (Scholtmeijer et al., 2001).

4. Antifungal Activity of Plants

The use of plants for medicinal purposes, for treatment or prevention of diseases, dates to antiquity. Even today, plant materials continue to play a major role in primary health care as therapeutic remedies in many developing countries.

A review of the literature on the evaluation of medicinal plant extracts shows that many studies of their antifungal activities have been performed in recent years. Various research groups have initiated antifungal screening programs for plants used all over the world as anti-infectious agents in traditional medicine (Abad et al., 2007).

A wide range of criteria are followed to select plants to be submitted to antimicrobial screening. Some studies focus on the ethnopharmacological uses; others investigate plants growing in a specific region or country, and others focus on the finding of inhibitors of one or more clinically important fungal species in randomly selected plants. Among the different recommendations, Ríos and Recio (2005) suggested that random selection be avoided and that plants be selected on the basis of ethnopharmacological perspective, because this approach appeared to enhance the probability of success in new drug-discovery efforts.

4.1. Medicinal Plants with Antifungal Activity

Among the different regions of the world, Latin America represents one of the wealthiest sources of flora with pharmacological activity due to its biodiversity (Brandão et al., 2008). It possesses a large number of vascular plants (around 85,000 plant species) (WCMC, 1992), and there is recent evidence that Neotropical forests located in Latin America possess the highest diversity of plants in the world. According to Svetaz et al. (2010), some factors critically distinguish the medicinal plants of Latin America, such as the huge, unexplored biodiversity of the region (Cruz et al., 2007); the rich tradition of the use of medicinal plants (Gupta, 1995); a wealth of ethnopharmacological knowledge that has been closely preserved or transmitted by the many indigenous populations still living in the region (Correia, 1984; Portillo et al., 2001; Souza et al., 2004); and the fact that these resources have been poorly studied.

Regarding the discovery of antifungal plants, the signs or symptoms that give support to a traditional antifungal use are mainly related to skin or mucosal conditions produced by dermatophytes (*Trichophyton, Microsporum* and *Epidermophyton* spp.). Less frequently described fungal-related conditions are leucorrhoea (mainly produced by *Candida* spp.) and respiratory diseases, which can be of fungal origin (caused by species of *Aspergillus* or *Cryptococcus* species) (Svetaz et al., 2010).

Svetaz et al. (2010) screened a group of 144 plants used medicinally for ailments related to fungal infections from Argentina, Bolivia, Brazil, Colombia, Costa Rica, Guatemala and Panama for antifungal activities. All plants selected were evaluated for antifungal properties with the Minimal Inhibitory Concentration (MIC) method following the guidelines of CLSI (Clinical Laboratory Standards Institute). These authors observed that, within Latin American plants, the probability of detecting antifungal activity (MIC≤1000 μg/mL) against at least one fungal species is increased when plants possess reported ethnopharmacological uses related to fungal infections.

According Svetaz et al. (2010), the families of plants with better results against pathogenic *Candida* genus (MIC≤250 μg/mL) are Fabaceae, Piperaceae, Sapindaceae, Winteraceae and Zygophyllaceae. The plants with the best antifungal activity against *C. albicans* and *C. tropicalis* (MIC 62.5 μg/mL) were *Zuccagnia punctata* (Fabaceae) from Argentina, *Serjania elongate* (Sapindaceae) from Bolivia and *Drimys winteri* (Winteraceae) from Brazil. For dermatophyte fungi, the better plant families were Amaranthaceae, Anacardiaceae, Arecaceae, Aristolochiaceae, Asteraceae, Bromeliaceae, Combretaceae, Euphorbiaceae, Fabaceae, Hydrophyllaceae, Lythraceae, Myristicaceae, Orchidaceae, Piperaceae, Polygonaceae, Ranunculaceae, Rubiaceae, Sapindaceae, Solanaceae, Winteraceae, Zingiberaceae, and Zygophyllaceae. The best results were shown for *Zuccagnia punctata* (Fabaceae) from Argentina, with MIC value of 15.6 μg/mL for *Microsporum gypseum, Trichophyton rubrum, Trichophyton mentagrophytes* and *Epidermophyton floccosum*. The species of plants from families Aristolochiaceae, Fabaceae, Piperaceae and Winteraceae present antifungal activity (MIC ≤ 250 μg/mL) against species of the genus *Aspergillus*. *Cryptococcus neoformans* was susceptible mainly to extracts of Aristolochiaceae, Fabaceae, Poaceae, Sapindaceae and Winteraceae. The best activity against *C. neoformans* was found for *Serjania elongata* (Sapindaceae), which had an MIC value of 31.2 μg/mL.

The methanolic extracts of 24 plants used medicinally in the treatment of skin infections in four different regions of Colombia were tested against *C. albicans*. The antifungal activity was determined using the disc diffusion assay. Extracts of *Piper lanceaefolium* (Piperaceae) and *Juglans neotropica* (Juglandaceae) presented activity against *C. albicans*. The decoction of the leaves of *P. lanceaefolium* is taken as a bath to treat skin infections. An infusion of the bark of *J. neotropica* is prepared and used as a bath to 'clean' the body or as an antiseptic (Lopez et al., 2001).

Fenner et al. (2006) observed that, in 98 families of plants from Brazil, Fabaceae and Asteraceae have a greater concentration of species of plants used for the treatment of the signs or symptoms that give support to a traditional antifungal use. In the state of Sergipe in Brazil, water infusion extracts of *Ziziphus joazeiro* (Rhamnaceae) and *Caesalpinea pyramidalis* (Caesalpinaceae) are used in traditional medicines. The aqueous extracts from these plants were tested and exhibited significant antifungal activity against *T. rubrum* and *C. guilliermondii,* with MIC values of 6.5 μg/mL.

Johann et al. (2007) demostrated the antifungal properties of extracts from eight plants traditionally used in popular Brazilian medicine. The plants were tested against five clinically relevant *Candida* species, *C. neoformans*, and *Sporothrix schenckii.* The ethanolic extract from the leaves of *Schinus terebinthifolius* (Anacardiaceae) exhibited potential antifungal activity against *C. glabrata* and *S. schenckii,* with MIC values of 125 and 15 μg/mL, respectively. The ethyl acetate fraction from the leaves of *S. terebinthifolius* presented activity inhibiting the growth of three *C. albicans* isolates at 7.8 μg/mL and significantly inhibiting of their adhesion (70%) in buccal epithelial cell (Johann et al., 2008a).

Several other medicinal plants from around the world were tested for their antifungal activity. The antifungal activity of extracts (hydro-alcohol (50 % v/v) and hexane) from Indian plants traditionally used as medicines as *Valeriana jatamansi* (Valerianaceae), *Coleus barbatus* (Lamiaceae), *Berberis aristata* (Berberidaceae), *Asparagus racemosus* (Asparagaceae), *Andrographis paniculata* (Acanthaceae), *Achyranthes aspera* (Amaranthaceae), *Tinospora cordifolia* (Menispermaceae), and *Plantago depressa* (Plantaginaceae) were evaluated against *C. albicans* and *A. niger*. Hydroalcoholic extracts of *A. paniculata* and *A. aspera* showed maximum potency against *A. niger* and *C. albicans* with the highest MIC values being 5 and 3 μg/mL, respectively. Hexane extracts of *A. paniculata* had the highest MIC value (7 μg/mL) against *A. niger* (Mathur et al., 2011).

Mentha longifolia L. (Lamiaceae) leaves have been traditionally used in the treatment of minor sore throat and minor mouth or throat irritation by the indigenous people of Iraq. A known compound was isolated from the essential oil of the plant and was identified as (-) menthol, one terpenoid. The menthol achieved considerable antifungal activity against the yeast *C. albicans,* with an MIC value of 125 μg/mL (Firas, 2009).

Numerous medicinal plants are commonly used in both urban and rural areas in Cameroon as well as in most African countries for the treatment of infectious diseases. The World Health Organization (WHO, 1996) estimates that 80% of the African population is affected by infectious diseases and argues that the use of medicinal plants, if applied appropriately, could significantly contribute to improving public health. Kuete (2010) summarized the activity of some extracts of plants against fungi and showed that plants with extracts that have better activity against *C. albicans, C. krusei, C. glabrata* and dermatophyte fungi include *Vismia guineensis* (Guttiferae), *Irvingia gabonensis* (Irvingiaceae), *Dorstenia angusticornis, D. barteri, D. elliptica, D. turbinate, Ficus cordata, F. ovata, Morus mesozygia, Treculia Africana, T. obovoidea* (Moraceae) and *Bersama engleriana* (Melianthaceae). Most of the antimicrobial substances isolated from Cameroonian medicinal plants belong to three main classes of secondary metabolites, terpenoids, phenolic compounds and alkaloids.

4.2. Compounds from Plants with Antifungal Activity

Plants have an almost limitless ability to synthesize aromatic substances, most of which are phenols or their oxygen-substituted derivatives. In many cases, these substances serve as plant defense mechanisms against herbivory by microorganisms, insects, and herbivores. Some, such as terpenoids, give plants their odors; others (quinones and tannins) are responsible for plant pigment. Many compounds are responsible for plant flavor (e.g., the terpenoid capsaicin from chili peppers), and some of the same herbs and spices used by humans to season food yield useful medicinal compounds (Cowan, 1999).

According to Saeelem et al. (2010), the alkaloid compounds have an interesting activity against fungi. Canthin-6-one (1), which is found in extracts of *Allium neapolitanum* and *Zanthoxylum chiloperone* var. *angustifolium*, exhibited a broad spectrum of activities against *A. fumigatus, A. niger, A. terreus, C. albicans, C. tropicalis, C. glabrata, C. neoformans, Geotrichum candidum, Saccharomyces cerevisiae, Trichosporon beigelii, T. cutaneum* and *T. mentagrophytes*, with MIC values between 1.66 and 10.12 μg/mL. In an experiment to determine the mechanism of action of canthin-6-one, Lagoutte et al.(2008) observed the accumulation of the derivatives of these compounds within lipid droplets and that the relative amount of unsaturated alkyl chain fatty acids is enhanced, which they propose is due to the stimulation of desaturase enzyme systems.

The triterpenes and phenols also present a good antifungal activity. Johann et al. (2010) isolated and identified the antifungal compounds schinol (tetracyclic triterpene) and a new biphenyl compound, namely, 4'-ethyl-4-methyl-2,2',6,6'- tetrahydroxy[1,1'-biphenyl]-4,4'-dicarboxylate, from the extracts of *S. terebinthifolius,* which has activity against clinical isolates of the pathogenic fungus *Paracoccidioides brasiliensis.* Schinol presented the better activity, with an MIC value of 7.8 μg/mL. The mechanism of action of terpenes is not fully understood but is speculated to involve membrane disruption by the lipophilic compounds (Mason et al., 1987, Cowan, 1999). The mechanisms thought to be responsible for phenolic toxicity to microorganisms include enzyme inhibition by the oxidized compounds, possibly through reaction with sulfhydryl groups or through more nonspecific interactions with the proteins.

Oenothein B, an ellagitannin, was isolated from extract of leaves of *Eugenia uniflora,* a Myrtaceae plant that is widely distributed in the tropical and subtropical areas of Brazil. This plant is well known in Brazilian savannah medicine and is used in folk practices against mycotic diseases (Holetz et al., 2002; Souza et al., 2002). Santos et al. (2007) evaluated oenothein B against *P. brasiliensis* yeast cells. The oenothein B dosage that most effectively inhibited the development (74%) of *P. brasiliensis* yeast cells *in vitro* was 500 μg/mL. The electron micrographs of oenothein B–treated yeast cells showed squashing, rough surface, cell wall rupture and cell membrane recess. The expression of *P. brasiliensis* genes was evaluated to investigate the action of oenothein B. The authors reported that oenothein B inhibited 1,3-β-glucan synthase (PbFKS1) transcript accumulation. The results indicated that oenothein B interferes with the cell morphology of *P. brasiliensis*, probably by inhibiting the transcription of 1,3-β-glucan synthase gene, which is involved in the cell wall synthesis.

Lignans were isolated from *Piper abutiloides*. This species of *Piper* (Piperaceae) is a traditional medicinal plant from Brazil and was evaluated against some opportunistic fungal species. The chromatographic fractionation of the bioactive part (hexanic part) was monitored with a bioautographic assay using *C. glabrata*, and this allowed the isolation of three antifungal compounds: pseudodillapiol (phenilpropanoid), eupomatenoid-6 and conocarpan (lignans). These compounds presented different potencies against the fungi tested, with the strongest effect being observed for eupomatenoid-6 against *C. glabrata*, which presented activity of 0.3 µg/spot in a bioautographic assay (Johann et al., 2008b).

A lignan, (+)-lyoniresinol-3a-O-b-D-glucopyranoside, isolated from the root bark of *Lycium chinense*, exhibited potent antifungal activity against *C. albicans, S. cerevisiae* and *T. beigelii* (MICs of 5.5 and 10 µg/mL, respectively). This compound induced the accumulation of intracellular trehalose in *C. albicans* as a stress response to the drug and disrupted the dimorphic transition that forms pseudo-hyphae caused by the pathogenesis (Lee et al., 2005).

Flavonoids have also been identified as potent antimicrobial agents. A dimeric amentoflavone from *Selaginella tamariscina* exhibited inhibitory activity (MIC value of 5 µg/mL) against *C. albicans, S. cerevisiae* and *T. beigelii* (Jung et al., 2007). The efficacy of flavonoids against a variety of pathogens can be attributed to cell-wall permeability, and the porins in the outer membrane present in microorganisms. Lipophilic flavonoids may also disrupt microbial membranes (Tsuchiya et al., 1996; Cowan, 1999, Alvarez et al., 2008).

Many phytochemicals compounds not mentioned above have been found to exert antifungal properties, including saponins (aginoside, dioscin), sulfoxides (ajoene) and quinones (3,4-Dihydroxy-1-methoxyanthraquinone-2-carboxaldehyde, ewbouldiaquinone A) (Cowan, 1999, Saleem et al., 2010). More of these compounds should be subjected to animal and human studies to determine their effectiveness in whole-organism systems, including, in particular, toxicity studies as well as an examination of their effects on beneficial normal microbiota. Disruption of adhesion is one example of an anti-infection activity that is currently not commonly screened for currently.

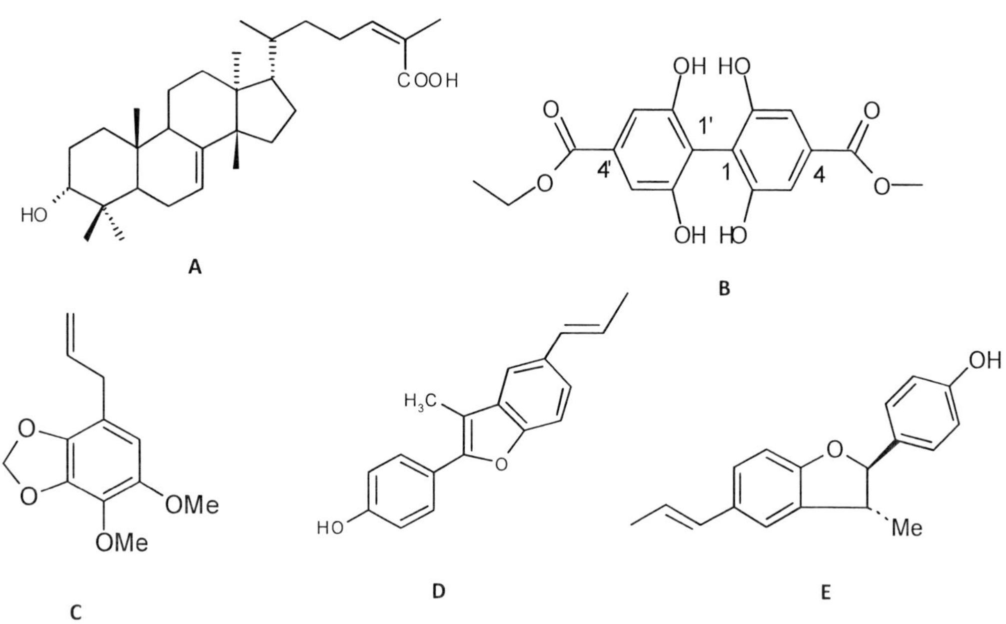

Figure 4. Compounds isolated from *Schinus terebinthifolius*. *A:* Schinol; *B*: 4'-ethyl-4-methyl-2,2',6,6'-tetrahydroxy[1,1'-biphenyl]-4,4'-dicarboxylate). Compounds isolated by hexane fraction from *Piper abutiloides*. *C:* Pseudodillapiol, *D:* Eupomatenoid 6 e *E:* Conocarpan.

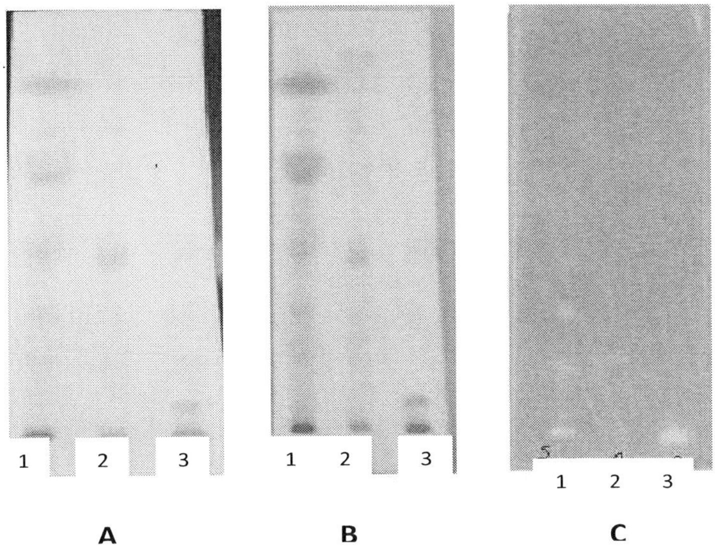

Figure 5. Thin-layer chromatography (TLC) and bioautography assay of extract and fractions of aerial parts from *Piper abutiloides*. A) TLC (silica gel plates 60F254) with 1: Crude extract; 2: Hexanic fraction; and 3: Dichloromethane fraction eluted with hexane-ethyl acetate (7: 3 v / v). B) TLC plate reveled with UV 366 and C) bioautographic assay with *Candida glabrata* were sprayed with p-iodonitrotetrazolium violet.

CONCLUSION

The massive emergence of fungal diseases associated with AIDS and the rising frequency of fatal mycoses associated with the increasing use of immunosuppressive medical therapies has stimulated research directed towards the discovery of novel antifungal agents. Thus, in recent decades, research on biologically active natural products has undergone a significant increase, particularly with the advent of the concept of diversity-oriented synthesis, which has been the basis for the development of new bioactive compounds as modified natural products. Microorganisms and plants are the sources of various known medicinal products and potential therapeutics. The compounds discussed in this review represent a subset of all the antifungal agents of natural origin that have been isolated, characterized and described in published literature to date. There are a huge number of other molecules with reported antifungal activity that have not been included here. Promising alternatives to produce potent biosurfactants with altered antimicrobial profiles and decreased toxicity against mammalian cells may be exploited by genetic alteration of biosurfactants. Furthermore, biosurfactants have the potential to be used as anti-adhesive biological coatings for medical insertional materials, thus reducing hospital infections and use of synthetic drugs and chemicals. In spite of the immense potential of biosurfactants in this field, their use still remains limited, possibly due to the high production and extraction costs and the lack of information concerning their toxicity in human systems. Further investigations on human cells and natural microbiota are needed to validate the use of biosurfactants in several biomedical and health-related areas. Nevertheless, there appears to be great potential for their use in the medical science arena waiting to be fully exploited.

Acknowledgment

We thanks the student Natália Priscila Silva de Oliveira for your help in the general organization of the references.

References

Abad, M.J.; Ansuategui, M.; Bermejo, P. (2007). Active antifungal substances from natural sources. *Arkivoc*, 7, 116-145.

Abalos, A.; Pinazo, A.; Infante, M.R.; Casals, M.; Garcia, F.; Manresa, A. (2001). Physicochemical and antimicrobial properties of new rhamnolipids produced by Pseudomonas aeruginosa AT10 from soybean oil refinery wastes. *Langmuir* 17, 1367–1371.

Abdel-Aegeed, A.; Al-Rahmah, A.N.; Mostafa, A.A, (2011). Biochemical characterization of anti-microbial activity of glycolipids produced by Rhodococcus erythropolis. *Pakistan Journal of Botany*, 43, 1323-1334.

Abdel-Mawgoud A.M.; Lépine, F.; Déziel, E. (2010). Rhamnolipids: diversity of structures, microbial origins and roles. *Applied Microbiology and Biotechnology*, 86, 1323-1336.

Al-Tahhan, R.A.; Sandrin, T.R.; Bodour, A.A.; Maier, R.M. (2000). Rhamnolipid-induced removal of lipopolysaccharide from Pseudomonas aeruginosa: effect on cell surface properties and interaction with hydrophobic substrates. Ap*plied and Environmental Microbiology*, 66, 3262–3268.

Alvarez, M. A.; Debattista, N. B.; Pappano, N. B. (2008). Antimicrobiol activity and synergism of some substituted flavonoids. *Folia Microbiologica*, 53, 23–28.

Andersen, G.R.; Nissen, P.; Nyborg, J. (2003). Elongation factors in protein biosynthesis. *TRENDS in Biochemical Sciences*, 23, 456-460.

Andes, D. (2003). Pharmacokinetics and pharmacodynamics in the development of antifungal compounds. *Current Opinion in Investigational Drugs*, 4, 991-998.

Arima, k.; Imanaka, H.; Kousaka, M.; Fukuda, A.; Tamura, G. (1964). Pyrrolnitrin, a new antibiotic substance, produced by *Pseudomonas. Agricultural and Biological Chemistry*, 28, 575-576.

Arutchelvi, J.I.; Bhaduri, S.; Uppara, P.V.; Doble, M. (2008). Mannosylerythritol lipids: a review. Journal of Industrial Microbiology and Biotechnology, 35, 1559–1570.

Baddly, J.W.; Moser, S.A. (2004). Emerging fungal resistance. *Clinics in Laboratory Medicine*, 24, 721–735.

Banat, I.M.; Makkar, R.; Cameotra, S. (2000). Potential commercial applications of microbial surfactants. *Applied Microbiology and Biotechnology*, 53, 495–508.

Bechard, J.; Eastwell, K.C.; Sholberg, P.L.; Mazza, G.; Skura, B. (1998). Isolation and partial chemical characterization of an antimicrobial peptide produced by a strain of Bacillus subtilis. *Journal of Agricultural and Food Chemistry*, 46, 5355–5361.

Benincasa, M.; Abalos, A.; Oliveira, I.; Manresa, A. (2004). Chemical structure, surface properties and biological activities of the biosurfactant produced by *Pseudomonas aeruginosa* LBI from soapstock. *Antonie Van Leeuwenhoek*, 85, 1–8.

Besson F.; Michel G. (1990). Mycosubtilins B and C: minor antibiotics from mycosubtilin producing *Bacillus subtilis*. *Microbios*, 62, 93-99.

Bills, G.; Platas, G.; Fillola, A.; Jiménez, M.R.; Collado, J.; Vicente, F.; Martín, J.; González, A.; Bur-Zimmermann, J.; Tormo, J.R.; Peláez, F. (2008). Enhancement of antibiotic and secondary metabolite detection from filamentous fungi by growth on nutritional arrays. *Journal of Applied Microbiology*, 104, 1644–1658.

Bills, G.F.; Platas, G.; Overy, D.P.; Fillola, A.; Jiménez, M.R.; Martín, J.; González, A.; Vicente, F.; Tormo, J.R.; Peláez, F. (2009). Discovery of the parnafungins, antifungal metabolites that inhibit mRNA polyadenylation, from the Fusarium larvarum complex and other Hypocrealean fungi. *Mycologia*, 101, 449-472.

Brandão, M.G.L.; Zanetti, N.N.S.; Oliveira, P.; Grael, C.F.F.; Santos, A.C.P.; Monte- Mór, R.L. (2008). Brazilian medicinal plants described by 19th century European naturalists in the Official Pharmacopoeia. *Journal of Ethnopharmacology* 120, 141–148.

Burkhead, K.D.; Schisler, D.A.; Slininger, P.J. (1994). Pyrrolnitrin production by biological control agent *Pseudomonas cepacia* B37w in culture and in colonized wounds of potatoes. *Applied and Environmental Microbiology*, 60, 2031-2039.

Burns, M.K.; Coffin, J.M.; Kurobane, I.; Vining, L.C. (1979). Biosynthesis of chlorflavonin in aspergillus candidus: a novel fungal route to flavonoids. *Journal of Chemical Society Chemical Commun*, 1, 426-427.

Burns, M.K.; Coffin, J.M.; Kurobane, I.; Vining, L.C.; McInnes, A.G.; Smith, D.G.; Walter, J.A. (1981). Biosynthesis of Chlorflavonin in Aspergillus candidus. ^{13}C- and ^{14}C-labelling evidence for a new route to the flavonoid structure. *Journal of Chemical Society Perkin Transation*, 1, 1411-1416.

Busscher, H.J.; Van de Belt-Gritter, B.; Westerhof, M.; Van Weissenbruch R.; Albers, F.W.; Van der Mei, H.C. (1999). Microbial interference in the colonization of silicone rubber implant surfaces in the oropharynx: *Streptococcus thermophilus* against a mixed fungal/bacterial biofilm. In: Rosenberg E, ed. *Microbial Ecology and Infectious Disease*. Washington, DC: American Society for Microbiology, 66–74.

Busscher, H.J.; Van Hoogmoed, C.G.; Geertsema-Doornbusch, G.I.; Van der Kuijl-Booij, M.; Van der Mei, H.C. (1997). *Streptococcus thermophilus* and its biosurfactants inhibit adhesion by *Candida* spp. on silicone rubber. *Applied and Environmental Microbiology*, 63, 3810–3817.

Cameotra, S.S.; Makkar, R.S. (2004). Recent applications of biosurfactants as biological and immunological molecules. *Current Opinion in Microbiology*, 7, 262-266.

Campos, F.F., Johann, S.; Cota, B.B.; Alves, T.M.A.; Rosa, L.H.; Caligiorne, R.B.; Cisalpino, P.S.; Rosa, C.A; Zani, C.L. (2010). Antifungal activity of trichothecenes from *Fusarium* sp. against clinical isolates of *Paracoccidioides brasiliensis*. *Mycoses*, 54, 127-136.

Céspedes, C.L.; Ávila, G.; Martinez, A.; Serrato, B.; Calderon-Mugica, J.C.; Salgado-Garciglia, R. (2006). Antifungal and antibacterial activities of Mexican tarragon (*Tagetes lucida*). *Journal of Agricultural Food Chemistry*, 54, 3521–3527

Chakrabarti, A.; Shivaprakash, M.R. (2005). Microbiology of systemic fungal infections. *Journal of Postgraduate Medicine*, 51, 16-20.

Chen J.; Song, X.; Zhang, H.; Qu, Y.B.; Miao, J.Y. (2006) Production, structure elucidation and anticancer properties of sophorolipid from Wickerhamiella domercqiae. *Enzyme and Microbial Technology*, 39, 501–506.

Cheng, Y.; McNally, D. J.; Labbe, C.; Voyer, N.; Belzile, F.; Belanger, R.R. (2003). Insertional mutagenesis of a fungal biocontrol agent led to discovery of a rare cellobiose lipid with antifungal activity. *Applied and Environmental Microbiology*, 69, 2595-2602.

Clardy, J.; Walsh, C. (2004). Lessons from natural molecules. *Nature*, 432, 829-837.

Clemons, K.V.; Sobel, R.A.; Williams, P.L.; Pappagianis, D.; Stevens, D.A. (2002). Efficacy of intravenous liposomal amphotericin B (AmBisome) against coccidioidal meningitis in rabbits. *Antimicrobial Agents Chemotherapy*, 46, 2420-2426.

Cooper, D.G.; Paddock, D.A. (1984) Production of a biosurfactant from *Torulopsis bombicola*. *Applied and Environmental Microbiology*, 47, 173–176.

Correia, P.M. (1984). Dicionário de plantas úteis do brasil e das exóticas cultivadas. Instituto Brasileiro de Desenvolvimiento Forestal, *Imprenta Nacional*, Rio de Janeiro.

Cowan, M.M. (1999). Plant products as antimicrobial agents. *Clinical Microbiology Reviews*, 12, 564–582.

Cruz, M.C.S.; Santos, P.O.; Barbosa Jr.A.; de Mélo, D.L.F.M.; Alviano, C.S.; Antoniolli, A.R.; Alviano, D.S.; Trindade, R.C. (2007). Antifungal activity of Brazilian medicinal plants involved in popular treatments of mycoses. *Journal of Ethnopharmacology*, 111, 409–412.

Delcambe, L.; Devignat, R. (1957). *L'iturine, nouvel antibiotique d'origine congolaise*. Académie Royale des Sciences Coloniales, 6, 1-77.

Deml, G.; Anke, T.; Oberwinkler, F.; Gianetti, B.M.; Steglich, W. (1980). Schizonellin A and B, new glycolipids from *Schizonella melanogramma*. *Phytochemistry*, 19, 83–88.

Denning, D.W. (2002). Echinocandins: a new class of antifungal. *Journal of Antimicrobial Chemotherapy*, 49, 889-891.

Denning, D.W. (2003). Echinocandin antifungal drugs. *Lancet*, 362, 1142–1151.

Desai, J.D.; Banat, I.M. (1997). Microbial production of surfactants and their commercial potential. *Microbiology and Molecular Biology Reviews*, 1, 47-64.

Di Santo, R. (2010). Natural products as antifungal agents against clinically relevant pathogens. *Natural Product Reports*, 27, 1084–1098.

Dias, L.C.; Oliveira, L.G.; Vilcachagua, J.D.; Nigsch, F. (2005). Total Synthesis of (+)-Crocacin D (I). *ChemInform*, 36, page no.

Earth's Living Resources. Chapman & Hall, London.

Endo, M.; Takesako, K.; Kato, I.; Yamaguchi, H. (1997). Fungicidal action of aureobasidin A, a cyclic depsipeptide antifungal antibiotic, against Saccharomyces cerevisiae. *Antimicrobial Agents and Chemotherapy*, 41, 672-676. *Ethnopharmacology*, 100, 80–84.

Ezra, D.; Castillo, U.F.; Strobel, G.A. (2004). Coronamycins, peptide antibiotics produced by a verticillate *Streptomyces* sp. (MSU-2110) endophytic on *Monstera sp. Microbiology*, 150, 785–793.

Feldmesser, M.; Kress, I.; Mednick, A.; Casadevall,A. (2000). The effect of the Echinocandin analogue Caspofunginon cell wall glucan synthesis by *Cryptococcus neoformans*. *The Journal of Infectious Diseases*, 182, 1791-1795.

Fenner, R.; Betti, A.H.; Mentz, L.A.; Rates, S.M.K. (2006). Plants with potencial antifungal activity employed in brazilian folk medicine. *Brazilian Journal of Pharmaceutical Sciences*, 42, 369-394.

Ficker, E.F.; Arnason, J.T.; Vindras, P.S.; Alvarez, L.P.; Akpagana, K.; Gbéassor, M.; Souza, C.; Smith, M.L. (2004). Inhibition of human pathogenic fungi by ethnobotanically selected plant extracts, *Mycoses*, 46, 29–37.

Firas, A.; Bayati, A. (2009). Isolation and identification of antimicrobial compound from *Mentha longifolia* L. leaves grown wild in Iraq. *Annals of Clinical Microbiology and Antimicrobials*, 8, 20.

Franzot SP, Casadevall A. (1997). Pneumocandin L-743,872 enhances the activities of amphotericin B and fluconazole against *Cryptococcus neoformans* in vitro. *Antimicrobial Agents Chemotherapy*, 41, 331–36.

Gharaei-Fathabad, E. (2011). Biosurfactants in pharmaceutical industry: A mini-review. *American Journal of Drug Discovery and Development*, 1, 58-69.

Gold, W.; Staut, H.A.; Pagano, J.F.; Donovick, R. (1956). Amphotericins A and B, antifungal antibiotics produced by a *Streptomycete*. I. I*n vitro studies*. Anibiotics Annual, Medical Ancycplopedia Inc, New York, 579-586.

Golubev, W.I.; Shabalin, Y. (1994). Microcin production in *Cryptococcus humicola*. FEMS Microbiology Letters, 119, 105–110.

Golubev, W.I.; Kulakovskaya, T.V.; Golubeva, E.W. (2001). The yeast *Pseudozyma fusiformata* VKM Y-2821 producing an antifungal glycolipid. *Microbiology*, 70, 553–556.

González, G.M.; Tijerina, R.; Najvar, L.K.; Bocanegra, R.; Rinaldi, M.G.; Graybill, J.R. (2004). Efficacies of amphotericin B (AMB) lipid complex, AMB colloidal dispersion, liposomal AMB, and conventional AMB in treatment of murine coccidioidomycosis. *Antimicrobial Agents and Chemotherapy*, 48, 2140-2143.

Gorin, P.A.J.; Spencer, J.F.T.; Tulloch, A.P. (1961). Hydroxy fatty acid glycosides of sophorose from *Torulopsis magnoliae*. *Canadian Journal of Chemistry*, 39, 846–855.

Gupta, M.P. (1995). 270 *Plantas medicinales Iberoamericanas*. CYTED-Convenio. Andrés Bello Ed., Bogotá.

Gupta, M.P.; Jones, A.; Solís, P.N.; Correa, M.D.A.; Galdames, C.; Guinneau-Sinclair, F. (1993). Medicinal plants inventory of Kuna Yala. Part I. *Journal of Ethnopharmacology*, 40, 77–109.

Haba, E. H; Pinazo, A.; Jauregui, O.; Espuny, M.J.; Infante, M.R.; Manresa, A. (20030. Physiochemical characterization and antimicrobial properties of rhamnolipids produced by *Pseudomonas aeruginosa* 47T2 NCBIM 40044. *Biotechnology and Bioengineering*, 81, 316–322.

Ha, K.C.; White, T.C. (1999). Effects of Azole Antifungal Drugs on the Transition from Yeast Cells to Hyphae in Susceptible and Resistant Isolates of the Pathogenic Yeast *Candida albicans*. *Antimicrobial Agents and Chemotherapy*, 43, 763–768.

Hadacek, F. (2002). Secondary metabolites as plant traits: current assessment and future perspectives. *Critical Reviews in Plant Science*, 21, 273-322.

Hamme, J.D.V.; Singh, A; Ward, O.P. (2006) Physiological aspects Part 1 in a series of papers devoted to surfactants in microbiology and biotechnology. *Biotechnology Advances*, 24, 604–620.

Harris, G.H.; Dai, P. (2001). *Journal of Liquid Chromatography & Related Technologies*, 24, 1775–1790.

Harrison, L.; Teplow, D.B.; Rinaldi, M.; Strobel, G. (1991). Pseudomycins, a family of novel peptides from *Pseudomonas syringae* possessing broad-spectrum antifungal activity. *Journal of General Microbiology*, 137, 2857-2865.

Hartmann, T. (1996). Diversity and variability of plant secondary metabolism: a mechanistic view. *Entomologia Experimentalis et Applicata*, 80, 177–188.

Haskins, R. H. (1950). Biochemistry of the Ustilaginales: I. Preliminary cultural studies of *Ustilago zeae. Canadian Journal of Research*, 28, 213-223.

Haskins, R.H.; Thorn, J.A.; Boothroyd, B. (1955). Biochemistry of the ustilagenales. XI. Metabolic products of *Ustilago zeae* in submerged culture. *Canadian Journal of Microbiology*, 1, 749–756.

Hauser, D.; Sigg, H.P. (1971). Isolation and decomposition of Sordarin. *Helvetica Chimica Acta*, 54, 1178–1190.

Hawser, S.P.; Douglas, L.J. (1995). Resistance of *Candida albicans* biofilms to antifungal agents in-vitro. *Antimicrobial Agents and Chemotherapy*, 39, 2128–2131.

Heidler, S.A.; Radding, J.A.(1995). The AUR1 gene in Saccharomyces cerevisiae encodes dominant resistance to the antifungal agent aureobasidin A (LY295337). *Antimicrobial Agents and Chemotherapy*, 39, 2765-2769.

Hektor, H.J.; Scholtmeijer, K. (2005). Hydrophobins: proteins with potential. *Current Opinion in Biotechnology*, 16, 434-439.

Henningsen, M. (2003). Pest Control in Agriculture. Fungicides up to Date. *ChemInform*, 34, 203-210.

Hino, M., Fujie, A.; Iwamoto, T.; Hori, Y.; Hashimoto, M.; Tsurumi, Y.; Sakamoto, K.; Takase, S.; Hashimoto, S. (2001). Chemical diversity in lipopeptide antifungal antibiotics. *Journal of Industrial Microbiology and Biotechnology*, 27, 157–162.

Holetz, F.B.; Pessini, G.L.; Sanches, N.R.Cortez, D.A.G.; Nakamura, C.V.; Dias Filho, B.P. (2002). Screening of some plants used in the Brazilian folk medicine for the treatment of infectious diseases. *Memórias do Instituto Oswaldo Cruz*, 97: 1027-1031.

Hossain, M.A.; Ghannoum, M.A. (2000). New investigational antifungal agents for treating invasive fungal infections. *Expert Opinion on Investigational Drugs*, 9, 1797-1813.

Hsieh, F.C.; Tsung-Chun, L.; Menhsiao, M.; Suey-Sheng, K. (2008). Comparing methods for identifying *Bacillus* strains capable of producing the antifungal lipopeptide iturin A. *Current Microbiology*. 56, 1-5.

Hutchison, M.L.; Gross, D.C. (1997). Lipopeptide phytotoxins produced by *Pseudomonas syringae* pv. *syringae*: comparison of the biosurfactant and ion channel-forming activities of syringopeptin and syringomycin, *Molecular Plant-Microbe Interactions*, 10, 347–354.

Ito, S.; Kinta, M.; Inoue, S. (1980). Growth of yeasts on normalalkanes – inhibition by a lactonic sophorolipid produced by *Torulopsis bombicola. Agricultural Biology and Chemistry*, 44, 2221–2223.

Ivanov, A.J.; Vagabov, V.M.; Fomchenkov, V.M.; Kulaev, I.S. (1996) Study of the influence of polyphosphates of cell envelope on the sensitivity of yeast *Saccharomyces carlsbergensis* to the cetyl-3-methylammonium bromide. *Microbiology*, 65, 611–616.

Jansen, R.; Washausen, P.; Kunze, B.; Reichenbach, H.; Höfle, G. (1999). The Crocacins, Novel Antifungal and Cytotoxic Antibiotics from *Chondromyces crocatus* and *Chondromyces pediculatus* (Myxobacteria): Isolation and Structure Elucidation. *European Journal of Organic Chemistry*, 1085–1089.

Jarvis, F.G.; Johnson, M.J. (1949) A glyco-lipid produced by *Pseudomonas aeruginosa. Journal of the American Chemical Society*, 71, 4124-4126.

Jekosch, K.; Kück, U. (2000). Glucose dependent transcriptional expression of the cre1 gene in *Acremonium chrysogenum* strains showing different levels of cephalosporin C production. *Current Genetics*. 37, 388-395.

Johann, S., Cisalpino, P.S., Watanabe, G.A., Cota, B.B., Siqueira, E.P., et al., 2010. Antifungal activity of extracts of some plants used in the Brazilian traditional medicine against the pathogenic fungus Paracoccidioides brasiliensis. *Pharmaceutical Biology*, 48, 388-396.

Johann, S.; Cota, B.B.; Souza-Fagundes, E.M.; Pizzolatti, M.G., Resende, M.A.; Zani, C. L. (2008b). Antifungal activities of compounds isolated from *Piper abutiloides* Kunth. *Mycoses*, 52, 499–506.

Johann, S.; Pizzolatti, M.G.; Donnici, C.L.; Resende, M.A. (2007) Antifungal properties of plants used in brazilian traditional medicine against clinically relevant fungal pathogens. *Brazilian Journal of Microbiology*, 38, 632-637.

Johann, S.; Silva, D.L.; Martins, C.V.B.; Resende, M.A. (2008) Inhibitory effect of extracts from Brazilian medicinal plants on the adhesion of *Candida albicans* to buccal epithelial cells. *World Journal of Microbiology and Biotechnology*, 24, 2459-2464.

Jones, J.B.G. (1994). Fungal adhesion. *Mycological research*, 98, 961–981.

Justice, M.C.; Hsu, M.J.; TSE, B.; Ku, T.; Balkovec, J.; Schmatz, D.; Nielsen, J. (1998). Elongation Factor 2 as a novel target for selective inhibition of fungal protein synthesis. *The journal of Biological Chemistry*, 273, 3148–3151.

Kaneko, H.; Hosohara, M.; Tanaka, M.; Itoh, T. (1976). Lipid composition of 30 species of yeast. *Lipids* 11, 837–840.

Kavamura, V.N.; Esposito, E. (2010). Biotechnological strategies applied to the decontamination of soils polluted with heavy metals. *Biotechnology Advances*, 28, 61-66.

Kearns, D.B; Losick, R. (2003). Swarming motility in undomesticated *Bacillus subtilis*. *Molecular Microbiology*, 49, 581–590.

Kim, B.S.; Lee, J.Y.; Hwang, B.K. (2000). In vivo control and in vitro antifungal activity of rhamnolipid B, a glycolipid antibiotic, against *Phytophthora capsici* and *Colletotrichum orbiculare*. *Pest Management Science*, 56, 1029–1035.

Kim, P.I.; Jaewon, R.; Young, H.K.; Youn-Tae, C. (2010). Production of biosurfactant lipopeptides iturin A, fengycin, and surfactin A from *Bacillus subtilis* CMB32 for control of *Colletotrichum gloeosporioides*. *Journal of Microbiology and Biotechnology*, 20, 138-145.

Kim, S.J.; Kwon-Chung, K.J.; Milne, G.W.A.; Hill, W.B.; Patterson, G. (1975). Relationship between polyene resistance and sterol compositions in *Cryptococcus neoformans*. *Antimicrobial Agents and Chemotherapy*, 7, 99–106.

Kitamoto, D. (2008). Naturally engineered glycolipid biosurfactants leading to distinctive self-assembling properties. *Yakugaku Zasshi*, 128, 695–670.

Kitamoto, D.; Haneishi, K.; Nakahara, T.; Tabuchi, T. (1990). Production of mannosylerythritol lipids by *Candida Antarctica* from vegetable-oils. *Agricultural and Biological Chemistry*, 54, 37-40

Kitamoto, D.; Yanagishita, H.; Shinbo, T.; Nakane, T.; Kamisawa, C.; Nakahara, T. (1993). Surface active properties and antimicrobial activities of mannosylerythritol lipids as biosurfactants produced by *Candida antarctica*. *Journal of Biotechnology*, 29, 91-96.

Knight, S.C.; Anthony, V.M.; Brady, A.M.; Greenland, A.J.; Heaney, S.P.; Murray, D. C.; Powell, K.A.; Schulz, M.A.; Spinks, C.A.; Worthington, P.A. Youle, D. (1997). Rationale and perspectives on the development of fungicides. *Annual Phytopathology*, 35, 349-372.

Konishi, M.; Nishio, M.; Saitoh, K.; Miyaki, T.; Oki, T.; Kawaguchi, H. (1989). Cispentacin, a new antifungal antibiotic. Production, isolation, physico-chemical properties and structure. *Jounal of Antibiotic,* 42, 1749-1755.

Konoshi, M.; Fukuoka, T.; Morita, T.; Imura, T.; Kitamoto, D. (2008). Production of new types of sophorolipids by *Candida batistae. Journal of Oleo Science*, 57, 359–369.

Kuete, V. (2010) Potential of Cameroonian Plants and Derived Products against Microbial Infections: A Review. *Planta Medica*, 76, 1479–1491.

Kulakovskaya, E.V.; Kulakovskaia, T.V.; Golubev, V.I.; Shashkov, A.S.; Grachev, A.A.; Nifant'ev, N.E. (2007). Fungicidal activity of cellobiose lipids from cultural fluid of yeast *Cryptococcus humicola* and *Pseudozyma fusiformata. Bioorganicheskaya Khimiya*, 33, 167-171.

Kulakovskaya, T.V.; Kulakovskaya, E.V.; Golubev, W.I. (2003). ATP leakage from yeast cells treated by extracellular glycolipids of *Pseudozyma fusiformata*. FEMS Yeast Research, 3, 401–404.

Kulakovskaya, T.V.; Shashkov, A.; Kulakovskaya, E.; Golubev, W., Zinin, A.; Tsvetkov, Y.; Grachev, A.; Nifantiev, N. (2009). Extracellular Cellobiose Lipid from Yeast and Their Analogues: Structures and Fungicidal Activities. *Journal of Oleo Science*, 58, 133-140.

Kulakovskaya, T.V.; Golubev, W.I.; Tomashevskaya, M.A.; Kulakovskaya, E.V.; Shashkov, A.S.; Grachev, A.A. Chizhov, A.S.; Nifantiev, N.E. (2010). Production of antifungal cellobiose lipids by *Trichosporon porosum. Mycopathologia*. 169, 117-123.

Kulakovskaya, T.V.; Shashkov, A.S.; Kulakovskaya, E.V.; Golubev, W. (2005). Ustilagic acid secretion by *Pseudozyma fusiformata* strains. *FEMS Yeast Research*, 5, 919-923.

Kurtzman, C. P.; Price, N. P.; Ray, K. J.; Kuo, T.-M. (2010). Production of sophorolipid biosurfactants by multiple species of the *Starmerella (Candida) bombicola* yeast clade. *FEMS Microbiology Letters,* 311, 140–146.

Kurz, M.; Eder, C.; Isert, D.; Li, Z.; Paulus, E.F.; Schiell, M.; Toti, L.; Vertesy, L.; Wink, J.; Seibert, G. (2003). Ustilipids, acylated beta-d-mannopyranosyl d-erythritols from *Ustilago maydis* and *Geotrichum candidum. Journal of Antibiotics*, 56, 91-101.

Lagoutte, D.; Nicolas, V.; Poupon, E.; Fournet, A.; Hocquemiller, R.; Libong, D.; Chaminade, P.; Loiseau, P.M. (2008). Antifungal canthin-6-one series accumulate in lipid droplets and affect fatty acid *Biomedicine and Pharmacotherapy*, 62, 99-103.

Lang S.; Wagner, G. (1987). Structure and properties of biosurfctants, p 21-47. In Kosaric, N.; Cairns, W.L.; Gray, N.C.C. (eds.) *Biosurfactants and biotechnology*. Marcel Dekker, Inc., New York.

Lang, S.; Katsiwela, E.; Wagner, F. (1989). Antimicrobial effects of biosurfactants. European *Journal of Lipid Science and Technology*, 91, 363–366.

Lang, S.;Wagner, F. In: Korasic N. (ed) (1993). Biosurfactants. *Surfactants Science series*, vol. 48, 251 -268.

Laniado-Laborın, R.; Cabrales-Vargas, M.N. (2009). Amphotericin B: side effects and toxicity. *Revista Iberoamericana de Micología*, 26, 223–227.

Lee, G.D.; Jung, H.J.; Woo, E-R. (2005). Antimicrobial property of (+)-lyoniresinol-3α- *O* -β-D -Glucopyranoside isolated from the root bark of *Lycium chinense* Miller against human pathogenic microorganisms. *Archives of Farmacal Research*, 28, 1031–1036.

Lee, S.C.; Ju-Soon, Y.; Sun-Hee, K.; Soo-Yeol, C.; Cher-Won, H.; Woo-Hong, J.; Yong-Lark, C. (2006). Production and characterization of lipopeptide biosurfactant from *Bacillus subtilis* A8-8. *Journal of Microbiology and Biotechnology*, 16, 716-723.

Liu, M.; Healy, M.D.; Dougherty, B.A.; Esposito, K.M.; Maurice, T.C.; Mazzucco, C.E.; Bruccoleri, R.E.; Davison, D.B.; Frosco, M.; Barrett, J.F.(2006). *Eukaryotic Cell*. 5, 638–649.

Lopez, A.; Hudson, J.B.; Towers, G.H.N. (2001). Antiviral and antimicrobial activities of Colombian medicinal Plants. *Journal of Ethnopharmacology*, 77, 189–196.

Mandala, S.M.; Harris, G.H. (2000). Isolation and characterization of novel inhibitors of sphingolipid synthesis: Australifungin, viridiofungins, rustmicin, and khafrefungin. *Methods in Enzymology*, 311, 335-348.

Mandala, S.M.; Thornton, R.A.; Milligan, J.; Rosenbach, M.; Garcia-Calvo, M.; Bull, H.G.; Harris, G.; Abruzzo, G.K.; Flattery, A.M.; Gill, C.J.; Bartizal, K.; Dreikorn, S.; Kurtz, M.B. (1998). Rustmicin, a potent antifungal agent, inhibits sphingolipid synthesis at inositol phosphoceramide synthase. *The Journal of Biological Chemistry*, 273, 14942–14949.

Mandala, S.M.; Thornton, R.A.; Rosenbach, M.; Milligan, J.; Garcia-Calvo, M.; Bull, H.G.; Kurtz, M.B. (1997). Khafrefungin, a novel inhibitor of sphingolipid synthesis. The *Journal of Biological Chemistry*, 272, 32709–32714.

Marybeth, A.; Maligie.; Selitrennikoff, C.P. (2005). Cryptococcus neoformans resistance to Echinocandins: (1,3)ß-Glucan Synthase activity is sensitive to Echinocandins. *Antimicrobial Agents and Chemotherapy*, 49, 2851-2856.

Mason, T. L.; Wasserman, B. P. (1987). Inactivation of red beet betaglucan synthase by native and oxidized phenolic compounds. *Phytochemistry*, 26, 2197–2202.

Mathur, A.; Singh, R.; Yousuf, S.; Bhardwaj, A.; Verma, S.K.; Babu, P.; Gupta, V.; Prasad, G.; Dua, V.K. (2011). Antifungal activity of some plant extracts against Clinical Pathogens. *Advances in Applied Science Research*, 2, 260-264.

Matsuyama, T.; Sogawa, M.; Nakagawa, Y. (1989). Fractal spreading growth of *Serratia marcescens* which produces surface active exolipids. *FEMS Microbiology Letters*, 61, 243-246.

Miller, C.M.; Miller, R.V.; Garton-Kenny, D.; Redgrave, B.; Sears, J.; Condron, M.M. (1998). Ecomycins, unique antimycotics from *Pseudomonas viridiflava*. *Journal of Applied Microbiology*, 84, 937–944.

Monteiro, A.S.; Coutinho, J.O.P.A.; Correa Junior, A.; Rosa, C.A.; Siqueira, E.P.; Saqntos, V.L. (2009). Characterization of new biosurfactant produced by *Trichosporon montevideense* CLOA 72 isolated from dairy industry. *Journal of Basic Microbiology*, 49, 553–563.

Monteiro, A.S.; Miranda, T.T.; Lula, I.; Denadai, A.M.L.; Sinisterra, R.D.; Santoro, M.M.; Santos, V.L. (2011).. Inhibition of *Candida albicans* CC biofilms formation in polystyrene plate surfaces by biosurfactant produced *by Trichosporon montevideense* CLOA72. Colloids and Surfaces B: *Biointerfaces*, 84, 467-476.

Moyne, A.L.; Shelby, R.; Cleveland, T.E.; Tuzun, S. (2001). Bacillomycin D: an iturin with antifungal activity against *Aspergillus flavus*. *Journal of Applied Microbiology*, 90, 622-629.

Munden, J.E.; Butterworth, D.; Hanscomb, G.; Verrall, M.S. (1970). Production of Chlorflavonin, an antifungal metabolite of *Aspergillus candidus*. *Applied Microbiology*, 20, 718-720.

Neu, T.R. (1996). Significance of bacterial surface-active compounds in interaction of bacteria with interfaces. *Microbiological Reviews*, 60, 151-166.

Nguyen, H.-V.; Panon, G. (1998). The yeast *Metschnikowia pulcherrima* has an inhibitory e¡ect against various yeast species. *Sciences des Aliments*, 18, 515-526.

Nyfeler, R.; Find all citations by this author (defaultOr filter your current search Keller-Schierlein, W. (1974). *Metabolites of microorganisms*. 143. Echinocandin B, a novel polypeptide-antibiotic from *Aspergillus nidulans* var. *echinulatus*: isolation and structural components. *Helvetica Chimica Acta*, 57, 2459-2477.

Overy, D.; Calati, K.; Kahn, J.N.; Hsu, M.; Martín, J.C.; Roemer, T.; Parish, C.A. (2009) Isolation and structure elucidation of parnafungins C and D, isoxazolidinone-containing antifungal natural products. *Bioorganic & Medicinal Chemistry Letters*, 19, 1224-1229.

Oxford, A.E.; Raistrick, H.; Simonart, P. (1939). Studies in the biochemistry of microorganisms. LX. Griseofulvin-a metabolic product of *Penicillium griseofulvum* Dierkx. *Biochemical Journal,* 33, 240-248.

Panda, D.; Rathinasamy, K.; Santra, M.K.; Wilson, L. (2005). Kinetic suppression of microtubule dynamic instability by griseofulvin: Implications for its possible use in the treatment of cancer. *Proceedings of the National Academy of Sciences*, 102, 28-33.

Peypoux, F.; Pommier, M.T.; Das, B.C.; Besson, F.; Delcambe, L.; Michel, G. (1984). Structures of bacillomycin D and bacillomycin L peptidolipid antibiotics from *Bacillus subtilis*. *Journal of Antibiotics*, 77, 1600-1604.

Portillo, A.; Vila, R.; Freixa, B.; Adzet, T.; Cãnigueral, S. (2001). Antifungal activity of Paraguayan plants used in traditional medicine. *Journal of Ethnopharmacology*, 76, 93–98.

Puchkov, E.O.; Wiese, A.; Seydel, U.; Kulakovskaya, T.V. (2001) Cytoplasmic membrane of a sensitive yeast is a primary target for *Cryptococcus humicola* mycocidal compound (microcin). *Biochimica et Biophysica Acta*, 1512, 239-250.

Puchkov, E.O.; Zahringer, U.; Lindner, B.; Kulakovskaya, T.V.; Seydel, U.; Wiese, A. (2002). The mycocidal, membrane-active complex of *Cryptococcus humicola* is a new type of cellobiose lipid with detergent features. *Biochimica et Biophysica Acta,* 1558, 161-170.

Rau, U.; Nguyen, L.A.; Schulz, S.; Wray, V.; Nimtz, M.; Roeper, H.; Koch, H.; Lang, S. (2005). Formation and analysis of mannosylerythritol lipids secreted by *Pseudozyma aphidis*. *Applied Microbiology and Biotechnology*, 66, 551–559.

Ríos, J.L.; Recio, M.C. (2005). Medicinal plants and antimicrobial activity. Journal of

Rodrigues L. R.; van der Mei, H.; Teixeira J.A.; Oliveira, R. (2004). Biosurfactant from Lactococcus lactis 53 inhibits microbial adhesion on silicone rubber. *Applied and Environmental Microbiology*, 66, 306–311(b).

Rodrigues L.R.; Banat, I.M., Teixeira, J.A.; Oliveira, R. (2006). Biosurfactants: potential applications in medicine. Journal Antimicrobial Chemotherapy, 57, 609-618 (a).

Rodrigues, L.R., Moldes, A.; Teixeira, J.A.; Oliveira, R. (2006). Kinetic study of fermentative biosurfactant production by *Lactobacillus strains*. *Biochemical Engineering Journal*, 28, 109-116 (b).

Rodrigues, L.R.; Banat, I.M.; Van der Mei, H.C.; Teixeira, J.A.; Oliveira, R. (2006). Interference in adhesion of bacteria and yeasts isolated from explanted voice prostheses to silicone rubber by rhamnolipid biosurfactants. *Journal of Applied Microbiology*, 100, 470–480 (e).

Rodrigues, L.R.; Teixeira, J.A.; van der mei, H.C.; Meib, Oliveira, R. (2006). Isolation and partial characterization of a biosurfactant produced by *Streptococcus thermophilus* A. *Colloids and Surfaces B: Biointerface*, 53, 105-111 (d).

Rodrigues, L.R.; Van der Mei, H.; Banat, I.M. (2006). Inhibition of microbial adhesion to silicone rubber treated with biosurfactant from *Streptococcus* thermophilus A. FEMS *Immunology and Medical Microbiology*, 46, 107–112(c).

Rodrigues, L.R.; van der Mei, H.; Teixeira, J.A.; Oliveira, R. (2004). Influence of biosurfactants from probiotic bacteria on formation of biofilms on voice prostheses. *Applied and Environmental Microbiology*, 70, 4408–4410 (a).

Ron, E.Z.; Rosenberg, E. (2001). Natural roles of biosurfactants. *Environmental Microbiology*, 3, 229–236.

Rothf, J.; Sallmabn.; Blankh. (1959). In vitro studies on the antifungal antibiotic griseofulvin. *Journal of Investigative dermatology,* 33, 403.

Saleem, M.; Nazir, M.; Ali, M.S.; Hussain, H.; Lee, Y.S.; Riaz, N.; Jabbar, A. (2010). Antimicrobial natural products: an update on future antibiotic drug candidates. *Natural Products Reporsts*, 27, 238–254.

Sanglard, D.; Odds, F.C. (2002). Resistance of *Candida* species to antifungal agents: molecular mechanisms and clinical consequences. *The Lancet infectious Diseases*, 2, 73–85.

Santos, G.D.; Ferri, P.H.; Santos, S.C.; Bao, S.N.; Soares, C.M.A.; Pereira, Maristela. (2007). Oenothein B inhibits the expression of PbFKS1 transcript and induces morphological changes in *Paracoccidioides brasiliensis. Medical Mycology*, 45, 7, 609 – 618.

Schmatz, D.M.; Powles, M.A.; McFadden, D.; Nollstadt, K.; Bouffard, F.A.; Dropinski, J.F.; Liberator, P.; Andersen, J. (1995). New semisynthetic pneumocandins with improved efficacies against *Pneumocystis carinii* in the rat. Antimicrobial Agents and *Chemotherapy*, 39, 1320-1323.

Scholtmeijer, K.; Wessels, J.G.H.; Wösten, H.A.B. (2001). Fungal hydrophobins in medical and technical applications. *Applied Microbiology and Biotechnology*, 56, 1–8.

Seneviratne, C.J.; Jin, L.; Samaranayake, L.P. (2008). Biofilm lifestyle of *Candida*: a mini review. Oral Diseases, 14, 582–590.

Shah, V.; Doncel, G.F.; Seyoum, T.; Eaton, K.M.; Zalenskaya, I.; Hagver, R.; Azim, A.; Gross, R. (2005). Sophorolipids, microbial glycolipids with anti-human immunodeficiency virus and sperm-immobilizing activities. *Antimicrobial Agents and Chemotherapy*, 49, 4093–4100.

Shastry, M..; Nielsen, J.; Ku, T.; Hsu, M.J.; Liberator, P.; Anderson, J.; Schmatz, D.; Justice, M.C. (2001). Species-specific inhibition of fungal protein synthesis by sordarin: identification of a sordarin-specificity region in eukaryotic elongation factor 2. *Microbiology,* 147, 383–390.

Sigg, H.P.; Stoll, C. (1969). Antibiotic SL 2266. U.S. Patent 3,432,598, March 11, 1969.

Singh, P.; Cameotra, S. (2004). Potential applications of microbial surfactants in biomedical sciences. *Trends in Biotechnology*, 22, 142–146.

Soll, D.R. Phenotypic switching. In: R. Calderone, Editor, *Candida and Candidiasis*, ASM Press, Washington (2002), pp. 123–142.

Soll, D.R.; Langtimm, C.J.; McDowell, J.; Hicks, J.; Galask, R. (1987). High-frequency switching in *Candida* strains isolated from vaginitis patients. *Journal of Clinical* Microbiology, 25, 1611–1622.

Souza, G.C.; Haas, A.P.S.; von Poser, G.I.; Schapoval, E.E.S.; Elisabetsky, E. (2004). Ethnopharmacological studies of antimicrobial remedies in the south of Brazil. *Journal of Ethnopharmacology*, 90, 135–143.

Souza, L.K.H.; Oliveira, C.M.A.; Ferri, P.H.; Santos, S.C.; Oliveira Junior, J.G.; Miranda, A.T.B.; Lião, L.M.; Silva, M.R.R. (2002). Antifungal properties of Brazilian Cerrado plants. *Brazilian Journal of Microbiology*, 33: 247-249.

Spencer, J.F.T.; Gorin, P.A.J.; Tulloch, A.P. (1970). *Torulopsis bombicola* sp. n. Antonie van Leeuwenhoek, 36, 129–133.

Stetter, J.; Lieb, F. (2000). Inovation in Crop Protection: Trends in Research. *Angew. Chem. Int*, 39, 1724-1744.

Strobel, G.A.; Daisy, B. (2003). Bioprospecting formicrobial endophytes and their natural products. *Microbiology and Molecular Biology Reviews*, 67, 491–502.

Strobel, G.A.; Miller, R.V.; Martinez-Miller, C.; Condron, M.M.; Teplow, D.B.; Hess, W.M. (1999). Cryptocandin, a potent antimycotic from the endophytic fungus *Cryptosporiopsis cf. quercina*. *Microbiology*, 145, 1919-1926.

Svetaz, L.; Zuljan, F.; Derita, M.; Petenatti, E.; Tamayo, G.; Cáceres,A.; Cechinel Filho, V.; Gimenez, A.; Pinzón, R.; Zacchino, s.A.; Gupta, M. (2010). Value of the ethnomedical information for the discovery of plants with antifungal properties. A survey among seven Latin American countries. *Journal of Ethnopharmacology* 127, 137–158.

Syldatk, C.; Lang, S.; Wagner, F.; Wray, V.; Witte, L. (1985). Chemical and physical characterization of 4 interfacial-active rhamnolipids from *Pseudomonas spec* Dsm 2874 grown on normal-alkanes. Zeitschrift fur Naturforschung C-*A Journal of Biosciences*, 40, 51-60.

Takesako, K.; Ikai, K.; Haruna, F.; Endo, M.; Shimanaka, K.S.; Nakamura, T.; Kato, I.; Yamaguchi, H. (1991). Aureobasidins, new antifungal antibiotics. Taxonomy, fermentation, and properties. *Journal of Antibiotics*, 44, 919–924.

Takesako, K.; Kuroda, H.; Inoue, T.; Haruna, F.; Yoshikawa, Y.; Kato, I.; Uchida, K.; Hiratani, T.; Yamaguchi, H. (1993). Biological properties of Aureobasidin A, a cyclic depsipeptide antifungal antibiotic. *Journal of Antibiotics,* 46, 1414–1420.

Tanaka, Y.; Tojo, T.; Uchida, K.; Uno, J.; Uchida, Y.; Shida, O. (1997). Method of producing iturin A and antifungal agent for profound mycosis. *Biotechnology Advances*, 15, 234–235.

Thimon, L.; Peypoux, F.; Wallach, J.; Michel, G. (1995). Effect of lipopeptide antibiotic, iturin A, on morphology and membrane ultrastructure of yeast cells. *FEMS Microbiology Letters*, 128, 101–106.

Thuuvenel, C.; Gantier, J.C.; Duret, P.; Fourneau, C.; Hocquemiller, M.E.F.; Arias, A.R. Fournet, A. Antifungal compounds from *Zanthoxylum chiloperone* var. *angustifolium*, 17, 678-680.

Tsuchiya, H.; Sato, M.; Miyazaki, T.; Fujiwara, S.; Tanigaki, S.; Ohyama, M.; Tanaka T.; Iinuma, M. (1996). Comparative study on the antibacterial activity of phytochemical flavanones against methicillin resistant *Staphylococcus aureus. Journal Ethnopharmacology*, 50, 27–34.

Tulloch, A.P.; Spencer, J.F.T.; Deinema, M.H. (1968). A new hydroxyl fatty acid sophoroside from Candida bogoriensis. *Canadian Journal of Chemistry*, 46, 345–348.

Tully, TP.; Bergum, J.S.; Schwarz, S.R.; Durand, S.C.; Howell, J.M.; Patel, R.N.; Cino, P.M. (2007). Improvement of sordarin production through process optimization: combining traditional approaches with DOE. *Journal of Industrial Microbiology and Biotechnology*, 34, 193-202.

Vicente, M.F.; Basilio, A.; Cabello, A.; Peláez, F. (2003). Microbial natural products as a source of antifungals. *Clinical Microbiology and Infection*, 9, 15–32.

WCMC, World Conservation Monitoring Center, (1992). Global diversity. Status of the WHO. (1996). WHO Guideline for the Assessment of herbal Medicines, WHO expert committee on specification for pharmaceutical preparation. *Technical Report series* No 863. Geneva.

Wicklow, D.T.; Poling, S.M. (2009). Antimicrobial activity of pyrrocidines from *Acremonium zeae* against endophytes and pathogens of maize. *Phytopathology*, 99, 109-115.

Wösten, H.A.B.; de Vocht, M.L. (2000). Hydrophobins, the fungal coat unravelled. *Biochimica et Biophysica Acta*, 1469, 79–86.

Wösten, H.A.B.; van Wetter, M.-A.; Lugones, L.G.; van der Mei, H.C.; Busscher, H.J.; Wessels, J.G.H. (1999). How a fungus escapes the water to grow into the air. *Current Biology*, 9, 85–88.

Wösten, H.A.B.; Wessels, J.G.H. (1997). Hydrophobins, from molecular structure to multiple functions in fungal development. *Mycoscience*, 38, 363–374.

Xu, D.; Jiang, B.; Ketela, T.; Lemieux, S.; Veillette, K.; Martel, N.; Davison, J.; Sillaots, S.; Trosok, S.; Bachewich, C.; Bussey, H.; Youngman, P.; Roemer, T. (2007). Genome-wide fitness test and mechanism-of-action studies of inhibitory compounds in *Candida albicans*. *PLoS Pathogens*, 3, e92.

Yoo, D.S.; Lee, B.S.; Kim, E.K. (2005). Characteristics of microbial biosurfactant as an antifungal agent against plant pathogenic fungus. *Journal of Microbiology and Biotechnology*, 15, 1164–1169.

Zhang, Y.; Lamm, R.; Pillonel, C.; Lam, S.; Xu, J.R. (2002). Osmoregulation and fungicide resistance: the *Neurospora crassa* os-2 *Gene encodes a HOG1 Mitogen-Activated Protein Kinase Homologue.*

Zhong, W.; Jeffries, M.W.; Georgopapadakou, N.H. (2000). Inhibition of Inositol Phosphorylceramide Synthase by Aureobasidin A in *Candida* and *Aspergillus* Species. Antimicrobial *Agents and Chemotherapy*, 44, 651-653.

INDEX

A

ABA, 157
absorption spectra, 134
access, 66, 68, 92, 112
accounting, 176
acetic acid, 19
acetone, 8, 40, 173
acetylation, 178, 180
achievement, 127
acid, 13, 14, 19, 20, 22, 24, 25, 26, 37, 40, 42, 43, 48, 50, 67, 68, 69, 79, 82, 85, 93, 94, 97, 99, 102, 104, 108, 111, 112, 113, 115, 117, 124, 127, 144, 151, 161, 162, 173, 174, 177, 178, 179, 180, 181, 183, 193, 196, 201
acidic, 179, 180
acidity, 108
acquired immunodeficiency syndrome, 13
active compound, 8, 59, 97, 117, 125, 162, 198
active site, 37, 45, 58, 66, 68
AD, 2, 31
adaptation, 3, 9
additives, 106, 159, 168
adenocarcinoma, 11
adenosine, 68
adhesion, 168, 182, 183, 184, 186, 188, 191, 195, 198, 199
adipocyte, 86, 89
adiponectin, 86
adsorption, 97, 102, 106, 114, 120, 121, 127, 128, 137, 141, 158
adverse conditions, 37
adverse effects, 66, 171
aerogels, 92
AFM, 150, 151, 152
agar, 8, 10
age, 83, 98, 161
ageing, 94, 98, 102, 127
aggregates, 113, 114, 115, 116, 118, 120, 134, 139, 149, 152
aggregation, 93, 94, 100, 102, 120, 121, 134, 146
aging process, 163
Agrobacterium, 8, 11
AIDS, 13, 144, 169, 170, 189
air temperature, 76
alanine, 177
albumin, 111, 124
alcohol, 92, 93, 94, 99, 113, 157, 158
alcoholism, 6
alcohols, 113, 121
aldehydes, 39
aldolase, 131
alkaloids, vii, viii, ix, 9, 12, 19, 20, 36, 39, 42, 45, 46, 50, 61, 64, 167, 186
aloe, 6
aluminium, 92
ambient air, 152
amine, 99, 113, 142
amines, 97, 165
amino, 10, 56, 121, 122, 144, 173, 177, 181
amino acid, 10, 56, 177
ammonium, 101, 113, 114, 117, 126, 135
ammonium salts, 114, 126, 135
amplitude, 150
analgesic, viii, 17, 75, 85, 86, 111
androgen, 65
angiogenesis, 85
ANOVA, 20
antibiotic, 3, 8, 9, 111, 170, 172, 173, 176, 179, 182, 190, 191, 192, 195, 196, 198, 199, 200
antibiotic resistance, 170
antibody, 110
anti-cancer, viii, 49, 56, 68, 73
anticancer activity, 181
anticancer drug, 112
antigen, 48

antioxidant, vii, viii, ix, 2, 7, 12, 14, 15, 16, 17, 20, 22, 25, 26, 27, 31, 32, 33, 75, 85, 86, 88, 91, 120, 124, 125, 144, 145, 152, 163
antioxidative activity, 33
antipyretic, 76
antitumor, viii, 11, 31, 75, 85, 125
antiviral agents, 12
apex, 3
apoptosis, viii, 12, 16, 31, 55, 56, 64, 67, 69
appetite, 6
apples, 181
applications, ix, 91, 95, 96, 97, 98, 99, 100, 103, 104, 106, 107, 108, 109, 110, 112, 114, 116, 121, 122, 123, 125, 126, 140, 143, 146, 150, 152, 153, 154, 155, 157, 159, 160, 161, 162, 165, 166
Argentina, 185
aromatic compounds, 130
arrest, 67
Artemia, 82
arthritis, 144
ascorbic acid, 19, 25, 26, 27
Asia, 31, 156
aspergillosis, 169
Aspergillus terreus, 9
assessment, v, vii, 35, 43, 193
asthma, 6
astringent, 6, 17
atherosclerosis, 14, 144
atomic force, 150, 151, 165
atomic force microscope, 150
atoms, 98, 113, 114, 121, 124, 127, 133, 140, 143, 148, 150, 165, 178
ATP, 57, 66, 68, 179, 196
attachment, 57, 58, 60, 182, 184
authentication, 143
authority, 38
authors, 92, 111, 112, 115, 122, 143
avoidance, 168
awareness, 36

B

bacillus, 52
Bacillus subtilis, 8, 84, 190, 191, 195, 197, 198
background, 144
bacteria, viii, 8, 9, 10, 11, 36, 37, 58, 75, 77, 84, 103, 168, 171, 172, 173, 177, 178, 180, 181, 183, 198, 199
bacterial artificial chromosome, 48
bacterial strains, 8
Bangladesh, 28
banks, 17
barley, 123, 160
base, 39, 85, 93, 95, 102, 105, 108, 112, 115
base catalysis, 95, 102
baths, 5, 17
beef, 175
behavior, 157
bending, 141
beneficial effect, 62, 125
benefits, 46, 124, 125
benzene, 45, 79, 113
beverages, 62
bilirubin, 124
binding, 95, 102, 110, 158
bioactive materials, 92
bioactive polyphenols, ix, 91, 125, 152
bioassay, 11
bioavailability, 47, 125, 170
biocatalysts, 97, 108
biochemistry, 51, 112, 140, 165, 198
biocompatibility, 97, 101, 110, 111, 112, 115, 122, 126
biodegradability, 110, 169
biodegradable materials, 111
biodegradation, 108
biodiversity, 2, 6, 28, 39, 45, 77, 185
bioflavonoids, 124
biogeography, 29
biological activities, vii, viii, ix, 7, 16, 28, 38, 39, 41, 48, 75, 84, 86, 88, 91, 124, 125, 153, 168, 172, 173, 176, 181, 190
biological control, 191
biological samples, 151
biological sciences, 149
biological stability, 101
biologically active compounds, 28
biomaterials, ix, 91, 96, 101, 103
biomedical applications, 97, 101, 154
biomolecules, 14, 56, 73, 101, 102, 103, 104, 106, 107, 118, 122, 144, 158, 161, 178, 180, 184
biosensors, 95, 97, 103, 110, 143, 158
biosurfactant, 168, 180, 182, 183, 190, 192, 194, 195, 197, 198, 199, 201
biosynthesis, 37, 48, 49, 50, 103, 170, 173, 175, 190
biotechnology, 100, 104, 110, 122, 193, 196
black tea, 125
bleaching, 15
blends, 105
blood, 17, 143
blood circulation, 17
body fluid, 144
body weight, 16, 101
Bolivia, 185
bonding, 96, 115, 119, 122, 126, 131, 140, 141, 182

bonds, 58, 68, 69, 92, 93, 98, 100, 105, 110, 114, 115, 117, 118, 120, 121, 123, 126, 127, 128, 130, 135, 140, 141, 142
bone, ix, 112, 167, 169
bone marrow, ix, 167, 169
bone marrow transplant, ix, 167
bottom-up, 113
brain, 111
branching, 100
Brazil, viii, 4, 17, 18, 75, 76, 77, 78, 82, 83, 88, 167, 185, 186, 187, 200
breakdown, 173
breast cancer, 11, 12, 37, 47, 50, 51, 52, 125, 163
brittleness, 104
bronchitis, 6
Brownian motion, 146, 165
buffer, 102, 103, 111
building blocks, 99, 101, 106, 126, 152, 160, 164
bulk materials, 96
burning, 108
Butcher, 51
by-products, 93, 168

C

calcium, 106
calibration, 19
Cameroon, 186
cancer, viii, 11, 12, 14, 38, 48, 50, 55, 56, 59, 62, 65, 66, 69, 84, 85, 88, 125, 144, 149, 163, 179, 198
cancer cells, viii, 12, 55, 65, 66, 69, 85, 125, 163, 179
cancer progression, 38
candidates, 53, 109, 123, 169, 199
candidiasis, 172, 173
capillary, 6, 111, 131
carbohydrate, 172, 180, 183
carbohydrates, 10, 146, 178
carbon, 15, 121, 124, 133, 152, 160, 168, 178, 181
carbon atoms, 124, 133, 178
carbon dioxide, 160
carbon tetrachloride, 15
carbonyl groups, 134, 135, 141
carboxyl, 179
carboxylic acid, 43, 46, 99, 173
carcinogen, 38, 50
carcinogenesis, 49
carcinoma, 11, 85
cardiovascular disease, 62
Caribbean, v, vii, viii, 35, 38, 39, 40, 42, 45, 47
carotene, 15
carotenoids, 124
carrier, 106, 112, 125, 155, 161
cascades, 66
case study, vii, 2
cast, 94, 95
catalysis, 94, 95, 100, 102, 107, 111, 158
catalyst, 92, 93, 94, 108
catalytic activity, 42, 62
catalytic properties, 152
cationic surfactants, 113, 117, 119, 135
CDC, 49
CEC, 42
cell, 97, 107, 108, 110, 125, 151, 163, 164
cell biology, 151
cell culture, 163, 172
cell cycle, 11, 12, 57, 67
cell death, 14, 179
cell division, 183
cell line, 11, 12, 46, 59, 61, 64, 85, 88
cell membranes, 173
cell signaling, 164
cell surface, 182, 183, 190
cellular homeostasis, viii, 55
Central Europe, 159
cephalosporin, 194
ceramic, 92, 94, 95, 96, 104
ceramic materials, 92, 96
ceramide, 173, 174, 175
Chagas disease, 14
challenges, 28, 36, 53, 152
channels, 96, 111, 115, 117, 122, 137
character, 113
chelates, 163
chemical bonds, 128
chemical inertness, 126
chemical interaction, 105, 106, 120, 128
chemical properties, 94, 95, 96, 112, 140, 196
chemical reactions, 94
chemical stability, 119
chemical structures, 137, 168
chemicals, ix, 40, 91, 125, 153, 189
chemiluminescence, 144, 145, 163, 165
chemoprevention, 11
chemopreventive agents, 48
chemotherapy, 11, 12, 36, 169
China, viii, 55, 75, 77, 86
Chinese medicine, 2, 12, 13
chlorination, 107
chlorine, 124
chloroform, 10, 19
chlorophyll, 123, 124, 160
cholesterol, 171
chromatography, 16, 40, 103, 178, 189
chromosome, 48
chymotrypsin, 58, 59, 61, 62, 64, 65

cigarette smoke, 38
city, 32, 73
civilization, 153
classes, ix, 3, 13, 19, 60, 62, 79, 167, 169, 186
classification, 3, 104, 123
clean technology, 159
cleaning, 114, 184
cleavage, 58, 174
climate, 83
climates, 4
clinical application, 155
clinical diagnosis, 143
clinical trials, viii, 11, 55
clustering, 106, 120
clusters, 95, 99, 105, 106, 116, 139, 150, 161
CMC, 185
coatings, 95, 96, 103, 104, 107, 109, 121, 151, 152, 155, 189
coccidioidomycosis, 193
coding, 37
collagen, 85, 164
collateral, 11
Colombia, 185
colon, 11, 50, 64, 125
colon cancer, 11, 50
colonization, 178, 191
color, 23, 77, 86
commercial, 116, 147, 184, 190, 192
communities, 169
community, 10, 76, 98
compatibility, 95, 97, 110, 141, 143, 169
competition, 178
competitive advantage, 178
competitive process, 118
complement, 140
complexity, ix, 109, 133, 153, 167
compliance, 36, 150
components, 92, 97, 98, 104, 105, 106, 107, 108, 109, 120, 123, 124, 125, 126, 134, 136, 141, 152, 153
composites, 96, 107, 109, 151, 156, 159
composition, viii, 19, 47, 75, 76, 77, 78, 82, 83, 84, 86, 88, 94, 95, 97, 100, 113, 116, 121, 133, 135, 140, 148, 152, 180, 195
ccomputer, 62, 68, 73, 148, 150
concentration, 94, 110, 111, 114, 121, 122, 134, 143, 144, 146, 147
concordance, 114
condensation, 4, 92, 93, 94, 95, 97, 102, 109, 116, 117, 118, 127, 141, 156
conditioning, 182
conductance, 181
conductivity, 95, 97, 150
configuration, 98, 108
confinement, 87, 134
Congress, 161
conjugation, 56, 58, 66, 67, 69, 127, 135
connectivity, 114
conservation, 20, 51
consolidation, 102
constituents, 12, 20, 30, 39, 49, 50, 62, 88, 125, 171
construction, 96, 123
consumption, 101, 131, 179
control, 95, 98, 109, 111, 112, 114, 119, 144, 148, 153, 156, 162
conversion, 110, 128
COOH, 129
cooking, 5
coordination, 98, 104, 117
copolymer, 100
copolymerization, 100, 127
copolymerization reaction, 127
copolymers, 113, 156, 157
coronary heart disease, 144
correlation, 85, 108, 122, 126, 146, 182
correlation function, 146
cosmetic, 77, 123, 144
cosmetics, 4
cost, 168
Costa Rica, 185
costs, 147
cough, 6, 17
coumarins, 52, 123
coupling, 97, 98, 109, 110, 122, 148
covalent bond, 64, 96, 105, 117, 119, 120, 122, 141
covalent bonding, 64, 96, 119
covering, 6, 17
CPC, 113
cracks, 76
crop, 169
crown, 11
cryptococcosis, 169, 174, 179
crystalline, 104, 116, 126
crystals, 121
Cuba, viii, 75, 77, 83, 88
culture, 10, 163, 171, 172, 176, 179, 180, 181, 191, 194
culture medium, 10
curcumin, 50, 64
cure, 6, 7, 28
cycles, 107
cyclicized side chain, viii, 36, 45
cyclodextrins, 115
cyclopentenone prostaglandins, 69
cysteine, 68
cytochrome, viii, 36, 38, 51, 158, 170, 173

cytochromes, 49, 51
cytometry, 122
cytoplasm, 58
cytotoxicity, 11, 46, 85, 122

D

dairy industry, 197
damages, 14
danger, 169
decay, 125
decomposition, 102, 131, 133, 194
decontamination, 96, 195
defects, 148
defence, 143
defense mechanisms, 9, 10, 14, 187
deficiency, 144
deformation, 178
degradation, viii, 55, 56, 57, 64, 65, 66, 68, 103, 106, 107, 108, 111, 120, 145, 173
degradation process, 108
degradation rate, 56
dehydration, 69, 106
delivery, 101, 110, 111
demulcent, 6
denaturation, 103, 111, 131, 134
density, 94, 95, 127, 147, 148
depolarization, 10
deposition, 151
depth, 12, 28, 152
derivatives, viii, 13, 36, 38, 45, 49, 59, 62, 64, 65, 99, 100, 102, 115, 121, 144, 171, 187
dermatitis, 77
dermatology, 199
dermatosis, 17
desorption, 182
detachment, 182
detection, 36, 42, 43, 50, 110, 143, 144, 146, 147, 160, 163, 169, 191
developing countries, vii, 13, 35, 170, 184
diabetes, 14, 144
diabetic cataract, 131
diabetic retinopathy, 85
dialysis, 169
diarrhea, 6
dielectric constant, 97
diet, 62, 86, 124, 144
diffraction, 148, 150
diffusion, 8, 9, 111, 146, 147, 185
digestion, 125
dimerization, 122, 134, 141
direct observation, 149
diseases, 6, 14, 27, 31, 38, 39, 56, 62, 70, 85, 111, 124, 125, 144, 169, 170, 171, 177, 184, 185, 186, 187, 189, 194
dispersion, 100, 139, 193
dissociation, 151
distillation, 47
distribution, 13, 17, 28, 108, 114, 118, 133, 137, 140, 146, 148, 151, 152, 153
distribution function, 147
diuretic, 6
diversity, ix, 3, 4, 83, 140, 167, 168, 185, 189, 190, 194, 201
DMF, 144
DNA, 3, 13, 14, 38, 110, 122, 144, 164, 170, 172
DNA polymerase, 13
docetaxel, 50
doctors, 2
donors, 85
dopants, 95, 96
doping, 106, 109, 122
dosage, 187
double bonds, 135
drawing, 45
drug delivery, 96, 100, 110, 111, 112, 155, 157, 161
drug design, vii
drug discovery, vii, ix, 38, 56, 73, 167
drug efflux, 174
drug interaction, 36, 170
drug metabolism, 38, 52
drug release, 111, 112
drug resistance, 36, 37, 38, 50, 51, 170, 182
drugs, vii, viii, ix, 7, 8, 9, 10, 11, 12, 13, 27, 28, 36, 37, 38, 49, 51, 53, 55, 56, 97, 110, 111, 155, 161, 167, 168, 170, 171, 178, 182, 189, 192
drying, 92, 94, 95, 97, 102, 106
DTA curve, 132
duodenal ulcer, 77
duration, 111, 147
dyes, 38, 108, 112, 119, 120, 121, 122, 134, 142, 159, 162
dynamics, 143

E

earth, 109
ecology, 29, 165
economic activity, 5
economic values, 2
economics, 169
eczema, 6
elastomers, 100, 158
electric field, 150
electrical conductivity, 95, 150

electron, 85, 121, 142, 146, 148, 150, 172, 173, 175, 181, 187
electron microscopy, 146, 148, 149
electronic structure, 148
electrons, 121, 127, 142, 143, 148
ELISA, 69, 122
elongation, 173, 175, 199
elucidation, 11, 28, 48, 70, 191, 198
emergency, 36
emission, 107, 108, 133, 134, 135, 136, 142, 143, 144, 145, 149
employment, 108
emulsions, 146
encapsulation, 95, 96, 97, 100, 101, 102, 103, 104, 112, 118, 119, 120, 121, 122, 123, 127, 128, 131, 133, 134, 137, 141, 143, 144, 145, 149, 151, 152, 158, 162
endothermic, 131, 133
energy, 58, 108, 109, 114, 116, 120, 131, 135, 141, 142, 144, 146, 151, 168, 178
engineering, ix, 91, 97, 100, 106, 110, 112, 154, 165, 169
entrapment, ix, 91, 92, 97, 102, 104, 107, 110, 113, 121, 122, 123, 126, 127, 134, 135, 141, 152, 163
entropy, 131
environment, 20, 95, 108, 114, 120, 127, 134, 141, 143, 145, 146, 147, 151, 152, 178, 182
environmental aspects, 96
environmental change, 168
environmental characteristics, 78
environmental protection, 145
environmental technology, 104
environments, 3, 4, 103, 165
enzymatic activity, 102
enzyme, vii, 11, 13, 35, 37, 38, 42, 44, 45, 47, 50, 52, 53, 57, 66, 67, 68, 97, 103, 107, 108, 110, 155, 170, 172, 173, 174, 178, 181, 187
enzyme inhibitors, 181
enzymes, viii, 11, 36, 37, 38, 39, 42, 44, 45, 46, 56, 57, 58, 66, 68, 69, 70, 92, 96, 102, 104, 157, 178
epidemiology, 50
epidermis, 144, 171
epilepsy, 111, 161
epithelial cells, 195
equilibrium, 14, 92
equipment, 147
erosion, 111
ESR, 87
ester, 50, 60, 62, 82, 85, 174, 180
estrogen, 37
ethanol, 8, 19, 20, 26, 27, 31, 127
ethers, 121
ethyl acetate, 11, 14, 19, 20, 22, 26, 27, 186, 189
ethylene, 113, 115
ethylene oxide, 113, 115
eukaryotic, viii, 55, 56, 57, 58, 169, 199
eukaryotic cell, viii, 55, 57, 58
Eurasia, 177
Europe, 76, 177
evaporation, 94, 107, 133
evidence, 15, 37, 38, 131, 134, 141, 148, 185, 191
evil, 164
evolution, 29, 36, 49, 92, 104, 113, 117, 118, 123, 126, 134
examinations, viii, 35
excitation, 134
expectorant, 6
experimental condition, 42, 103, 128
exploitation, ix, 91, 109, 110
exposure, 49
expulsion, 128
external environment, 145
extinction, 4
extraction, 22, 40, 103, 117, 174, 181, 189
extracts, ix, 7, 8, 9, 10, 11, 15, 17, 19, 20, 22, 25, 27, 28, 30, 31, 32, 33, 47, 52, 68, 83, 89, 91, 101, 113, 123, 125, 126, 128, 129, 131, 134, 136, 137, 138, 143, 152, 164, 172, 184, 185, 186, 187, 192, 195, 197
extrusion, 119

F

fabrication, ix, 91, 116, 121, 152
failure, 140
families, vii, viii, 2, 3, 5, 10, 45, 65, 75, 86, 185, 186
family, 93, 98, 124, 126, 162
family members, 59
farmers, 4
fatty acids, 168, 179, 180, 187
FDA, 38, 170
FDA approval, 170
feedback, 150, 159
fermentation, 69, 173, 174, 200
ferrous ion, 15
fever, 6, 17
fibers, 143, 155
fibrin, 168, 181
fibroblasts, 184
fibrosarcoma, 85
filament, 163
films, 85, 92, 94, 95, 96, 107, 115, 126, 142, 151, 154
filters, 121
filtration, 147
financial, 47

financial support, 47
fitness, 201
flavonoids, ix, 9, 11, 12, 14, 15, 17, 19, 20, 22, 24, 25, 27, 32, 33, 84, 85, 113, 123, 124, 125, 126, 129, 131, 134, 163, 164, 167, 188, 190, 191
flavor, 187
flexibility, ix, 91, 98, 104, 109, 120
flora, 4, 82, 83, 185
flowers, 3, 62, 124
fluctuations, 146
fluid, 106, 146, 157, 182, 196
fluorescence, ix, 42, 91, 97, 101, 106, 108, 112, 118, 119, 120, 121, 122, 123, 125, 126, 127, 131, 134, 135, 136, 137, 142, 143, 152, 159, 160, 163, 164
focusing, 100
folklore, 32, 39
food, 2, 3, 6, 29, 62, 77, 97, 101, 143, 145, 165, 168, 187
food additives, 165
Food and Drug Administration, 38
food products, 143
force, 114, 115, 150, 151, 165, 166
formation, 11, 20, 23, 38, 40, 57, 61, 66, 67, 68, 92, 100, 102, 105, 106, 114, 115, 116, 117, 118, 121, 122, 123, 124, 126, 127, 134, 140, 141, 143, 145, 148, 154, 168, 179, 181, 182, 183, 197, 199
formula, 98
fractures, 6
fragility, 28, 131
fragments, 58
free energy, 131
free radical copolymerization, 157
free radicals, vii, 2, 14, 15, 20, 23, 27, 124, 125, 144, 145
FRP, 15
fructose, 115
fruits, 33, 40, 62
functionalization, 156
funding, 36
fungal infections, vii, ix, 10, 84, 167, 169, 171, 182, 185, 191, 194
fungi, 9, 10, 77, 84, 103, 168, 169, 170, 171, 172, 173, 175, 176, 177, 179, 180, 181, 185, 186, 187, 188, 191, 192
fungus, 67, 69, 173, 175, 176, 177, 187, 195, 200, 201

G

gametophyte, 10
gas sensors, 121
gases, 107, 150
gastrointestinal tract, 5
gel formation, 141
gelation, 94, 95, 111, 118
gene expression, 65
generation, 96, 119, 144, 153, 157
genes, 3, 37, 51, 52, 172, 187
genetic alteration, 189
genetics, 50
genome, 37, 38, 48, 49, 52
genomics, 48
genus, 8, 10, 12, 13, 14, 22, 31, 60, 83, 173, 179, 182, 185
genus Candida, 182
geographical origin, 77, 83, 84
geometry, 161
Germany, 161
germination, 39, 180
gland, 163
glass transition, 151
glass transition temperature, 151
glasses, 94, 95, 96, 98, 102, 103, 109, 110, 119, 120, 121, 126, 137, 140, 153, 157, 159, 160
global HIV cases, vii, 35
glucose, 115, 127, 128, 130, 131, 135, 137, 138, 139, 172, 183
glucoside, 15, 115
glue, 76
glutathione, 124
glycerol, 102, 115, 173
glycine, 66, 177
glycoside, 16, 113, 141
gonorrhea, 6
grain boundaries, 148
grains, 124
granules, 111
grass, 177
groups, 92, 93, 98, 99, 100, 104, 105, 107, 108, 109, 113, 115, 116, 117, 121, 122, 126, 127, 128, 129, 130, 131, 134, 135, 140, 141, 142, 156, 157, 159
growth, 3, 8, 9, 10, 11, 12, 45, 46, 47, 56, 76, 85, 92, 110, 111, 124, 125, 163, 172, 173, 174, 180, 181, 184, 186, 191, 197
growth factor, 85, 111
Guatemala, 185
guidelines, 185
Gulf of Mexico, 32
Guyana, 39

H

habitat, 9
habitats, 168
hair, 169
halitosis, 6

hallucinations, 5
halogen, 99
hardness, 98, 108, 119
healing, 76
health, vii, 3, 28, 35, 36, 47, 77, 86, 97, 123, 124, 125, 169, 184, 186, 189
health care, 36, 184
health care professionals, 36
heart disease, 144
heat, 94, 98, 103, 108, 123, 127, 135, 144
heavy metals, 195
height, 17, 150, 152
Helicobacter pylori, 11
heme, 38
hemodialysis, 169
hepatitis, 86, 89
hepatoma, 11
hepatotoxicity, 12, 16
herbal medicine, 64
herpes, 6
hexane, 19, 20, 40, 186, 188, 189
HHS, 31
histidine, 40
history, 2, 86
HIV, vii, 13, 35, 36, 39, 170
HIV/AIDS, 36
HIV-1, 13
hives, 76, 82
homeostasis, 28, 56, 70
homogeneity, 95, 96, 128, 137
homovanillic acid, 79
honey bees, 86
Hong Kong, 55, 70, 73
host, 37, 38, 95, 101, 104, 117, 119, 120, 121, 126, 134, 137, 141, 151, 153, 168, 182
hotspots, 51
human, vii, viii, 1, 2, 7, 10, 11, 14, 27, 28, 31, 36, 37, 38, 39, 42, 44, 45, 46, 49, 50, 52, 62, 85, 88, 111, 123, 124, 144, 153, 163, 169, 171, 172, 173, 176, 177, 184, 188, 189, 192, 196, 199
human health, 7, 123, 169
human immunodeficiency virus, 199
human skin, 144, 169
humidity, 76, 155
Hunter, 157
hydrazine, 36
hydrocarbons, viii, 36, 38
hydrogels, 102
hydrogen, 62, 68, 100, 105, 114, 115, 120, 121, 126, 127, 128, 131, 139, 141, 155, 182
hydrogen atoms, 121, 127
hydrogen bonds, 62, 68, 100, 105, 114, 115, 121, 127, 128, 141
hydrogen peroxide, 155
hydrolysis, 40, 43, 92, 93, 94, 95, 96, 102, 106, 109, 115, 116, 117, 118, 127
hydrophilicity, 99
hydrophobicity, 112, 118, 153, 182
hydroxyethyl methacrylate, 115
hydroxyl, 59, 107, 109, 121, 124, 126, 127, 128, 135, 141, 142, 144, 174, 179, 180, 201
hydroxyl groups, 109, 121, 126, 127, 128, 135, 142
hypertension, 144
hypotensive, 6
hypothesis, 39

I

ideal, 120, 121, 144
identification, 7, 19, 45, 142, 165, 168, 193, 199
identity, 40, 140, 148
immersion, 111
immobilization, 101, 103, 107, 108, 126, 127, 128, 131, 134, 138, 140, 163
immune system, 14, 169
immunocompromised, ix, 167
immunodeficiency, vii, 35, 199
implants, 96, 97, 101, 157
impotence, 6
impregnation, 106
imprinting, 110, 153, 154
improvements, 100
in vitro, ix, 7, 11, 12, 13, 25, 27, 47, 62, 64, 68, 69, 111, 112, 124, 155, 161, 167, 173, 174, 187, 193, 195
in vivo, ix, 7, 11, 12, 14, 15, 16, 27, 97, 100, 111, 122, 144, 163, 167, 170, 176
incidence, 10, 125, 169
income, 4
India, 3, 5, 14, 29, 30, 32, 64
Indians, 32
indication, 119, 134, 141
individuals, 17, 36
inducer, 59, 64
induction, 12, 38, 64
industrial sectors, 153
industries, 98
industry, 45, 52, 101, 168, 193, 197
ineffectiveness, 169
infection, 17, 84, 123, 169, 173, 182, 188
infectious agents, 12, 184
inflammation, 62, 64
information technology, 160
infrared spectroscopy, 140, 142
ingredients, 125, 135

inhibition, viii, 8, 10, 11, 12, 15, 20, 27, 35, 38, 40, 42, 43, 45, 46, 47, 49, 60, 61, 62, 64, 65, 66, 68, 69, 171, 173, 174, 175, 180, 181, 183, 187, 194, 195, 199
inhibitor, viii, 42, 50, 55, 56, 59, 62, 67, 68, 69, 131, 173, 197
initiation, 11
injury, 171
inositol, 172, 173, 174, 175, 197
insects, 187
instability, 98, 135
integrated optics, 109
integration, 108, 128
integrity, 37, 39, 173
interaction, 114, 121, 122, 123, 126, 131, 133, 134, 140, 148, 150, 159
interactions, 103, 104, 105, 110, 112, 114, 115, 117, 120, 127, 128, 131, 141, 143, 148, 151, 152, 164, 166
interface, 66, 98, 113, 152, 168, 182, 183
interference, 19, 183, 191
interval, 92, 118
intervention, 51, 56
intoxication, 6
intravenously, 169
ionization, 179
ions, 103, 109, 124, 146, 159
IR spectra, 129, 141, 142
IR spectroscopy, 140, 164
Iraq, 186, 193
Ireland, 73
iron, 124
irradiation, 144, 145
islands, 39
isoflavonoids, 77, 83, 84, 85, 88
isolation, 11, 14, 33, 39, 40, 114, 134, 176, 187, 196, 198
isomerization, 109
isomers, 42, 181
isoniazid, vii, 35, 37, 40, 42, 43, 52, 53
issues, 36, 110, 153, 168

J

Jamaica, 35, 39, 40, 47, 51
Japan, 29, 170, 173

K

kaempferol, 11, 12, 13, 15, 19, 27, 31, 62
keratin, 171
kidneys, 97
kill, 76
kinetics, 42, 112

L

labeling, 12
Lactobacillus, 183, 198
lasers, 123, 159
lateral roots, 3
Latin America, 32, 185, 200
leaching, 106, 122
lead, ix, 11, 14, 39, 46, 68, 70, 95, 100, 102, 118, 119, 121, 126, 127, 128, 150, 167, 172, 173, 175
leakage, 103, 178, 179, 181, 196
learning, 87
learning task, 87
legs, 76
leishmaniasis, 13
lens, 131
leprosy, 6
lesions, 53, 163
leukemia, 11, 62
lice, 6
life cycle, 3
life expectancy, 125
life sciences, 110
lifetime, 121, 184
ligand, 93, 166
light, 77, 118, 128, 135, 137, 142, 144, 146, 147, 148, 150, 165, 173
light beam, 146
light scattering, 118, 137, 146, 165
lignans, 123, 188
limitation, 120
lipid oxidation, 124
lipid peroxidation, 15
lipids, 14, 48, 113, 144, 163, 168, 178, 179, 180, 190, 195, 196, 198
lipoproteins, 168
liposomes, 163
liquid chromatography, 20
liquid phase, 94, 102, 119
liquids, 150
liver, 11, 101, 124
liver cells, 11
localization, 51
luminescence, 109, 121, 159
Luo, 163
lysine, 57
lysis, 172, 175, 178, 180

M

macromolecules, 38, 103

magnesium, 157
magnetic field, 151
magnetic materials, 150
majority, 5, 7, 9, 10, 20, 169, 176
maltose, 115
mammalian cells, 171, 189
man, 36, 52
management, 36, 49
manufacturing, 102
mapping, 48, 151
marine environment, 165
marrow, ix, 167, 169
Marx, 154
mass, 94, 110, 131, 133, 147, 181
mass loss, 131, 133
material sciences, 96
materials science, 92, 100, 104, 140, 149
matrix, 92, 97, 98, 100, 102, 103, 105, 106, 107, 109, 110, 111, 117, 119, 122, 123, 126, 127, 128, 129, 130, 131, 133, 134, 135, 137, 139, 141, 143, 149, 153, 158, 159, 160, 163, 164, 179
matrixes, 158
matter, 143
MB, 12
measurement, 134, 140, 143, 146, 147, 151
measurements, 95, 119, 123, 131, 137, 138, 144, 146, 147, 149, 150, 164
measures, 146, 150
mechanical properties, 97, 98, 100, 101, 108, 109, 149
media, 97, 104, 122, 144, 151, 184
medical, ix, 2, 70, 91, 96, 97, 106, 107, 125, 153, 167, 169, 178, 184, 189, 199
medical science, 189
medicine, vii, ix, 2, 4, 6, 8, 10, 14, 17, 30, 39, 53, 87, 91, 96, 97, 100, 104, 123, 125, 143, 145, 153, 184, 186, 187, 192, 194, 195, 198
melanoma, 85
melt, 132
melting, 133, 151
melting temperature, 151
membrane permeability, 178
membranes, 42, 95, 102, 124, 125, 143, 144, 173, 181, 188
meningitis, 192
mesoporous materials, 111, 112, 114, 115, 161
Metabolic, 194
metabolic changes, 171
metabolic pathways, 9
metabolism, vii, 35, 37, 38, 42, 46, 47, 53, 193
metabolites, ix, 7, 9, 10, 11, 12, 19, 40, 51, 59, 68, 167, 168, 169, 172, 176, 186, 191, 193
metabolizing, vii, 35, 37, 50
metal complexes, 100
metal ion, 73, 117, 124
metal oxides, 102
metal salts, 92
metals, 109, 151, 195
methanol, 8, 19, 22, 40
methodology, 107, 110
methyl methacrylate, 157
methylation, 174
Mexico, viii, 32, 75
mice, 38, 50, 53, 64, 111, 163, 172, 173
microbial cells, 183
microbiota, 188, 189
microemulsion, 121
micrometer, 92, 122, 150
microorganisms, ix, 9, 28, 53, 167, 168, 169, 171, 178, 181, 182, 183, 184, 187, 188, 196, 198
microparticles, 102
microscope, 148, 150
microscopy, 122, 137, 138, 146, 148, 149, 150, 151, 152, 165, 166
microspheres, 102, 111, 121, 155, 161
microstructure, 94, 112, 140, 148
migration, 36
MIP, 110
mitogen, 31
mitosis, 171, 175
mixing, 102, 118
mobility, 127, 137
model, 111, 131
models, 38
modifications, 19, 100, 126, 184
modulations, 148
molecular mass, 147, 181
molecular mobility, 143
molecular sensors, 143
molecular structure, 113, 140, 201
molecular weight, 60, 147, 177
monolayer, 120
monomers, 57, 93, 100, 110, 126
Moon, 163
morphology, 114, 132, 137, 148, 149, 151, 176, 187, 200
motion, 146
MR, 29, 32
mRNA, 173, 174, 175, 191
mucous membrane, 144
mucous membranes, 144
multidrug-resistant tuberculosis, 51
multiple sclerosis, 14
multiplication, 8
mustard oil, 6
mutagenesis, 68, 192

mutant, 68
mutations, 14
mycobacteria, 36, 39, 45, 50, 52
mycotoxins, 176

N

N- acetyltransferase (NAT), vii, 35
NADH, 164
nanocomposites, 96, 100, 101, 116, 155, 157, 159, 160
nanocrystals, 149, 159
nanomaterials, ix, 91, 92, 97, 98, 99, 100, 101, 103, 107, 112, 113, 116, 123, 125, 126, 134, 139, 140, 142, 149, 152, 153, 161, 164
nanometer, 101, 113, 116, 140, 148, 150, 151
nanometer scale, 116, 151
nanometers, 118, 127, 137, 139, 150
nanoparticles, 96, 109, 116, 122, 146, 147, 148, 156
nanostructured materials, ix, 91, 111, 153, 163, 164
nanostructures, 95, 112, 116, 140, 149, 154, 157, 160
nanotechnology, ix, 91, 96, 101, 112, 146, 149, 150, 151, 152
nanowires, 149
naphthalene, 82
National Academy of Sciences, 52, 198
natural compound, ix, 12, 38, 39, 42, 43, 44, 45, 88, 153, 167
natural evolution, 10
natural isolates, 39
neodymium, 109, 159
nerve, 6
nervous system, 124
network, 92, 93, 94, 95, 98, 102, 105, 106, 108, 109, 117, 118, 119, 126, 127, 128, 129, 131, 133, 135, 137, 138, 139, 141, 144, 145, 146, 149, 151
network polymers, 102
neurodegenerative diseases, 59, 85
neurological disease, 111
neuroprotection, 64
neutral, 96, 114, 126, 143, 168
neutral lipids, 168
New England, 49, 51
NH2, 31, 121
Nicaragua, 5, 30
Nile, 121, 155
NIR, 128, 130, 131, 134
nitric oxide, 11
nitrogen, 14, 37, 124
NMR, 13, 40, 88
nonlinear optics, 108
North America, 2, 28
novel materials, 107, 119
novelty, 111
nuclear magnetic resonance, 50
nuclei, 127
nucleic acid, 53, 143
nucleotides, 164
nucleus, 58, 123
null, 11
nutrient(s), 85, 88, 168
nutrition, 124, 169

O

obesity, viii, 75, 85, 86
observations, 128, 129, 130, 134
OH, 93, 97, 99, 100, 107, 115, 118, 124, 126, 128, 130, 131, 134, 135, 141
oil, 6, 33, 116, 121, 182, 186, 190
oligomers, 93, 102, 128, 156
opportunities, ix, 91, 109, 125, 152, 153
optical fiber, 143
optical microscopy, 122, 148
optical properties, 100, 101, 103, 108, 110, 119, 121, 126, 127, 134, 137, 140, 159
optimization, 70, 126, 149, 169, 201
optoelectronics, 100
order, ix, 91, 96, 101, 106, 108, 109, 110, 111, 114, 115, 121, 126, 127, 131, 134, 137, 140, 142, 143, 144, 146, 149, 150, 152, 153
ores, 103, 115
organ, ix, 92, 113, 167, 169
organelles, 181
organic compounds, 92, 121, 137, 141
organic polymers, 95, 105, 106, 140, 160
organic solvents, 20, 95
organism, vii, 97, 124, 168, 188
organize, 116
organs, 103
orientation, 115, 148
ornamental plants, 4
oscillation, 150
ovulation, 6
oxidation, 20, 61, 85, 100, 124, 131, 133, 144
oxidative damage, 15, 144
oxidative stress, 14, 16, 28, 124, 144
oxides, 92, 103, 116
oxygen, 14, 85, 92, 98, 121, 124, 127, 133, 143, 144, 187

P

p53, 66, 68, 69
pain, 6
pairing, 110

Pakistan, 190
Panama, 185
parallel, 183
parameter, 146
parameters, 94, 106, 111, 121, 123, 134, 143, 147, 151, 155, 166
parasite, 13, 84, 176
parasites, viii, 75, 84, 86
particles, 92, 94, 95, 99, 102, 105, 122, 143, 146, 147, 164, 165
partition, 14, 182
pathogenesis, 188
pathogens, 52, 169, 173, 177, 178, 180, 188, 192, 195, 201
pathways, vii, viii, 55, 56, 58, 66, 68, 70
penicillin, 8
peptide, 58, 176, 177, 190, 192
peptides, 57, 172, 181, 193
perchlorate, 122
performance, 97, 98, 103, 127, 131
periodontal, 125
periodontal disease, 125
permeability, 131, 170, 171, 178, 188
permit, 76, 115, 169
peroxidation, 27
peroxide, 155
pH, 94, 96, 102, 107, 118, 122, 127, 143, 144, 179, 180
pharmaceutical, vii, 2, 30, 45, 52, 106, 123, 144, 178, 193, 201
pharmacology, 73
phenol, 123, 124, 174
phenolic compounds, vii, 2, 8, 10, 11, 15, 17, 19, 20, 22, 79, 82, 83, 84, 85, 125, 186, 197
phenotype, 182
phenytoin, 111, 161
phosphatidylserine, 12
phospholipids, 168
phosphorylation, 65, 173
photochemical degradation, 120
photoluminescence, 120, 121, 123
photonics, 120
photons, 148
physical and mechanical properties, 149
physical characteristics, 94
physical chemistry, 114
physical environment, 143
physical interaction, 106, 127, 131, 141
physical properties, 140, 149
physicians, 76
physicochemical properties, 161
physics, 114, 126, 143, 153, 165
Physiological, 193
physiology, 37, 51, 56, 165, 178
phytotherapy, 5
plants, vii, viii, ix, 1, 2, 3, 4, 5, 7, 8, 9, 10, 11, 13, 15, 20, 25, 27, 28, 29, 31, 32, 33, 35, 39, 40, 42, 47, 62, 75, 76, 77, 78, 83, 86, 91, 123, 124, 125, 131, 134, 152, 167, 168, 184, 185, 186, 187, 189, 191, 192, 193, 194, 195, 198, 200
plasma membrane, 10, 181
plasticity, 140
plastid, 3
platform, 110
playing, 143
polar, 20, 27, 113, 115, 117, 182
polarity, 143, 182
pollen, 76, 77, 78, 87
pollutants, 108
pollution, 149
poly(methyl methacrylate), 157
polycondensation, 92, 94, 96, 102, 106, 118, 127
polycondensation process, 102, 118
polycyclic aromatic hydrocarbon, 50
polydimethylsiloxane, 112, 161
polydispersity, 137, 146
polymer, ix, 91, 100, 102, 105, 106, 110, 111, 116, 117, 151, 158, 160
polymer matrix, 106
polymer properties, 100
polymer synthesis, 100
polymerase, 13, 174, 175
polymeric chains, 92, 100
polymeric materials, 100, 120, 137, 138
polymeric matrices, 128, 138, 152
polymerization, 92, 93, 94, 100, 103, 109, 110, 128, 156
polymers, 95, 98, 100, 102, 103, 104, 105, 106, 110, 120, 121, 123, 135, 140, 146, 151, 154, 156, 158, 160, 165, 172
polypeptide, 56, 173, 175, 198
polyphenols, ix, 91, 123, 125, 130, 135, 152, 163
polyphosphates, 194
polysaccharide, 168, 172
polystyrene, 183, 197
poor, 125
population, 2, 10, 13, 186
porosity, 94, 95, 97, 102, 103, 106, 112, 117, 118, 127, 137
porous materials, 96, 100, 103, 111, 114, 116, 121, 126
porphyrins, 154
potassium, 143
potato, 11
poverty, 36
power, 148

precipitation, 4, 76, 94, 128
preparation, ix, 4, 5, 17, 19, 28, 91, 92, 95, 96, 97, 103, 104, 109, 114, 116, 125, 126, 127, 137, 140, 152, 160, 180, 201
preservation, 103, 104, 121
pressure, 159
prevention, 6, 20, 62, 86, 125, 131, 184
principles, ix, 3, 10, 91, 97, 107, 112, 114, 125, 126, 133, 148, 152
probability, 184, 185
probe, 150
probiotic, 183, 199
process control, 146
prodrugs, 46
production, 93, 102, 114
prokaryotes, 37
proliferation, 10, 12, 56, 66, 183
proline, 177
propane, 79
properties, ix, 91, 94, 96, 97, 98, 100, 101, 103, 104, 106, 107, 108, 109, 110, 112, 114, 115, 116, 119, 120, 121, 123, 124, 125, 126, 127, 131, 136, 137, 140, 141, 142, 144, 148, 149, 150, 152, 153, 155, 158, 159, 160, 161, 162, 164
prophylaxis, 170
propylene, 113
prostaglandins, 69
prostate cancer, 12, 31, 64, 65
prostheses, 183, 199
prosthesis, 155
proteasome, vii, viii, 55, 56, 57, 58, 59, 60, 61, 62, 63, 64, 65, 66, 68, 70
protection, 27, 145, 168, 173
protective coating, 107
protein conformations, 178
protein synthesis, 173, 195, 199
proteins, viii, 14, 16, 55, 56, 57, 58, 66, 92, 102, 104, 111, 143, 144, 146, 151, 157, 161, 164, 184, 187, 194
proteolysis, viii, 55, 58
protocol, 95
Pseudomonas aeruginosa, 8, 84, 190, 193, 194
public health, 13, 186
purification, 173
purity, 40, 93, 95
pyrimidine, 170
pyrolysis, 94

Q

quality assurance, 146
quality of life, vii, 1, 4
quantum dot, 96
quantum dots, 96
quassinoids, viii, 35, 52
quaternary ammonium, 101, 114
quercetin, 15, 19, 20, 22, 24, 27, 62, 80, 164
quinolone antibiotic, 49
quinone, 11, 65
quinones, 187, 188

R

radiation, 76, 123, 133, 141, 142
radical copolymerization, 157
radical formation, 124
radicals, 14, 15, 124, 125, 143, 145
radius, 146, 147, 150, 165
radius of gyration, 147
rain forest, 83
range, 94, 95, 96, 104, 111, 114, 121, 122, 124, 126, 131, 140, 143, 145, 146, 149
raw materials, ix, 91, 114
RE, 29, 31
reactants, 108
reaction mechanism, 94
reactions, 42, 58, 61, 84, 92, 93, 94, 96, 97, 102, 104, 106, 109, 115, 118, 120, 127, 137, 140, 141, 149, 157, 163
reactive oxygen, 124, 143
reactivity, 99, 120, 158, 163
reading, 16, 20, 40
reagents, 40, 93
real time, 148
reason, 120, 131
recognition, 49, 58, 110, 151
recommendations, 184
recovery, 111
recrystallization, 132
recycling, 69
red propolis samples, viii, 75, 77, 78, 79, 82, 87
reducing sugars, 9
refractive index, 120, 147
refractive indices, 147
region, 131, 135, 141
regions of the world, 6, 185
regression, 42
regression analysis, 42
relationship, 146
relatives, 29
relaxation, 109
relevance, 123
repair, 76
repetitions, 20
replication, 13
repulsion, 116

requirements, 103, 109, 110, 113, 146
researchers, 11, 12, 92, 96
residues, 40, 57, 58, 68
resins, viii, 75, 78, 79, 82, 86
resistance, ix, 7, 9, 12, 13, 47, 49, 52, 53, 77, 84, 100, 103, 120, 131, 167, 169, 170, 179, 182, 190, 194, 195, 197, 201
resolution, 142, 148, 149, 150
resorcinol, 82
resource utilization, 169
resources, 36, 185
respect, ix, 91, 101, 121, 123, 125, 148, 153
respiration, 172, 175
respiratory syncytial virus, 12, 13
response, 64, 65, 85, 97, 123, 155, 168, 188
retention, 102, 120, 162
retinol, 165
reverse transcriptase, 13
rhamnolipid, 183, 195, 199
rheumatoid arthritis, 144
rhizome, 6, 8, 13
Rhizopus, 10, 171
ribosome, 173, 175
rings, 58, 124
RNA, 13, 172
rods, 114
rolls, 116
Romania, 91
room temperature, 94, 96, 97, 106, 107, 108, 120, 144
root, 51, 188, 196
roots, 5, 6, 11, 40
roughness, 109, 150
roughness measurements, 150
rubber, 183, 191, 198, 199
rural areas, 186

S

safety, 28, 145
Salmonella, 8, 37, 84
salts, 92, 113, 114, 126, 135
sampling, 142
savannah, 187
scaling, 149
scattering, 118, 137, 146, 165
scavengers, 145
science, ix, 2, 91, 92, 96, 97, 98, 100, 101, 102, 104, 140, 141, 149, 153, 154, 189
scientific investigations, 39
scientific knowledge, 28
searching, 111
secondary metabolism, 193
secretion, 76, 196
security, 96
sediments, 78
seed, 3, 29
selectivity, 45, 58, 110, 143
self-assembly, 107, 112, 113, 116, 184
self-organization, 113
semiconductor, 149, 154
sensing, 73, 107, 143, 158, 160, 168
sensitivity, 143, 145, 171, 173, 180, 194
sensors, 96, 104, 107, 110, 120, 121, 143
separation, 100, 102, 116, 128, 152, 163
sepsis, 169
sequencing, 48
serine, 68, 177
serum, 111, 174
serum albumin, 111
shape, 99, 111, 113, 116, 117, 140, 149, 153
shaping, 98
showing, 3, 14, 15, 17, 27, 36, 61, 69, 76, 85, 111, 133, 172, 194
shrubs, 17
side chain, viii, 36, 45, 46, 181
side effects, 168, 170, 196
signal transduction, 56, 66, 68
signaling pathway, 70
signalling, 143
signals, 131, 137
signal-to-noise ratio, 142
signs, 101, 185, 186
silane, 98, 123
silanol groups, 93, 117, 121, 127, 128, 133, 141
silicon, 92, 97, 98, 100, 107, 110, 117, 133, 150, 162
Singapore, 32
sintering, 94
SiO2, 100, 108, 111, 122, 126, 136, 139, 155, 158, 159, 161, 162
skin, ix, 6, 50, 91, 101, 124, 125, 144, 153, 169, 177, 185
skin diseases, 6, 144
smart materials, 101
society, ix, 91
sodium, 113, 122
sodium dodecyl sulfate (SDS), 113
sol-gel, ix, 91, 92, 93, 94, 95, 96, 97, 101, 102, 103, 104, 105, 106, 107, 108, 109, 110, 111, 112, 113, 115, 116, 117, 118, 119, 120, 121, 123, 126, 127, 131, 134, 140, 141, 143, 144, 145, 152, 153, 154, 155, 156, 157, 158, 159, 160, 161, 162, 164
solid state, 123
solid surfaces, 143, 150, 183
solubility, 100, 113, 122, 123, 171

solution, 19, 20, 23, 59, 92, 94, 102, 104, 106, 107, 108, 114, 115, 117, 118, 120, 121, 122, 127, 134, 144, 146, 160, 165, 184
solvent molecules, 146
solvents, 95, 97, 107
South America, 4, 30, 86
Southeast Asia, 4
SP, 29, 32, 193
space, 109, 114
specialization, 76, 87
specific surface, 101
spectroscopy, 50, 122, 128, 129, 133, 134, 140, 142, 143, 146, 164
spectrum, ix, 91, 104, 119, 125, 133, 136, 140, 153
speed, 122, 143
sperm, 199
spin, 94
spindle, 171, 175
sponge, 65, 114
spore, 180, 184
sporophyte, 10
SS, 33
stability, 39, 45, 95, 97, 98, 100, 103, 104, 107, 108, 110, 112, 115, 119, 120, 122, 123, 126, 128, 131, 133, 140, 141, 143, 144, 145, 146, 157, 162, 178, 179
stabilization, 122, 131
standard deviation, 20, 24, 25
standardization, 76
starvation, 85
state, 4, 61, 82, 103, 122, 127, 135, 142, 154, 184, 186
states, 77, 78, 82, 121, 144
sterile, 8, 175
steroids, 9, 159
sterols, 170
stigma, 36
stimulant, 6, 17, 39
stimulation, 173, 187
stomach, 6
storage, 103, 109, 131, 144, 151, 168
storage media, 151
strategies, 107, 112, 113, 120, 152, 153
strategy, 109
strength, 94, 95, 97, 98, 150
stress, 124, 144, 165, 188
stress response, 188
stretching, 141
strong interaction, 105, 115, 126
structural changes, 132
structural characteristics, 103
structure formation, 154
structuring, 169
styrene, 99, 157
substitution, 59, 107
substrate, 38, 42, 57, 58, 66, 94, 110, 124, 168
substrates, 40, 48, 58, 66, 68, 103, 107, 168, 190
sugar, 115
sulfate, 113
sulfur, 14
Sun, 53, 73, 197
supervision, 73
supplementation, 163
suppression, 62, 169, 180, 198
surface area, 95, 96, 111, 116, 120
surface properties, 157, 190
surface reactions, 149
surface structure, 146
surfactant, 108, 113, 114, 115, 118, 119, 121, 122, 126, 133, 135, 155, 160, 161, 183
surfactants, 112, 113, 114, 116, 117, 118, 119, 126, 135, 143, 161, 162, 165, 178, 182, 184, 190, 192, 193, 199
surveillance, 51
survival, vii, 9, 28, 35, 37, 66, 168, 171
survival rate, vii, 35
susceptibility, 88, 120, 169, 171, 179
suspensions, 92, 118
Sweden, 47
swelling, 95
switching, 109, 122
Switzerland, 173
symmetry, 98
symptomology, 14
symptoms, 185, 186
synergistic effect, 123, 125, 136
synthesis, vii, ix, 7, 14, 35, 45, 48, 50, 53, 91, 92, 95, 96, 98, 100, 101, 104, 106, 107, 108, 109, 111, 112, 114, 115, 116, 117, 120, 122, 125, 139, 140, 143, 148, 152, 153, 154, 155, 157, 160, 161, 167, 170, 172, 173, 174, 175, 187, 189, 192, 195, 197, 199
synthetic analogues, 169
synthetic methods, 107
synthetic polymers, 158

T

tannins, 9, 11, 19, 20, 123, 187
target, viii, 12, 35, 37, 38, 45, 48, 56, 57, 59, 61, 64, 65, 66, 68, 69, 104, 110, 163, 164, 169, 172, 173, 174, 195, 198
techniques, 13, 42, 47, 53, 62, 70, 94, 103, 107, 116, 122, 140, 141, 142, 148, 149, 152, 153
technology, ix, 96, 100, 101, 104, 108, 114, 154, 159, 160, 167

TEM, 137, 138, 139, 148, 149
temperature, 19, 92, 94, 95, 96, 97, 100, 103, 104, 106, 107, 108, 109, 114, 120, 131, 133, 144, 151
template molecules, 117
temporal lobe, 111, 161
temporal lobe epilepsy, 111, 161
tension, 113
TEOS, 93, 102, 111, 117, 118, 121, 127, 128
terpenes, 65, 125, 187
testing, 33
tetraethoxysilane, 121
texture, 76, 149
therapeutic agents, 2, 13, 56, 110
therapeutic approaches, 56
therapeutic benefits, 39
therapeutic effects, 86
therapeutic targets, 53
therapeutic use, 7
therapeutics, 45, 56, 70, 189
therapy, viii, 35, 36, 46, 51, 53, 68, 120, 169, 171, 173
thermal analysis, 131, 133
thermal decomposition, 133
thermal degradation, 106, 120
thermal energy, 146
thermal properties, 100, 150
thermal stability, 95, 97, 98, 100, 104, 110, 115, 120, 126, 128, 133, 140, 141, 157, 179
thermal treatment, 92, 123
thermoplastics, 100
thin films, 94, 95, 126, 151, 154
threats, 28
three-dimensional representation, 150
threonine, 58, 59, 60, 61, 62, 64, 65, 177
threshold, 97, 111
time periods, 147
time-frame, 111
tissue, 38, 97, 101, 111, 112, 155, 164, 182
tissue engineering, 112
titanium, 156, 161
TMC, 61
TNF, 11
TNF-α, 11
tobacco, 5
toluene, 14
tonic, 6
tooth, 125
toxic effect, 30
toxicity, 12, 13, 29, 85, 89, 100, 111, 157, 168, 170, 172, 179, 184, 187, 188, 189, 196
tradition, 123
traits, 193
transcription, 57, 187
transduction, 110
transesterification, 102
transformation, 14, 92, 102, 127, 146, 149
transition, 94, 102, 109, 118, 124, 131, 154
transition metal, 109, 124
transition metal ions, 124
transition temperature, 151
transitions, 129, 131, 163
translocation, 173, 175
transmission, 107, 127, 137, 138, 148, 149
transmission electron microscopy, 148, 149, 165
transparency, 97, 108, 119, 126, 140
transport, 103, 168, 172, 173, 175, 178
transportation, 58
treatment, vii, 1, 6, 13, 17, 28, 36, 39, 46, 51, 56, 70, 84, 86, 92, 94, 101, 108, 123, 125, 127, 144, 161, 169, 171, 173, 184, 185, 186, 193, 194, 198
trypsin, 58, 61, 62
tuberculosis, vii, 35, 36, 37, 38, 39, 45, 46, 47, 48, 49, 50, 51, 52, 53
tumor, 10, 11, 12, 62, 64, 66, 85, 163
tumor cells, 10, 11, 62, 85
tumor growth, 64, 163
tumorigenesis, 50
tumors, 5, 11, 64
tumour growth, 124
tumours, 38
tyramine, 53

U

ubiquitin, vii, viii, 55, 56, 57, 58, 62, 64, 66, 67, 68, 69, 70
ubiquitin-proteasome system, viii, 55, 56, 57, 66, 68, 70
UK, 160, 164
ultrasound, 19
ultrastructure, 3, 200
uniform, 94, 115, 116, 121, 148, 151
unique features, 103
United, 48, 163
United States, 48, 163
urban, 16, 186
urban areas, 16
urea, 101, 115
uric acid, 124
USA, 40, 50, 52, 71, 73
uterus, 101
UV, 19, 40, 123, 128, 129, 130, 133, 134, 189
UV radiation, 123
UVB irradiation, 144, 145

V

vaccine, 36
vaginitis, 200
valuation, 155
vanadium, 157
variables, 76, 94
variations, 83, 86, 180
varieties, 5, 28, 39, 107
vascular plants, vii, 1, 3, 4, 7, 185
vascularization, 3
vegetables, 6, 38, 62, 124
vegetation, 78
vein, 3
velocity, 146
Venezuela, viii, 75, 77
versatility, 8
vessels, 131
vibration, 141
viruses, 13, 28
viscosity, 92, 94, 102, 118, 127, 147
vitamin C, 125
vitamins, ix, 91, 124, 144, 152

W

Washington, 191, 200
water, 6, 9, 19, 20, 47, 76, 92, 93, 94, 102, 103, 111, 113, 114, 116, 121, 133, 144, 182, 186, 201
water evaporation, 133
wavelengths, 128, 129, 131, 142
weak interaction, 105, 127, 128, 134, 137, 141
wealth, 185
web, 165, 166
weight gain, 101
weight ratio, 122
well-being, vii, 1, 3, 28, 125
West Indies, 35, 40, 47, 51
Western blot, 67, 68
WHO, 31, 36, 53, 186, 201
wilderness, 51
wind speed, 76
wood, 17
workers, 36, 62, 96, 100, 102, 106, 108, 111
World Health Organisation, 53
World Health Organization (WHO), 13, 31, 36, 186
worldwide, 4, 7, 28, 169
worms, 6
wound healing, 85

X

xylem, 3

Y

Yale University, 73
yeast, viii, 61, 62, 75, 84, 172, 179, 180, 181, 182, 183, 186, 187, 193, 194, 195, 196, 198, 200
yield, 45, 134, 173, 187

Z

zoospore, 180
zygote, 10